bracket—

Bracket is an almanac that highlights emerging critical issues at the juncture of architecture, environment, and digital culture. The series looks at thematics in our age of globalization that are shaping the built environment in unexpected yet radically significant ways.

—on farming

Farming harnesses the efficiency of collectivity and community. Whether cultivating land, harvesting resources, extracting energy or delegating labor, farming reveals the interdependencies of a globalized world. Simultaneously, farming represents the local gesture, the productive landscape, and the alternative economy. The processes of farming are mutable, parametric, and efficient.

bracket — on farming — almanac 1

www.brkt.org

collaborators —

Archinect

www.archinect.com

InfraNet Lab

www.infranetlab.org

sponsor —

Graham Foundation for Advanced Studies in the Fine Arts

www.grahamfoundation.org

designer —

Thumb

www.thumbprojects.com

published by —

Actar

Barcelona — New York
Part of ActarBirkhäuser
www.actar.com

distributed by —

ActarBirkhäuserD

Barcelona — Basel — New York

Roca i Batlle 2
E-08023 Barcelona
T +34 93 417 49 93
F +34 93 418 67 07
salesbarcelona@actarbirkhauser.com

Viaduktstrasse 42
CH-4051 Basel
T +41 61 5689 800
F +41 61 5689 899
salesbasel@actarbirkhauser.com

151 Grand Street, 5th floor
New York, NY 10013
T +1 212 966 2207
F +1 212 966 2214
salesnewyork@actarbirkhauser.com

isbn —

978-84-92861-21-7

Printed and bound in the European Union.

Bracket "On Farming" is set in Akzidenz Grotesk Old Face, Dolly, and Sauna.

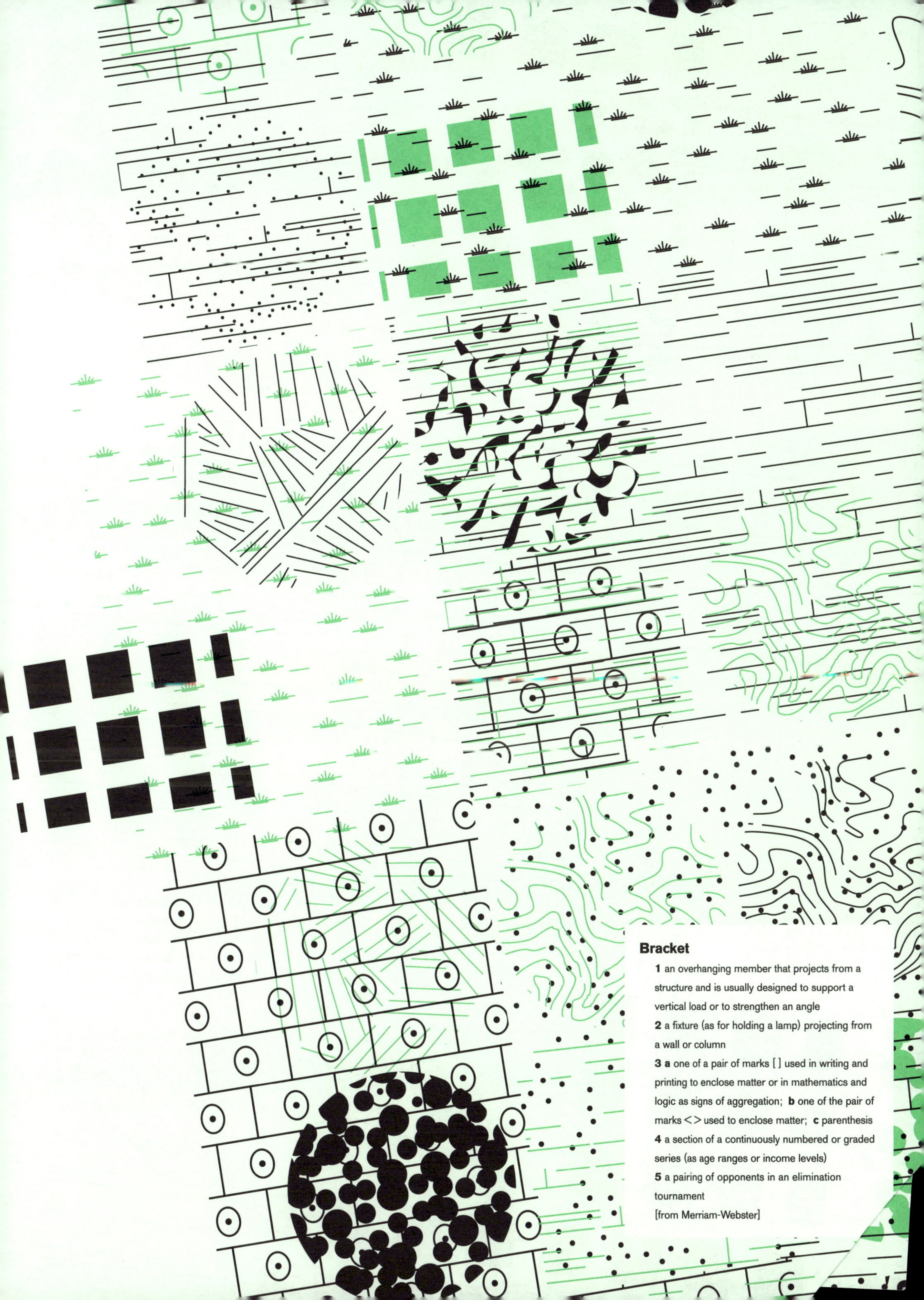

Bracket

1 an overhanging member that projects from a structure and is usually designed to support a vertical load or to strengthen an angle

2 a fixture (as for holding a lamp) projecting from a wall or column

3 a one of a pair of marks [ ] used in writing and printing to enclose matter or in mathematics and logic as signs of aggregation; **b** one of the pair of marks <> used to enclose matter; **c** parenthesis

4 a section of a continuously numbered or graded series (as age ranges or income levels)

5 a pairing of opponents in an elimination tournament

[from Merriam-Webster]

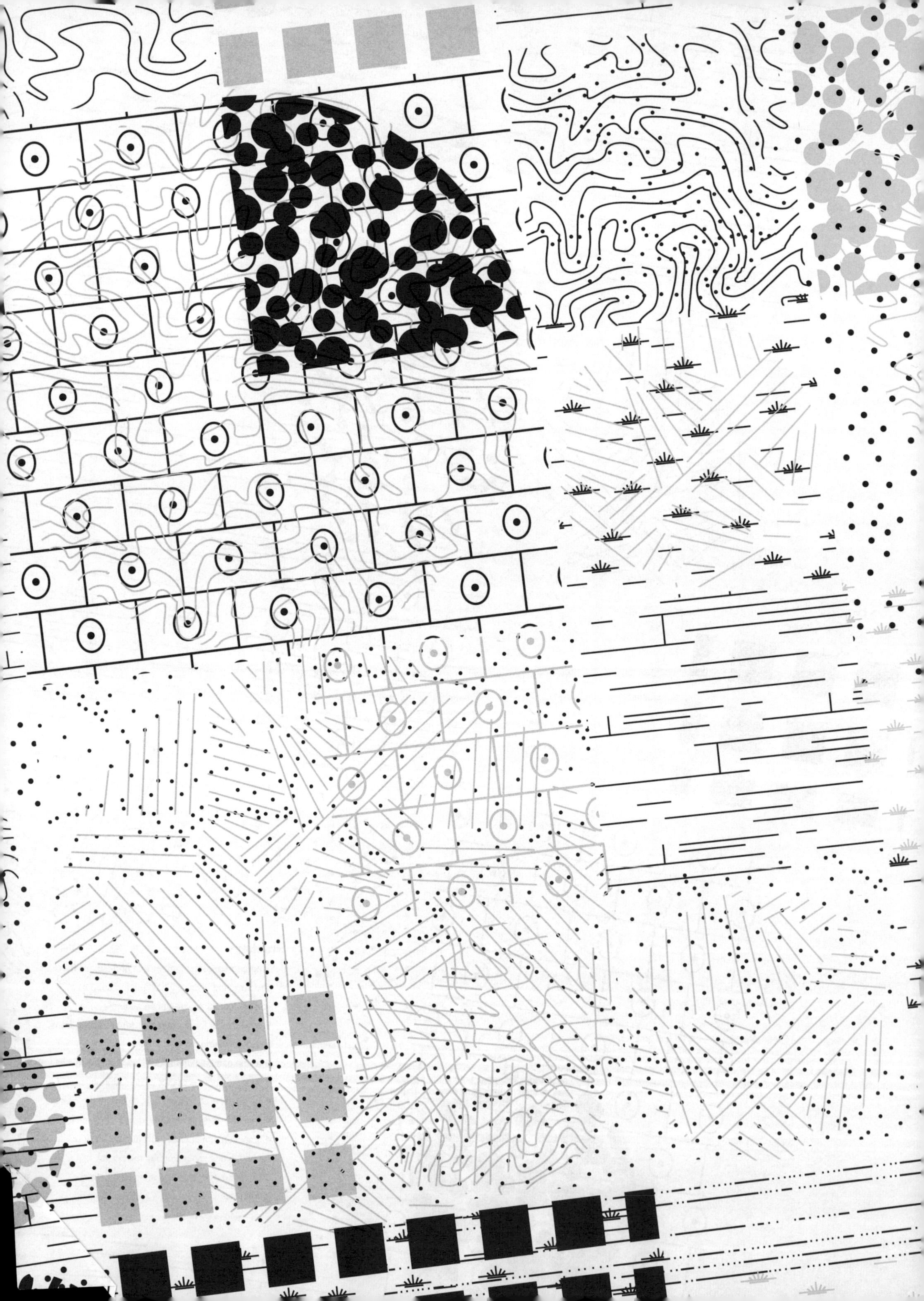

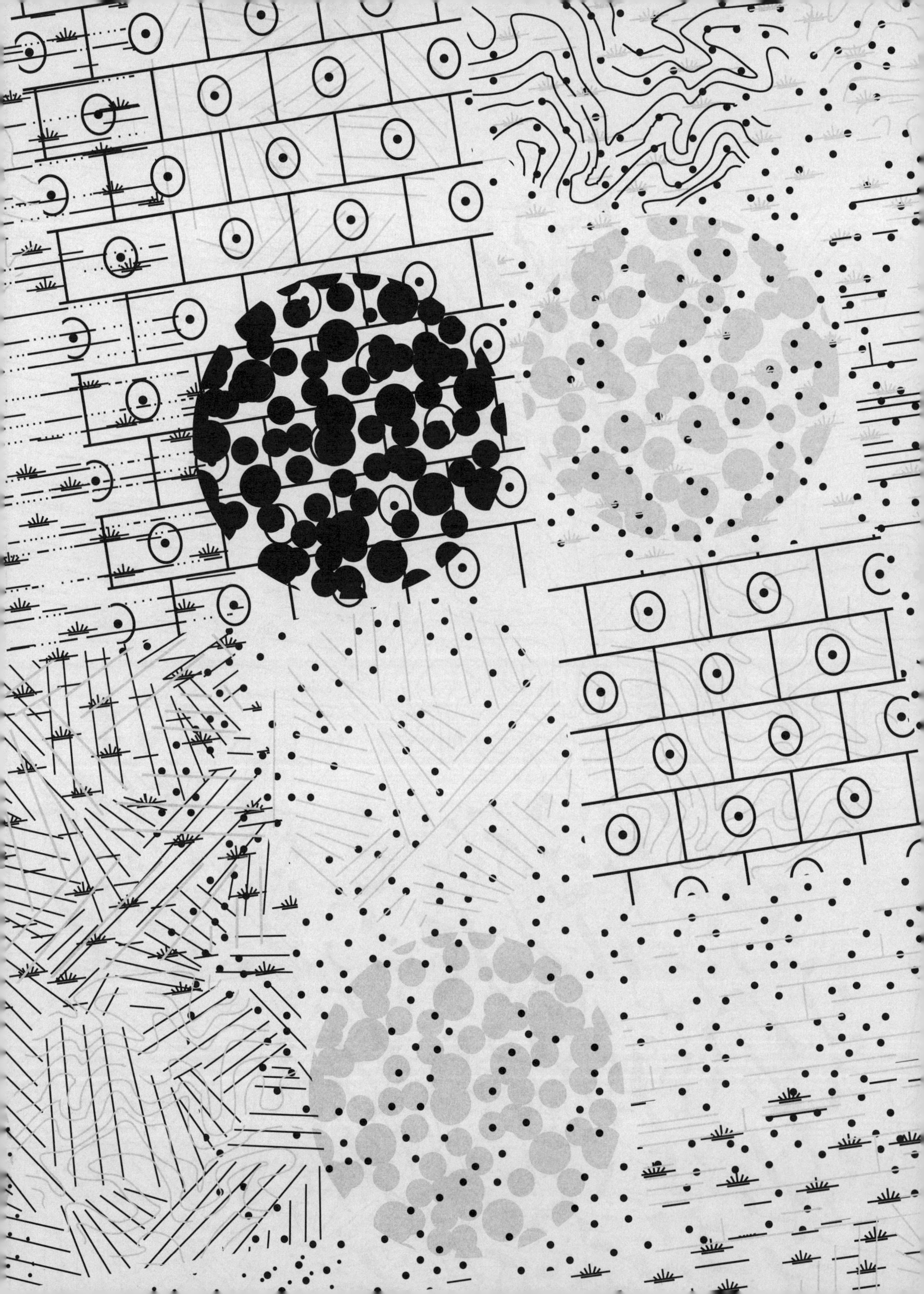

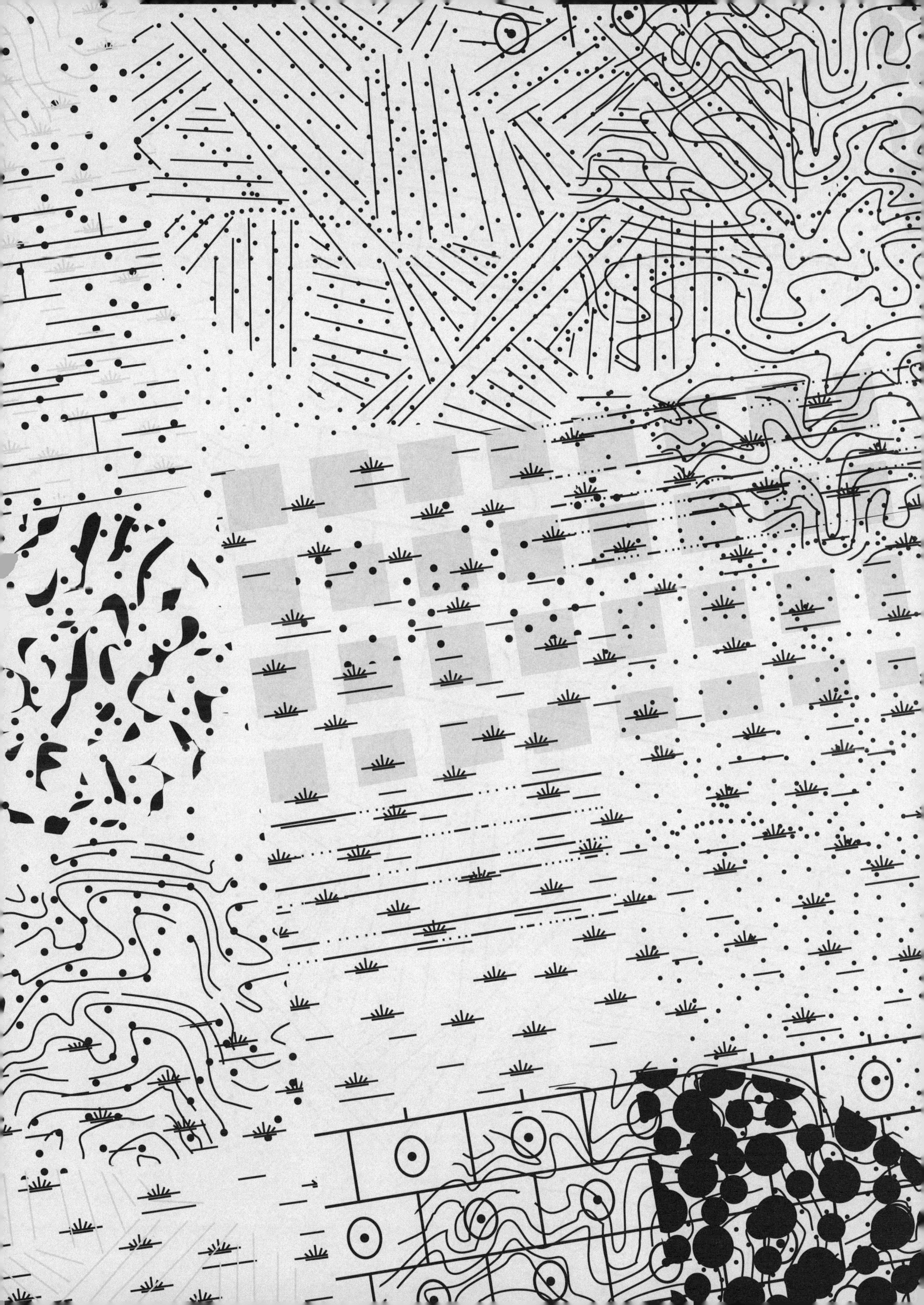

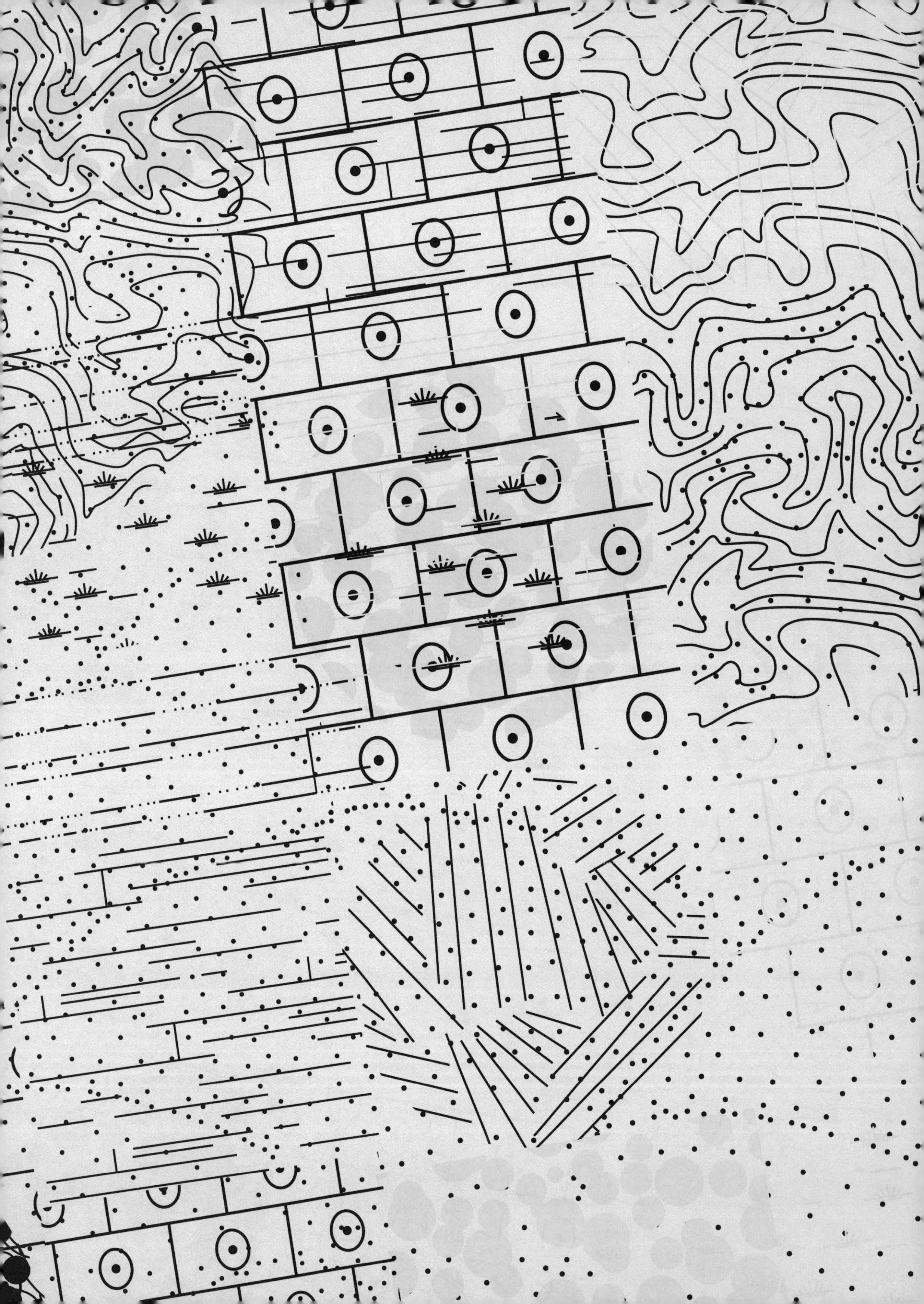

We would like to acknowledge the contributions, energy, dialogue, and support of the following people in the sphere of Archinect and InfraNet Lab: Paul Petrunia, John Jourden, Heather Ring, Javier Arbona, Orhan Ayyüce, Nicole Schiller, Lola Sheppard, Neeraj Bhatia, Joshua Cohen, Geoff Manaugh, and Alexander Trevi. At the Graham Foundation we would like to thank Sarah Herda, Carolyn Kelly, and Stephanie Whitlock. At Actar, we would like to thank Ramon Prat, Albert Ferré, Anna Tetas, and Dolors Soriano for believing in this project. At Thumb, we would like to thank Luke Bulman and Jessica Young for their patience and vision in helping us shape the identity of Bracket.

bracket — on farming — almanac 1

on farming

mason white & maya przybylski

NOVEMBER

The Eleventh Month • 2010

CALENDAR

SKY WATCH ☆ *During the first six hours after sunset, Jupiter, having retrograded to the Aquarius–Pisces border, remains brilliant and dominant in the south even as it fades ever so slightly to magnitude –2.6. The Moon floats to the right of Jupiter on the 15th. In the predawn insomniac hours, Saturn and Venus speedily return, with Venus rivetingly brilliant as it explosively brightens from magnitude –4.1 to –4.9. At midmonth, 40 minutes before sunrise, UFO-like Venus stands 15 degrees high, with Virgo's blue star Spica just above it and Saturn higher still. The group's brightness range is enormous. While the Ringed Planet barely exceeds Spica's so-so magnitude 1, Venus is 250 times more brilliant than the other two.*

●	**New Moon**	6th day	0 hour	52nd minute
◐	**First Quarter**	13th day	11th hour	39th minute
○	**Full Moon**	21st day	12th hour	27th minute
◑	**Last Quarter**	28th day	15th hour	36th minute

After 2:00 A.M. on November 7, Eastern Standard Time is given.

Purchase these pages with times set to your zip code at MyLocalAlmanac.com.

Day of Year	Day of Month	Day of Week	☼ Rises h. m.	Rise Key	☼ Sets h. m.	Set Key	Length of Day h.	m.	Sun Fast m.	Declination of Sun °	'	High Tide Times Boston		☾ Rises h. m.	Rise Key	☾ Sets h. m.	Set Key	☾ Place	☾ Age
305	1	M.	7:17	D	**5:37**	B	10	20	32	14 s.	32	7	**7½**	1:43	D	**2:49**	D	LEO	25
306	2	Tu.	7:18	D	**5:36**	B	10	18	32	14	51	8	**8½**	2:58	D	**3:17**	C	LEO	26
307	3	W.	7:19	D	**5:35**	B	10	16	32	15	10	9	**9½**	4:13	E	**3:45**	C	VIR	27
308	4	Th.	7:21	E	**5:33**	B	10	12	32	15	29	10	**10½**	5:29	E	**4:17**	B	VIR	28
309	5	Fr.	7:22	E	**5:32**	B	10	10	32	15	47	10¾	**11¼**	6:45	E	**4:52**	B	VIR	29
310	6	Sa.	7:23	E	**5:31**	B	10	08	32	16	05	11½	—	8:00	E	**5:34**	B	LIB	0
311	7	**C**	6:24	E	**4:30**	B	10	06	32	16	23	12¼	11¼	8:11	E	**5:23**	B	LIB	1
312	8	M.	6:26	E	**4:29**	B	10	03	32	16	40	12	**12¼**	9:14	E	**6:19**	B	OPH	2
313	9	Tu.	6:27	E	**4:28**	B	10	01	32	16	57	12¾	**1**	10:08	E	**7:20**	B	OPH	3
314	10	W.	6:28	E	**4:27**	B	9	59	32	17	14	1¼	**1¾**	10:52	E	**8:24**	B	SAG	4
315	11	Th.	6:29	E	**4:26**	B	9	57	32	17	31	2½	**2¾**	11:28	E	**9:27**	B	SAG	5
316	12	Fr.	6:31	E	**4:25**	B	9	54	32	17	47	3½	**3½**	11:58	E	**10:30**	C	CAP	6
317	13	Sa.	6:32	E	**4:24**	B	9	52	31	18	03	4¼	**4½**	**12:24**	D	**11:30**	C	AQU	7
318	14	**C**	6:33	E	**4:23**	B	9	50	31	18	19	5¼	**5½**	**12:47**	D	—	–	CAP	8
319	15	M.	6:34	E	**4:22**	B	9	48	31	18	34	6	**6¼**	**1:09**	C	12:30	D	AQU	9
320	16	Tu.	6:36	E	**4:21**	B	9	45	31	18	49	7	**7¼**	**1:31**	C	1:29	D	PSC	10
321	17	W.	6:37	E	**4:20**	B	9	43	31	19	04	7¾	**8**	**1:54**	C	2:29	E	PSC	11
322	18	Th.	6:38	E	**4:19**	B	9	41	31	19	18	8½	**9**	**2:19**	B	3:30	E	PSC	12
323	19	Fr.	6:39	E	**4:19**	B	9	40	30	19	32	9	**9¾**	**2:48**	B	4:33	E	PSC	13
324	20	Sa.	6:41	E	**4:18**	B	9	37	30	19	46	9¾	**10¼**	**3:22**	B	5:37	E	ARI	14
325	21	**C**	6:42	E	**4:17**	A	9	35	30	19	59	10½	**11**	**4:04**	B	6:42	E	ARI	15
326	22	M.	6:43	E	**4:16**	A	9	33	30	20	12	11	**11¾**	**4:54**	B	7:44	E	TAU	16
327	23	Tu.	6:44	E	**4:16**	A	9	32	29	20	25	11¾	—	**5:53**	B	8:41	E	TAU	17
328	24	W.	6:45	E	**4:15**	A	9	30	29	20	37	12½	**12½**	**6:58**	B	9:32	E	GEM	18
329	25	Th.	6:46	E	**4:15**	A	9	29	29	20	49	1¼	**1½**	**8:09**	C	10:15	E	GEM	19
330	26	Fr.	6:48	E	**4:14**	A	9	26	28	21	00	2	**2¼**	**9:21**	C	10:51	E	CAN	20
331	27	Sa.	6:49	E	**4:14**	A	9	25	28	21	11	3	**3**	**10:34**	D	11:23	D	CAN	21
332	28	**C**	6:50	E	**4:13**	A	9	23	28	21	22	3¼	**4**	**11:46**	D	11:52	D	LEO	22
333	29	M.	6:51	E	**4:13**	A	9	22	27	21	32	4¼	**5**	—	–	**12:19**	D	SEX	23
334	30	Tu.	6:52	E	**4:12**	A	9	20	27	21 s.	42	5¼	**6¼**	12:59	E	**12:46**	C	VIR	24

134 To use this page, see p. 106; for Key Letters, see p. 234. ☞ **Bold = P.M.** ☞ Light = A.M. 2010

NOVEMBER

November hath 30 days • 2010

The autumn is old,
The sere leaves are flying. –Thomas Hood

Day of Month	Day of Week	Dates, Feasts, Fasts, Aspects, Tide Heights	Weather
1	M.	**All Saints'** • Michelangelo's fresco on Sistine Chapel ceiling unveiled, 1512 • { 9.7 / 10.1	*Skim*
2	Tu.	**All Souls'** • **Election Day** • ☾ ON EQ. • Tides { 10.2 / 10.3	*ice*
3	W.	☾ AT PERIG. • Mary Jacobs granted patent for first modern bra, 1914 • { 10.8 / 10.4	*cracking,*
4	Th.	☌♄☾ • (Royal) Montreal Golf Club, oldest in North America, founded, 1873 • { 11.4 / 10.6	*snow*
5	Fr.	☌♀☾ • *Keep your shop and your shop will keep you.* • Tides { 11.7 / 10.6	*for*
6	Sa.	Sadie Hawkins Day • **New ●** • Tides { 11.9 / —	*tracking.*
7	**C**	**Daylight Saving Time ends, 2:00 A.M.** • ☌☿☾ • ☌♂☾ • ♆ STAT.	*Sodden:*
8	M.	☾ RUNS LOW • Black bears head to winter dens now. • Tides { 10.3 / 11.5	*Snow's*
9	Tu.	☾ AT ☊ • ☌♇☾ • Worst day of lethal Great Lakes storm, 1913 • { 9.9 / 11.1	*gone*
10	W.	Continental Marines (now U.S. Marine Corps) established, 1755	*but*
11	Th.	**St. Martin of Tours** • **Veterans Day** • Tides { 9.2 / 10.0	*not*
12	Fr.	Indian Summer • Lobsters move to offshore waters. • Tides { 8.9 / 9.5	*forgotten.*
13	Sa.	Ground-breaking ceremony for Martin Luther King Jr. Memorial, Wash., D.C., 2006 • { 8.7 / 9.1	*Glory*
14	**C**	**25th S. af. P.** • ☌♆☾ • Actress Veronica Lake born, 1919	*days*
15	M.	☾ ON EQ. • ☾ AT APO. • Crab apples are ripe now. • { 8.7 / 8.7	*for*
16	Tu.	☌♃☾ • ☌♅☾ • ♀ STAT. • The Philadephia Orchestra debuted, 1900	*football*
17	W.	**St. Hugh of Lincoln** • First U.S. postage stamp with American eagle issued, 1851 • { 9.2 / 8.7	*stars,*
18	Th.	Botanist Asa Gray born, 1810 • Desoto car discontinued, 1960 • Tides { 9.5 / 8.8	*heavy*
19	Fr.	♃ STAT. • *Old friends, old wine, and old gold are best.*	*snow*
20	Sa.	☌☿♂ • Skunks hibernate now. • Tides { 10.1 / 9.1	*buries*
21	**C**	**26th S. af. P.** • **Full Beaver ○** • Tides { 10.4 / 9.2	*our*
22	M.	☾ RIDES HIGH • Santa Ana winds in parts of southern Calif. made airborne rubble a hazard, 1957	*cars.*
23	Tu.	**St. Clement** • U.S. president Franklin Pierce born, 1804 • Tides { 10.7 / —	*Thanks*
24	W.	☾ AT ☋ • *The belly carries the legs and not the legs the belly.* • Tides { 9.3 / 10.8	*for*
25	Th.	**Thanksgiving Day** • National Independent Party organized, 1874 • Tides { 9.3 / 10.7	*turkey,*
26	Fr.	France's first satellite, *Astérix 1*, launched 1961 • { 9.3 / 10.6	*bless*
27	Sa.	0.5" snow began falling in northern Fla., 1912 • Tides { 9.4 / 10.3	*the*
28	**C**	**1st S. of Advent** • Basketball inventor James Naismith died, 1939	*sage:*
29	M.	☾ ON EQ. • Richard Byrd's expedition first to fly over South Pole, 1929 • Tides { 9.8 / 9.9	*ice*
30	Tu.	**St. Andrew** • ☾ AT PERIG. • Deadly tornado, Simsboro, La., 1996 • { 10.1 / 9.7	*age!*

I never think of the future—it comes soon enough. –Albert Einstein

Farmer's Calendar

■ The thing to remember about weather signs in nature is that they are highly reliable. In fact, many are never wrong. Consider the celebrated woolly bear caterpillar, larva of the Isabella tiger moth *(Pyrrharctia isabella)*. It's a plump, 2-inch worm covered in soft bristles—black at the creature's ends and in its middle the color of a red fox. Everybody knows that the ratio of black bristles to rust on the woolly bear foretells the coming winter: More black bristles mean a hard winter; more rust, a mild one.

Or maybe it's the other way around. It scarcely matters. This is because woolly bears have been predicting the weather for a long time, and they know how to cover themselves. Some years ago, for example, we had an unusual spate of woolly bears in this neighborhood. They appeared late in the fall, after the frosts had set in, and they appeared in numbers. In a half-hour's walk, you'd see hundreds of woolly bears creeping across the road. On examination, they showed a great variety of color proportions, from nearly all-black to rust with the least dip of black at the ends. Evidently, then, there could be no kind of winter that some of them would not accurately predict.

The benefit for the would-be weather-wise is obvious. If you distrust the augury of a given woolly bear, then you simply ignore that caterpillar, take a step, and believe another.

CALENDAR

2010 *Listen to the Farmer's Calendar at* Almanac.com. 135

^ **Calendar of November 2010 from *The Old Farmer's Almanac*, 2010.**

"[I]t is the territory that becomes the privileged protagonist of the post-industrial economy, acting as a place for working out the weak and diffuse energies of a powder-fine productivity."
—Andrea Branzi, "Architecture and Agriculture," *Weak and Diffuse Modernity*, 114.

^ **Robert B. Thomas' *The Farmer's Almanack* (1810) and *The Old Farmer's Almanac* (1910).**

In publication continuously since 1792, *The Old Farmer's Almanac* contains weather forecasts, tide tables, planting charts, astronomical data, and articles on environmental data and trends. Originally founded by Robert B. Thomas, the *Almanac* predicts weather up to 18 months in advance using a combined study of solar activity, weather patterns, and atmospherics. The mathematics behind their high success rate has been stored in a black box at the headquarters in Dublin, New Hampshire. The competitive market of predictive weather has always been essential to a farmer's operational successes, and this is the context in which the *Almanac* emerged. Combined with Thomas's original "secret formula," the *Almanac* today uses advanced enviro-veillance technologies in solar science, climatology, and meteorology.

Originally intended solely as a calendar, competition encouraged the almanac's supplementary material to ultimately occupy a majority of the published material. The 2010 Canadian Edition, for example includes only 40 pages of calendar from its total 256 pages. This same 2010 edition featured supplementary material such as an article about green manure titled "The Old and New Farmer's Essential Manure Manual," another article titled "What is Normal Weather?" by Peter Spotts, and an article on botanical nomenclature titled "Ghosts in your Garden?" by Cynthia van Hazinga. Early almanacs, preceding the *Farmer's Almanac* were centered on predicting the position of the sun (sometimes up to four years in advance), while later editions expanded to include supplementary projections, such as horoscopic divinations. Almanacs have served as an essential medium of common cultural understanding of the future. In the 1600s, many English-language almanacs were bestsellers and had almost 400,000 other almanacs to compete with for such a title.

Rather than as a journal, a magazine, or an annual, *Bracket*'s preferred publishing format takes its cues from the *Almanac*. Instead of a calendar, tabular charts, or data sheets, *Bracket* offers a design almanac—predictions through writing and design. As a predictive medium, *Bracket* seeks to interrogate the fertile territory where architecture, environment, and digital culture collide; these transdisciplinary alignments themselves a prediction of emerging collaborations between the analog and the digital. Architecture is not only a byproduct of predictions, but Architecture itself is a prediction machine. Its inherent slowness necessitates a *Farmer's Almanac*-like forecasting of its site's future: future weather, future society, future politics, future economics, and future ecologies, to name a few.

The first issue of *Bracket* is titled "On Farming" partially in homage to the long-standing immediacy of the *Farmer's Almanac*, and also to further isolate and interrogate a contemporary phenomenon that has managed to proliferate within both the wider environment and digital culture, which has made it an ideal initial platform for *Bracket*'s inaugural issue. Toward a broader understanding of its theme, *Bracket* assembles an issue-specific editorial board to review submitted projects and texts. For "On Farming," we have engaged Fritz Haeg, Heather Ring, Michael Speaks, Charles Waldheim, Nathalie de Vries, as well as this issue's editors, Maya Przybylski and Mason White.

Through the selection of the projects and texts featured here, farming emerges less as a conscious practice than a collective behavior across mediums. Agriculture, but also information, energy, and labor can be farmed. Farming harnesses the efficiency of collectivity and community. Whether cultivating land, harvesting resources, extracting energy or delegating labor, farming reveals the interdependencies of our globalized world. Simultaneously, farming represents the local gesture, the productive landscape, and the alternative economy. The processes of farming are mutable, parametric, and efficient. From terraforming to foodsheds to crowdsourcing, farming often involves the management of the natural mediated by the technologic. Farming, beyond its most common agricultural understanding is the modification of infrastructure, urbanisms, architectures, and landscapes toward a privileging of production.

With a global food crisis looming, even the traditional farm's impact on land, resources, and economics is in need of re-visioning. Innovations have led to a growing number of people investing in shares of a local farmer's crop, reducing trips to the supermarket and the cost of shipping food. Energy farming has seen immense diversification in the last decade with essential innovations in renewable energies such as wave farms, wind, tidal, solar, among others. Investment in wind power alone rose from an $8 billion per year plateau from 2002-2004 to more than $18 billion annually on average for 2008-2010, with most of the growth in North America and Asia. Information farming has also experienced considerable rise in the last 5 years and, in fact, could be architecture's newest building typology, the "data campus." Google completed a 30-acre server farm in Oregon on converted agricultural land in 2006. The town of Quincy, Washington has seen Yahoo, Microsoft, and Intuit establish large internet data-processing facilities.

Fish farms, server farms, energy farms, information farms, Wikipedia, Twitter; our contemporary daily life owes so much to the resourceful, convenient intelligence of farming. How is it shaping or how could it shape our design environments? How are these developments shaping our larger environment? And what are new potentials for these typologies? These are the issues and questions that designers and writers have been asked to respond to in the first edition of *Bracket*, "On Farming."

notes towards a history of agrarian urbanism

charles waldheim

The agrarian and the urban are two categories of thought that have more often than not been opposed to one another. Across many disciplines, and for many centuries, the city and the country have been called upon to define one other through binary opposition. Contemporary design culture and discourse on cities are, by contrast, awash in claims of the potential for urban agriculture. Enthusiasm for agricultural production in and around cities has grown through an increased environmental literacy on behalf of designers and scholars. Equally this renewed interest in the relation of food production to urban form has been made possible by increased public literacy about food and the forms of industrial food production and distribution that characterize globalization. This renewed interest in food production and consumption has been shaped by a variety of authors and interests, but has been most forcefully felt as a call for more renewable or sustainable agricultural practices associated with local food production, reduced carbon footprint, increased public health, and the associated benefits of pre-industrial farming techniques including increased biodiversity and ecological health. These tendencies have been most clearly articulated through the so-called 'slow food' and 'locavore' movements. While much has been written on the implications of these tendencies for agricultural production, public policy, and food as an element of culture, little has been written on the potentially profound implications of these transformations for the shape and structure of the city itself. Much of the enthusiasm for slow and local food in the context of urban populations has been predicated on the assumption that abandoned or underused brownfield sites could be remediated and repurposed with productive potential. Equally, this enthusiasm for urban agriculture has been based on the rededication of greenfield sites peripheral to the city, focusing valuable ecological assets on food production rather than suburban sprawl. While both of these remain viable and laudable goals, they shed little light on the implications of such transformations on the shape and the structure of urban form. For those interested in the city as an object of study and subject of design, this suggests the need for further inquiry into the possibilities for an agricultural urbanism. This essay proposes a history of urban form conceived through the spatial, ecological, and infrastructural implications of agricultural production. In the projects that form this tentative counter-history, agricultural production is conceived as a formative element of the city's structure, rather than being considered adjunct to, outside of, or inserted within traditional urban forms. While this may remain an alternative or even marginal counter-history, it may be useful as architects and urbanists grapple with the implications for urban form attendant to their renewed interest in the agricultural. This alternative history of the city seeks to construct a useful past from three urban projects organized explicitly around agricultural production as inherent to the economic, ecological, and spatial order of the city.

Many projects of 20th-century urban planning explicitly aspired to construct an agrarian urbanism. Often these agrarian aspirations were an attempt to reconcile the seemingly contradictory impulses of the industrial metropolis with the social and cultural conditions of agrarian settlement. In many of these projects, agrarianism offered an alternative to the dense metropolitan form of industrial arrangement that grew from the great migrations from farm village to industrial city in the 19th- and early 20th-century cities of Western Europe and North America. The agrarian aspirations of many modernist urban planning proposals originate in the relatively decentralized model of industrial order favored by Henry Ford and other industrialists as early as the 1910s and 20s. Following Ford's organizational preference for spatial decentralization, industrial organizations tended to spread horizontally and abandon the traditional industrial city. In part as a response to the social conditions of the Depression era, agrarianism came to be seen as a form of continuity between formerly agrarian populations based on subsistence farming and the relatively vulnerable industrial workforce of the modern metropolis. By mixing industry with agriculture, many modernist urban planners imagined a rotational labor system in which workers alternated between factory jobs and collective farms. Most often these new spatial orders were understood as vast regional landscapes, and their representation conflated aerial view and orthographic map.

The emergence of these tendencies in the twentieth century might be read through three unbuilt projects advocating a decentralized agrarian urbanism: Frank Lloyd Wright's "Broadacre City" (1934-35), Ludwig Hilberseimer's "New Regional Pattern" (1945-49), and Andrea Branzi's "Agronica" (1993-94) or "Territory for the New Economy" (1999).[1] While these projects were produced decades apart by three very different authors, taken collectively they illustrate the implications for urban form of agricultural production as inherent to the structure of the city. These projects also form a

1. Frank Lloyd Wright, *The Living City* (New York: Horizon Press, 1958); Ludwig Hilberseimer, *The New Regional Pattern: Industries and Gardens, Workshops and Farms* (Chicago: Paul Theobald & Co., 1949); Andrea Branzi, D. Donegani, A. Petrillo, and C. Raimondo, "Symbiotic Metropolis: Agronica" *The Solid Side*, ed. Ezio Manzini and Marco Susani (Netherlands: V+K Publishing / Philips, 1995), 101-120; and Andrea Branzi, "Preliminary Notes for a Master Plan," and "Master Plan Strijp Philips, Eindhoven 1999." *Lotus*, no. 107 (2000): 110-123.

^ Ludwig Hilberseimer, "The City in the Landscape" (1944). Reprinted from Ludwig Hilberseimer, *The New City* (Chicago: Paul Theobald, 1944) p. 55, ill. 33, Ludwig Hilberseimer Papers, Ryerson & Burnham Library Archives, The Art Institute of Chicago.

coherent genealogy of thought on the subject of agricultural urbanism as Branzi explicitly references Hilberseimer's urban proposals, and Hilberseimer's work was informed by familiarity with Wright's urban project. Each of the projects presented their audiences with a profound reconceptualization of the city, proposing radical decentralization and dissolution of the urban figure into a productive landscape. The dissolution of figure into field rendered the classical distinction between city and countryside irrelevant in favor of a conflated condition of suburbanized regionalism. From the perspective of contemporary interests in urban agriculture, both tendencies offer equally compelling alternatives to the canonical history of urban form.

Implicit in the work of these three urbanists was the assumption of an ongoing process of urban decentralization led by an industrial economy. For Wright, Hilberseimer, and Branzi, the decreased density produced through the new industrial logic of decentralization came to depend upon landscape as the primary medium of urban form. These suburban landscapes were embodied and fleshed out with agricultural lands, farms, and fields. These projects proposed large territorial or regional networks of urban infrastructure bringing existing natural environments into relationship with new agricultural and industrial landscapes.

Broadacres / Usonia

In the depths of the Depression, lacking reasonable prospects for a recovery of his once towering stature as the dean of American architects, Frank Lloyd Wright persuaded his lone remaining patron to fund a traveling exhibition of Wright's conception of an organic American urbanism. Broadacre City, as it was referred to, consisted of a large model and supporting materials produced by student apprentices at Taliesin in the winter of 1934-35. While the premises underpinning the project were evident in Wright's lectures as early as the 1920s and fully informed Wright's 1932 publication *The Disappearing City*, the Broadacre model and drawings were not debuted until a 1935 New York City exhibition. Subsequently, the traveling exhibition toured extensively and the remarkably durable project was further disseminated in subsequent publications including *When Democracy Builds* (1945) and *The Living City* (1958).[2]

2. The principles underpinning Wright's Broadacre project were published in 1932 in Frank Lloyd Wright, *Disappearing City* (New York: W. F. Payson, 1932); and subsequently reformulated as *When Democracy Builds* (Chicago: University of Chicago Press, 1945). For an historical overview of Broadacre's influences and contemporary reception, see Peter Hall, *Cities of Tomorrow* (Oxford: Blackwell, 1996), 285-90.

Broadacre City offered American audiences the clearest crystallization of Wright's damning critique of the modern industrial city, positing Broadacre as an autochthonous organic model for North American settlement across an essentially boundless carpet of cultivated landscape. Eschewing traditional European distinctions between city and countryside, Broadacre proposed a network of transportation and communication infrastructures using the Jeffersonian grid as its principal ordering system. Within this nearly undifferentiated field, the county government (headed by the county architect) replaced other levels of government administering a population of landowning citizen-farmers. Wright was clearly conversant with and sympathetic to Henry Ford's notion of a decentralized settlement pattern for North America and the closest built parallel for Wright's work on Broadacre can be found in Ford's instigation of what would become the Tennessee Valley Authority (TVA). As an autonomous public agency, TVA was charged with the construction of hydro-electric dams and highways along the Tennessee River in the electrification of an entire region as a seeding process for future urbanization.[3]

Enjoying ownership of one acre of land per person as a birthright, residents of Broadacre (or Usonia, as Wright would come to refer to it) were to enjoy modern houses set in relation to ample subsistence gardens and small-scale farms. This basic pattern of variously scaled housing and landscape types was interspersed with light industry, small commercial centers and markets, civic buildings, and of course the ubiquitous highway. In spite of the project's extremely low density, most of the ground was cleared and cultivated. Occasionally this constructed and maintained landscape relented in favor of extant waterways, topographic features, or other pre-existing ecologies. Presumably the extrapolation of Broadacre City from its chiefly middle-western origins to the margins of the continent would have been accomplished with varying degrees of accommodation to local climate, geography, and geology, if not cultural or material history. The status of previously urbanized areas existing outside of Wright's Broadacre remained an open question; presumably these would be abandoned in place, again following Ford's lead in this regard.

Wright's critique of private ownership, conspicuous consumption, and accumulation of wealth associated with cities was no small part of the explicit social critique offered by Broadacre, as the worst of the Depression forced bankrupt family farmers to flee their mortgaged farms in the midwest for protest in the east or California in the west. Ironically, given his anxiety over the corrosive effects of accumulated wealth and speculative capital, Wright found in Ford's notion of regional infrastructure the basis for an American pattern of organic urban development. Wright's Broadacre provided a respite from the relentless demands of profit associated with the industrial city, even as the American city was well on a course toward decentralization, driven by the tendencies of Fordist production.

The New Regional Pattern / The New City

Another modernist architect/urbanist grappling with the impacts of decentralization on urban form was Ludwig Hilberseimer. Born and educated in Karlsruhe, Germany, Hilberseimer worked with Mies van der Rohe at the Bauhaus until the rise of fascism precipitated their emigration to Chicago and the Armour Institute of Technology (later IIT) in 1938. While Hilberseimer is most notoriously known for his earlier studies for totalizing rationally-planned schemes of modern urbanism from the 1920s such as *Hochhausstadt* (Highrise City, 1924), Hilberseimer quickly abandoned those schemes in favor of projects that explored decentralization and landscape as remedies to the ills of the industrial city. This was evident as early as 1927 in a sketch titled "The Metropolis as a Garden-City."[4] Hilberseimer's work over the course of the 1930s was clearly influenced by European precedents for the garden-city and evidenced a strategy for the use of landscape and mixed-height housing in a low-density pattern. This is a pattern that would continue to appear in his work in the U.S. over the ensuing decades. Particularly formative in this regard was Hilberseimer's project for *Mischbebauung* (Mixed-height Housing, c. 1930), the principles of which would inform the balance of his career. Hilberseimer during this period was committed to the inevitable decentralization of the traditional city as the resultant of industrial policy. This tendency was evident to Hilberseimer as early as the 1920s in Henry Ford's decision to relocate industrial production outside the city of Detroit in the previous decade.

By the 1940s, Hilberseimer's notion of the "settlement unit" took clearer form through anticipating the development of an interstate highway system and articulating precise relationships between transportation networks, settlement units, and the regional landscape. Hilberseimer's interest in an organic urbanism for North America was further fueled by civil defense imperatives encouraging decentralization in the years following the war.[5] In the wake of Hiroshima, Hilberseimer adapted his proposals to anticipate the construction of the interstate highway system as a civil defense infrastructure and an extension of Fordist production logics. In this context—and conversant with Wright's Broadacre City as well as the progressive TVA project and its proponents in the Regional Planning Association of America—Hilberseimer developed his "New Regional Pattern"

3. For an overview of the Tennessee Valley Authority, see Walter Creese, *TVA's Public Planning* (Knoxville: University of Tennessee Press, 1990); and Peter Hall, *Cities of Tomorrow*, 161-3.

4. For an overview of the origins of Hilberseimer's interpretation of the garden city, see David Spaeth, "Ludwig Hilberseimer's Settlement Unit: Origins and Applications," *In the Shadow of Mies: Ludwig Hilberseimer, Architect, Educator, and Urban Planner*, ed. Richard Pommer, David Spaeth, and Kevin Harrington (New York/Chicago: Rizzoli/Art Institute of Chicago, 1988), 54-68.

5. Hilberseimer and Caldwell advocated for decentralization as a civil defense strategy in the wake of Hiroshima. See Caldwell, "Atomic Bombs and City Planning," *Journal of the American Institute of Architects*, vol. 4 (1945: 289-299); and also Hilberseimer, "Cities and Defense," (c. 1945) reprinted in: *In the Shadow of Mies: Ludwig Hilberseimer, Architect, Educator, and Urban Planner*, ed. Richard Pommer, David Spaeth, and Kevin Harrington (New York/Chicago: Rizzoli/Art Institute of Chicago, 1988), 89-93.

^ Ludwig Hilberseimer, "Bird's-eye view of commercial area and settlement unit" (1944). Reprinted from Ludwig Hilberseimer, *The New City* (Chicago: Paul Theobald, 1944) p. 121, ill. 92, Ludwig Hilberseimer Papers, Ryerson & Burnham Library Archives, The Art Institute of Chicago.

as a strategy for the urbanization of a low-density North American settlement pattern based on regional highway systems and natural environmental conditions. Hilberseimer disseminated his proposals through a publication: *The New Regional Pattern: Industries and Gardens, Workshops and Farms* (1949). The principles and analysis informing Hilberseimer's project was published prior to the project itself in *The New City: Principles of Planning* (1944) and was disseminated a decade later in *The Nature of Cities* (1955).[6]

As with Broadacre, the "New Regional Pattern" was organized around the distribution of transportation and communication networks across an essentially horizontal field of landscape. Within this extensive horizontal territory, housing, farms, light industry, commercial buildings, and civic spaces formed variously scaled networks across a field of decentralized distribution. The organizational pattern of "New Regional Pattern" did not defer to the abstraction of the grid, but was informed by the natural environment; topography, hydrology, vegetation, wind patterns, among others. It conflated infrastructural systems with built landscapes and found environmental conditions to produce a radically reconceived settlement pattern for North America. While Hilberseimer's exquisite drawings (many are the uncredited work of IIT colleague Alfred Caldwell) did not make an explicit case for the kind of ecological awareness apparent in contemporary landscape urbanism, they clearly inflected urban infrastructure to ambient environmental conditions.[7] In this regard, the project offers a profound critique of traditional nineteenth-century urban form, as well as the architectural and urban practices associated with that that persisted into the twentieth century.

Agronica / Territory for the New Economy

The work of the Italian architect and urbanist Andrea Branzi might be found equally relevant to the emergent discourse on agrarian urbanism. Branzi's work reanimates a long tradition of using the urban project as a social and cultural critique. This form of urban projection deploys a project not simply as an illustration or 'vision,' but rather as a demystified distillation

6. Ludwig Hilberseimer, *The New City: Principles of Planning* (Chicago: Paul Theobald & Co., 1944); *The Nature of Cities: Origin, Growth, and Decline, Pattern and Form, Planning Problems* (Chicago: Paul Theobald & Co., 1955).

7. For a detailed account of Hilberseimer's professional relationship with Caldwell, see Caroline Constant, "Hilberseimer and Caldwell: Merging Ideologies in the Lafayette Park Landscape," *CASE: Lafayette Park Detroit*, ed. Charles Waldheim (Cambridge/Munich: Harvard/Prestel, 2004), 95-111. On Caldwell's life and work, see Dennis Domer, *Alfred Caldwell: The Life and Work of a Prairie School Landscape Architect* (Baltimore: Johns Hopkins University Press, 1997).

and description of our present urban predicaments. In this sense, Branzi's urban projects can be read less as a utopian future possible world, and more as a critically engaged and politically literate delineation of the power structures, forces, and flows shaping the contemporary urban condition. Over the past four decades Branzi's work has articulated a remarkably consistent critique of the social, cultural, and intellectual poverty of laissez-faire urban development and the realpolitik assumptions of much urban design and planning. As an alternative, Branzi's projects propose urbanism in the form of an environmental, economic, and aesthetic critique of the failings of the contemporary city.

Born and educated in Florence, Branzi studied architecture in the cultural milieu of the Operaists and the scholarly tradition of Marxist critique. Branzi first came to international visibility as a member of the collective Archizoom Associati (mid-1960s) based in Milano but associated with the Florentine Architettura Radicale movement. Archizoom's project and texts for "No-Stop City" (1968-71) illustrate an urbanism of continuous mobility, fluidity, and flux. While "No-Stop City" was received on one level as a satire of the British technophilia of Archigram, on another level it was an illustration of an urbanism without qualities, a representation of the 'degree-zero' conditions for urbanization.[8]

Archizoom's use of typewriter keystrokes on A4 paper to represent a non-figural planning study for "No-Stop City" anticipated contemporary interest in indexical and parametric formulations of the city. Their work prefigured the current interest in describing the relentlessly horizontal field conditions of the modern metropolis as a surface shaped by the strong forces of economic and ecological flows. Equally, these drawings and their texts anticipate current interest in infrastructure and ecology as non-figurative drivers of urban form. As such, a generation of contemporary urbanists has drawn from Branzi's intellectual commitments. This diverse list of influence ranges from Stan Allen and James Corner's interest in field conditions to Alex Wall and Alejandro Zaera-Polo's interest in logistics.[9] More recently Pier Vittorio Aureli and Martino Tattara's project "Stop-City" directly references Branzi's use of non-figurative urban projection as a form of social and political critique.[10] Branzi's urban projects are equally available to inform contemporary interests within architectural culture and urbanism on a wide array of topics as diverse as animalia, indeterminacy, and genericity, among others.

Branzi's "No-Stop City" proposed an explicitly non-figurative urbanism. In so doing, it renewed a longstanding tradition of non-figurative urban projects as a form of social critique. In this regard, Branzi's "No-Stop City" draws upon the urban planning projects and theories of Ludwig Hilberseimer, particularly Hilberseimer's "New Regional Pattern" and that project's illustration of a proto-ecological urbanism.[11]

Not coincidentally, both Branzi and Hilberseimer chose to illustrate the city as a continuous system of relational forces and flows, as opposed to a collection of objects. In this sense, the ongoing recuperation of Hilberseimer, and Branzi's renewed relevance for discussions of contemporary urbanism render them particularly relevant to discussions of ecological urbanism. Andrea Branzi occupies a singular historical position as a hinge figure between the social and environmental aspirations of modernist planning of the post-war era and the politics of 1968 in which his work first emerged for English-language audiences. As such, his work is particularly well suited to shed light on the emergent discussion around ecological urbanism.

Branzi's "Agronica" project (1993-94) illustrated the relentlessly horizontal spread of capital across thin tissues of territory, and the resultant 'weak urbanization' that the neo-liberal economic paradigm affords. Agronica embodies the potential parallels between agricultural and energy production, new modalities of post-Fordist industrial economy, and the cultures of consumption that they construct.[12] More recently in 1999, Branzi (with the Domus Academy, a post-graduate research institute founded in the 1980s) executed a project for Philips in Eindhoven. These projects returned to the recurring themes in Branzi's oeuvre with typical wit and pith, illustrating a "Territory for the New Economy" in which agricultural production was instrumental in deriving urban form.[13]

Branzi's 'weak work' maintains its critical and projective relevance for a new generation of urbanists interested in the economic and agricultural drivers of urban form. His long-standing call for the development of weak urban forms and non-figural fields has already influenced the thinking of those

8. Archizoom Associates, "No-Stop City. Residential Parkings. Climatic Universal Sistem," ***Domus*** **496 (March 1971): 49-55. For Branzi's reflections on the project, see Andrea Branzi, "Notes on No-Stop City: Archizoom Associates 1969-1972,"** ***Exit Utopia: Architectural Provocations 1956-1976,*** **ed. Martin van Schaik and Otakar Macel, (Munich: Prestel, 2005), 177-182. For more recent scholarship on the project and its relations to contemporary architectural culture and urban theory, see Kazys Varnelis, "Programming After Program: Archizoom's No-Stop City,"** ***Praxis,*** **no. 8 (May 2006): 82-91.**

9. On field conditions and contemporary urbanism, see James Corner "The Agency of Mapping: Speculation, Critique and Invention," ***Mappings*** **ed. Denis Cosgrove (London: Reaktion Books, 1999), 213-300; and Stan Allen, "Mat Urbanism: The Thick 2-D,"** ***CASE: Le Corbusier's Venice Hospital and the Mat Building Revival,*** **ed. Hashim Sarkis (Munich: Prestel, 2001), 118-126. On logistics and contemporary urbanism, see Susan Nigra Snyder and Alex Wall, "Emerging Landscape of Movement and Logistics,"** ***Architectural Design Profile,*** **no.134 (1998): 16-21; and Alejandro Zaera Polo, "Order out of Chaos: The Material Organization of Advanced Capitalism,"** ***Architectural Design Profile,*** **no. 108 (1994): 24-29.**

10. See Pier Vittorio Aureli and Martino Tattara, "Architecture as Framework: The Project of the City and the Crisis of Neoliberalism," ***New Geographies,*** **no. 1 (September 2008): 38-51.**

11. Ludwig Hilberseimer, ***The New Regional Pattern: Industries and Gardens, Workshops and Farms*** **(Chicago: Paul Theobald, 1949).**

12. Andrea Branzi, D. Donegani, A. Petrillo, and C. Raimondo, "Symbiotic Metropolis: Agronica" ***The Solid Side,*** **ed. Ezio Manzini and Marco Susani (Netherlands: V+K Publishing / Philips, 1995), 101-120.**

13. Andrea Branzi, "Preliminary Notes for a Master Plan," and "Master Plan Strijp Philips, Eindhoven 1999" ***Lotus,*** **no. 107 (2000): 110-123.**

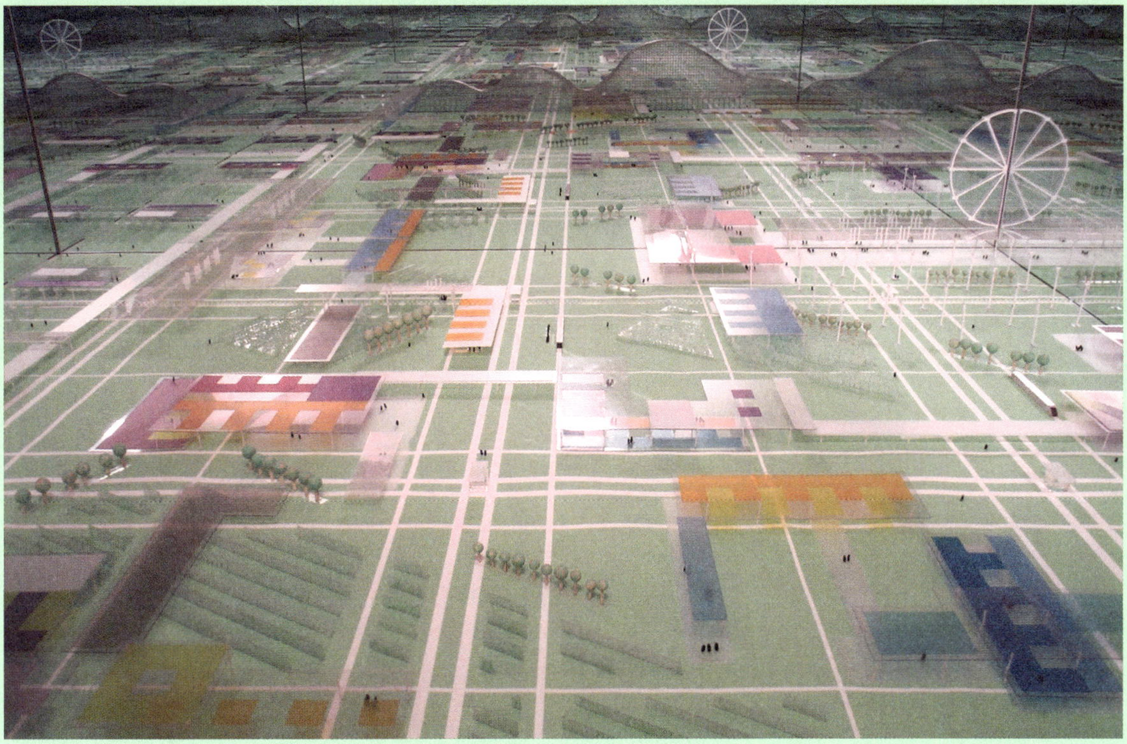

^ Andrea Branzi, et. al., "Masterplan Strijp Philips, Eindhoven," model view (1999-2000).

^ Andrea Branzi, et. al., "Agronica," model view (1993-94).

who articulated landscape urbanism over a decade ago and promises to reanimate emergent discussions of ecological urbanism.[14] Equally, Branzi's projective and polemic urban propositions promise to shed light on agrarian urbanism, and its potential for shaping the contemporary city and the disciplines that describe it.

While this brief pre-history of agricultural urbanism raises more questions than it answers, and may do little to convince contemporary readers of the efficacy of organizing the city in this way, it seems a useful (if not necessary) exercise in understanding the broader implications of contemporary food culture for the design disciplines. In this regard, it is significant that each of the three architect/urbanists presented here as pursuing an explicitly agricultural urbanism did so as part of a broader critical position engaged with economic inequality, social justice, and environmental health. Wright, Hilberseimer, and Branzi, each in their own way, embodied a longstanding tradition of using the urban project as a form of social critique in which the production and consumption of the city, its economy and ecology, are available as tools of analysis and critique. While Wright, Hilberseimer, and Branzi were responding to different economic and ecological contexts, each of them found the urban project an effective vehicle for critiquing the form of their contemporary cities, and the economic, social, and political orders that produced them.

14. Andrea Branzi, "The Weak Metropolis," Ecological Urbanism Conference, Harvard Graduate School of Design, April 4, 2009.

tilling / seeding

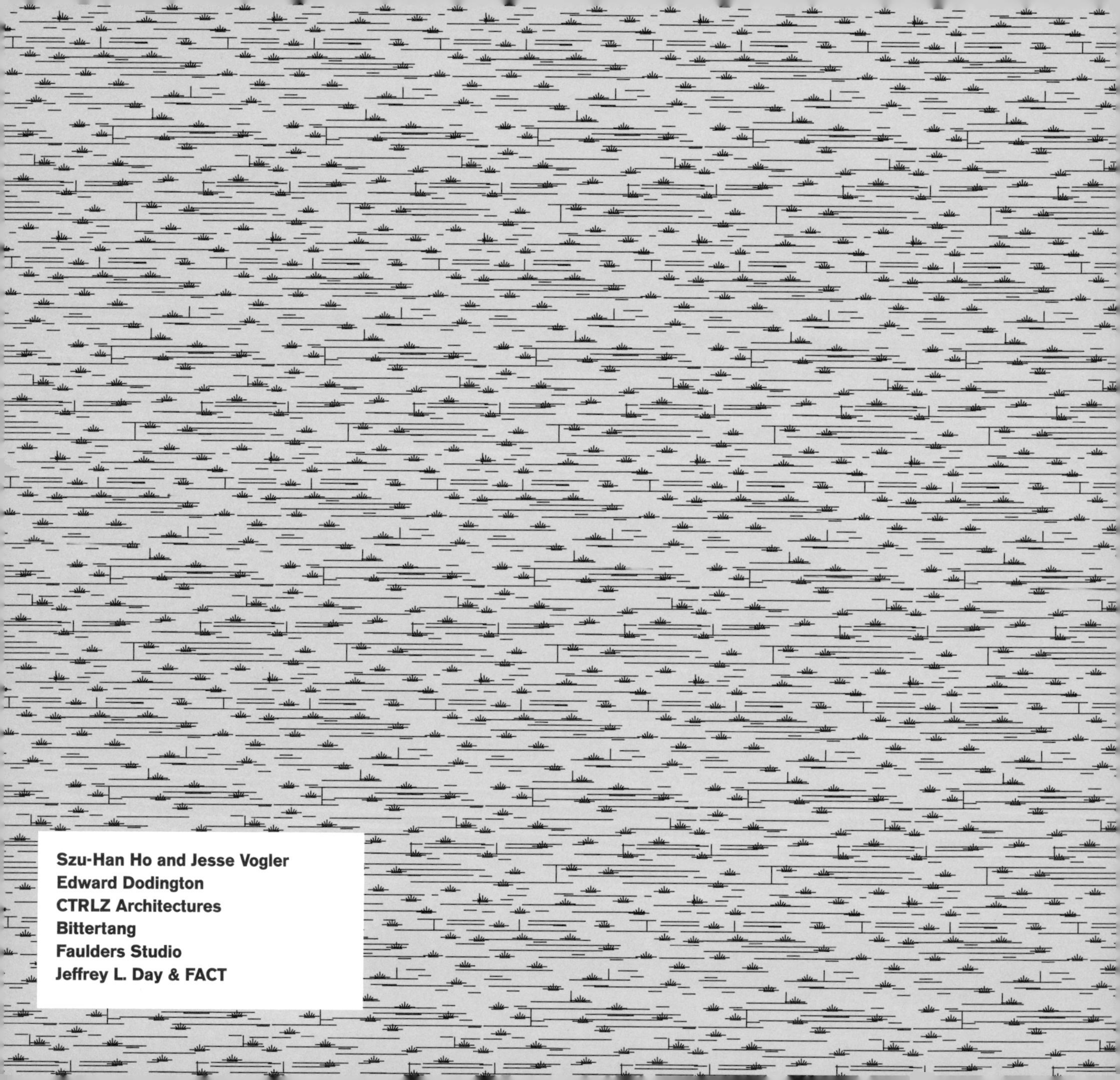

Szu-Han Ho and Jesse Vogler
Edward Dodington
CTRLZ Architectures
Bittertang
Faulders Studio
Jeffrey L. Day & FACT

butter in the mail: experiments in an epistolary economy

szu-han ho & jesse vogler

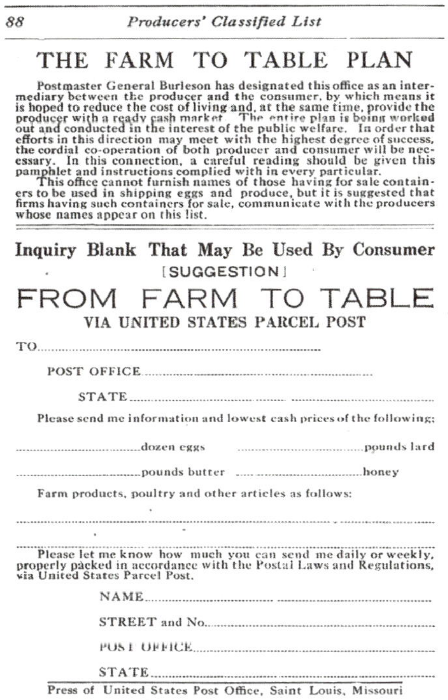

88 *Producers' Classified List*

THE FARM TO TABLE PLAN

Postmaster General Burleson has designated this office as an intermediary between the producer and the consumer, by which means it is hoped to reduce the cost of living and, at the same time, provide the producer with a ready cash market. The entire plan is being worked out and conducted in the interest of the public welfare. In order that efforts in this direction may meet with the highest degree of success, the cordial co-operation of both producer and consumer will be necessary. In this connection, a careful reading should be given this pamphlet and instructions complied with in every particular.

This office cannot furnish names of those having for sale containers to be used in shipping eggs and produce, but it is suggested that firms having such containers for sale, communicate with the producers whose names appear on this list.

Inquiry Blank That May Be Used By Consumer

[SUGGESTION]

FROM FARM TO TABLE

VIA UNITED STATES PARCEL POST

TO..........

POST OFFICE..........

STATE..........

Please send me information and lowest cash prices of the following;

..........dozen eggs ....pounds lard

..........pounds butter ....honey

Farm products, poultry and other articles as follows:

....

....

Please let me know how much you can send me daily or weekly, properly packed in accordance with the Postal Laws and Regulations, via United States Parcel Post.

NAME..........

STREET and No..........

POST OFFICE..........

STATE..........

Press of United States Post Office, Saint Louis, Missouri

^ **Form: Correspondence relating to the shipment of farm produce by postal motor truck service. National Archives and Records Administration, 1917.**

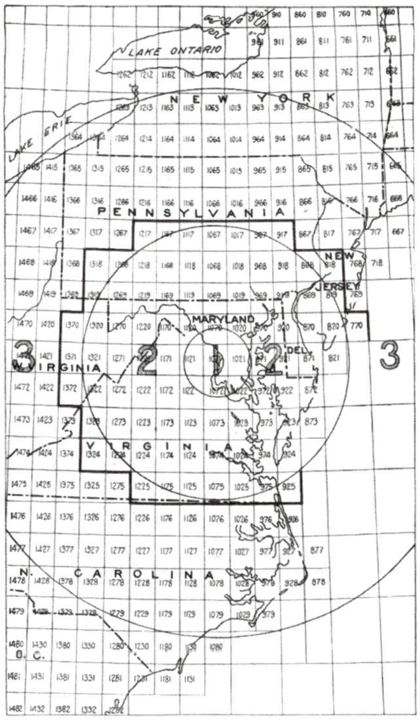

^ **Postal zone map: Post Office Department Annual Report, 1917.**

Beginning in 1914, the United States Post Office Department initiated a little known, and short-lived, program to ship produce directly from rural farmers to urban consumers through the mechanisms of the postal system. Promoted as "Farm-to-Table" (FTT), this program sought to enmesh the protocols of agricultural production with the convenience of a daily delivery system. Simultaneously a plan to manage excess within the economy of farming and an effort to spark rural commerce, the FTT plan fundamentally relied on the system of equivalences that was at the core of postal logistics. As every household was formatted within a grid of equal addressability, it was a short leap to conceptualize producers and consumers as mere variables in a postal matrix—with each as near-at-hand as the corner store. The local postmaster solicited information on goods and pricing from farmers along rural delivery routes and disseminated this information through posted advertisements, published brochures, or bulk mailings. All that remained was for shoppers to mail their order through the local post office and await delivery of farm-fresh foodstuffs at their doorstep. The promise of this model was that agriculture, infrastructure, and communication could be collapsed into a seamless system of exchange. What resulted was the nascent outline of an epistolary economy in which farming protocols and delivery infrastructures were married through the logistics of the postal system.

While the practices of farming and mail delivery might seem unlikely bedfellows, their underlying spatial tendencies and historically conditioned trajectories bracket a specific early 20th-century attitude towards the nature of geographical difference and infrastructural coherence. The emerging practices of industrialized production and commodities trading, on the one hand, and the dream of ubiquitous interconnectivity afforded by uniform postal rates and comprehensive home delivery, on the other, define a particular universalizing, abstracting tendency across the space of the nation. To be sure, neither the postal service nor industrialized agriculture alone could collapse regional or social difference, yet in each can be found the material ambitions of coordinated nation-building, as well as the discursive sympathies of a broad-based, if heterogeneous, public. As both urban and agricultural patterns were undergoing the paradoxical processes of consolidation and fragmentation, norms dictating community and propinquity were recalibrated according to a postal logic of diffusion. Historians have located an identifiable "postal culture" that grew in response to the practices of the late 19th-century postal system, and by the initial adoption of the FTT plan in the second decade of the 20th century, few branches of national life were to be excluded as constituent elements in a looming agro-postal empire.[1]

While it is tempting to consider the FTT program in relation to contemporary farm-to-market alternatives such as the CSA (community-supported agriculture) model or even internet grocery delivery, it is perhaps more accurate to read this program in relation to contemporaneous attitudes towards the expanding role of the federal government, a strong anti-middleman populism, and changing priorities accompanying the rise in industrial production. Seen in this way, the FTT plan can be read as a test-case to measure conflicting sentiments in regard to issues of commodity standardization, shifting territorial coherences, expanding infrastructural rollout, and uncertain patterns of belonging and community—in a word, as a measure of the spaces and textures of modernity. The butter and eggs sent by mail in the early years of this experiment did more than sate the needs of an urban "housewife aristocracy" and provide additional income to farmers. Butter and eggs were additionally, and importantly, signifiers for the essential and the familiar within a changing territorial and social framework.

When first introduced, enthusiasm for the program quickly spread among members of the press, city governments, and the public. The program's inaugural trial took place in ten cities nationwide, including Washington, DC, St. Louis, Boston, Baltimore, Atlanta, Birmingham, and San Francisco. A 1915 newspaper column entitled "Butter from the Post Office" alleged that 250 parcels a day were exchanged through the FTT program in Washington, DC, boasting: "These have traveled direct from the farm to kitchen—with no middleman save Uncle Sam."[2] Promoters hailed the direct marketing of food products through the postal system as a way to profit both producer and consumer, to relieve them from the "tender mercies of an unscrupulous middleman."[3] Architects of the plan hoped that FTT would correct the imbalance between surplus on the farm and demand in the city, at a time of expanding transportation and communication networks, rising cost of living, and anticipation of wartime food shortages. FTT eventually grew to include twenty-eight participating cities; and while the plan was from its conception a nationwide initiative, the programs of each participating city differed slightly in operation with control ceded to local postmasters.

For this new program to gain traction, its success would rely on shifts in the management of spatial proximities that were at the center of postal strategies—a shift toward a unified administration across cultural and geographical difference. Simple bureaucratic tools such as standardized forms for requesting goods and the standardized containers used in their shipment were the ambassadors of exchange within this new postal regime. The range of perishable goods that could be sent by mail was impressive, including: apples, asparagus, beans, butter, corn, dairy products, eggs, fish, flour, fowl, fruits, honey, lettuce, meat products, nuts, oysters, potatoes, sugar, vegetables, and yams, among others. These goods were subject to new standards of consistency and uniformity, as the postmaster took on the dual role of product list distributor as well as de facto agricultural inspector. And while the proliferation of diagrams showing proper packaging techniques of such diverse foodstuffs underscores the U.S. Post Office Department's concern with efficiency and hygiene, the number of complaint letters in the archives serves as a measure of their troubled adoption.[4]

Seen within the history of the American postal system, the FTT program was reliant on two relatively recent innovations that greatly extended the administrative and geographical reach of the U.S. Post Office Department: Rural Free Delivery (RFD) and parcel post service. By the end of the 19th century, City Free Delivery—the practice of delivering mail directly to each urban residence—had not yet been extended to rural citizens. And while some vocal urban constituents actively opposed the idea of Rural Free Delivery as an expensive subsidy, farmers organized under the National Grange associations and successfully petitioned their right to have "everyman's mail to everyman's door."[5] Beginning in 1896 and extending over the next five years, Rural Free Delivery brought daily letter service to individual, isolated farms. This was no small benefit to those who previously had to travel into town for their mail and could only do so intermittently

1. David M. Henkin, *The Postal Age: The Emergence of Modern Communications in Nineteenth-Century America.* (Chicago: The University of Chicago Press, 2006). See also: Bernhard Siegert, *Relays: Literature as an Epoch of the Postal System*, trans. Kevin Repp. (Stanford: Stanford UP, 1999).

2. "Butter from the Post Office," *The Independent*, vol. 82, (April 19, 1915), p. 112-13.

3. "The Farm Products Post," *Literary Journal*, vol. 48, (March 28, 1914), p. 685.

4. *Correspondence Relating to the Shipment of Farm Produce by Postal Motor Truck Service*, 1919-1929. National Archives and Records Administration (NARA).

5. Wayne E. Fuller, *RFD: The Changing Face of Rural America*, (Bloomington: Indiana University Press, 1964), p. 53.

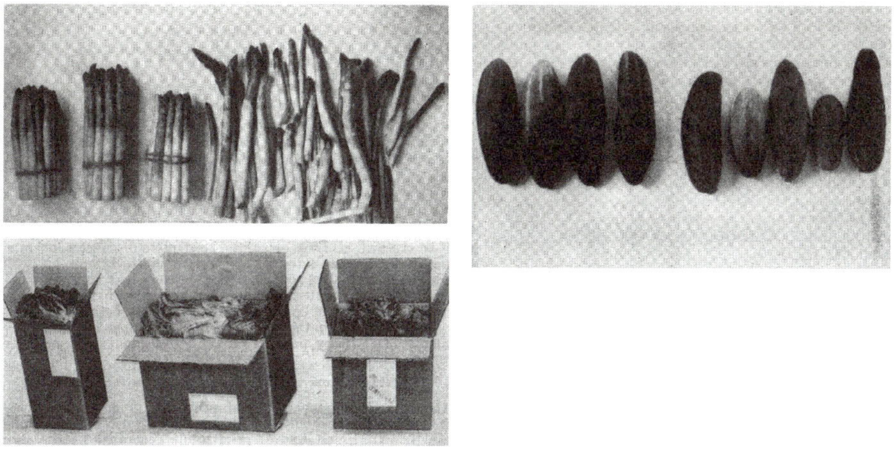

^ **Asparagus, cucumbers, lettuce. Suggestions for parcel post marketing. USDA Farmer's Bulletin No. 703, 1916.**

when roads and weather permitted. A daily visit from the postman brought farmers not only letters from loved ones but also newspapers with weather forecasts and daily market prices. The niceties brought about by this service were reliant upon an adequate infrastructure network, and during these years the extension of postal services and road improvements developed in mutually reaffirming cycles. The success of this project was enough to persuade President McKinley of its merits, as he addressed the Congress in 1900: "[Rural Free Delivery service] ameliorates the isolation of rural life, conduces to good roads and quickens and extends the dissemination of general information."[6] Such big-tent rhetoric is representative of much postal-political discourse at the time, as mail delivery, infrastructure rollout, and progressive-era reforms were framed as promoting the social and geographical concerns of a heterogeneous constituency.

Where Rural Free Delivery brought the furthest geographical reaches into the postal fold, parcel post delivery service extended this universalizing model by opening the mails to increasingly large and complex transactions. Adopted as late as 1913, parcel post was, from its inception, explicitly championed as a means for farmers to send their produce through the mail, and nearly all stories that circulate regarding the first package to be sent include foodstuffs of some sort. Prior to its adoption, items weighing more than four pounds had to be sent by one of the private express companies whose rates were unregulated and often prohibitively high. Rather than foster the flow of goods from country to city, however, parcel post service ultimately encouraged a reverse flow through the widespread and successful adoption of another marketing concept: the mail-order catalog. Montgomery Ward and Sears, Roebuck & Co. extended their domestic dominance as this service invigorated advertising and production of affordable manufactured goods while transforming mail service from a channel dominated by news, business, and personal correspondence, to one literally choking with material goods.

With these innovations newly in place, America's entry into World War I set the stage for a particularly patriotic drama involving farmers and consumers as the FTT program took on a pedagogical role in a broader mobilization of materiel for the wartime effort. In a precursor to Roosevelt's Victory Gardens during World War II, President Wilson issued a patriotic appeal for farmers to increase their food production, warning that "Upon the Farmers of the country, in a large measure, rests the fate of the war and the fate of the nations."[7] Following Wilson's lead, the postmaster of St. Louis, Colin M. Delph, called for the "harmonious" linking of producer and consumer through the FTT plan, calling upon each farmer to become "an agricultural soldier" with a plea towards "making heroes on the Farms as well as on the battlefield."[8] Wilson and Delph had reason to sound the rallying cry in the countryside. American contributions to the Great War, even prior to the direct mobilization of troops, consisted to a large extent in the successful supply and transport of subsistence goods to the Allied front lines. Provisioning, distribution, and warehousing of wartime foodstuffs all contributed to the logistical suite that became known as the "service of supply,"[9] which lent imaginative and calculable argument to the planned expansion of the FTT program after the War.

6. Ibid, p. 60.

7. *C.M. Delph to the Farmer and Producer of Fruits, Vegetables, Dairy Products and Other Foodstuffs*, July 1, 1917, NARA.

8. Ibid.

9. Congressional Record. 65th Congress, 3rd sess. 1918, December 18, p. 636.

With this successful mobilization came the dual challenge of how best to reframe wartime protocols to address a peacetime, domestic audience and the related organizational question of which mechanisms—fiscal, political, and infrastructural—were required to carryout this transition. One argument focused on the social cause of returning veterans, as an army of returning troops (and their fleet of returning motor vehicles) was imagined as a ready-made workforce to bolster new FTT efforts to provide a dedicated trucking fleet for the delivery of agricultural products. A second, related argument foresaw the predominantly urban in-migration that the War would only accelerate, and accordingly promoted the FTT plan as a potential counteragent against the disassociation and alienation of the new metropolis. Toward this end, planners attempted to retool the FTT to link with already existing public buildings and gathering places such as schools and libraries, outlining a collective, agro-postal infrastructure in support of the practices of everyday life. The program began to conceive of these public spaces as the kernel of a neighborhood unit that could double as a centralized distribution center for farm products and serve as an organizational center on the city end of the FTT trajectory. While such experiments eventually proved ill-equipped at tackling food distribution on such a large scale, the far-reaching imperatives of the FTT plan fleetingly held the position of a comprehensive panacea to pressing post-war challenges.[10]

At a time when the Post functioned not only as a letter and parcel delivery service but also as a trusted banking firm for savings, there seemed no reason why it could not also become a low-cost food supplier. But for a nation that would "harness its energies into an invincible fighting machine within a single year,"[11] the ability of the government to link the farmer and the consumer proved limited. By placing the Department between individual transactions, FTT created cumbersome accounting for farmers and postmasters, while a 1916 economic study cites the high cost of packaging relative to contents, inconsistent quality and supply, lack of a pricing structure, enduring consumer habits, and the difficulty of maintaining good customer relations among the many challenges to direct marketing through the Post. Numerous attempts to improve the process arose, yet the logistics of handling, shipping, and purchasing small lots proved too great an obstacle, and the nation's system of food distribution tended toward the trucking of commodities on a large scale. This would help to fuel the rise of new self-service grocery store chains to replace smaller, localized middlemen. Meanwhile, postal trucks broke down, goods spoiled in transit, consumers' expectations were unmet, and public enthusiasm waned. Many of these factors led Congress to cut funding for the program in July 1920, but at least one historian of the postal system attributes the collapse in part to a pervasive "specter of Bolshevism" and the tendency for opponents to equate the FTT program with an overly active central government. Struggling farms caused a resurgence of interest in the marketing program into the 1920s and 30s, but these attempts also either failed or never took off.

In piggybacking on existing infrastructures and daily cultural practices, the ambition of the FTT program was to imagine-away all intermediary frictions within the farm-to-table vector, positing a seamless topological surface between town and country. The audacity of this remains astonishing particularly when, in the present, any perceived encroachment of government into private industry continues to be highly contentious. Supporting this desired continuity were two sometimes contradictory notions of infrastructure—one that prioritized the spatial logistics of delivery and another that remained attuned to the informational content and signal-to-noise ratio of the postal channels. Efforts to ensure the legibility and connectivity of so many diffuse exchanges constituted part of a larger idea of linking heterogenous publics within the national landscape. By leveraging expanding notions of statecraft and adjusting the micro-processes of these infrastructures, the hopes and failures of the FTT plan highlight the ways in which the leveling effects of agricultural and postal protocols met the tensions and disruptions of the terrain, both profoundly shaped by and giving rise to the manifold of American spaces, natures, and socialities.

>>> Szu-Han Ho is an artist whose work and research looks at economic language and public sentiment.

>>> Jesse Vogler's work and teaching investigates the territorial and geographic ambitions of architecture.

10. Fuller, *RFD*, 228-258.

11. Congressional Record. 65th Congress, 3rd sess. p. 634-36.

factory-farmed architecture: you are how you eat

edward dodington

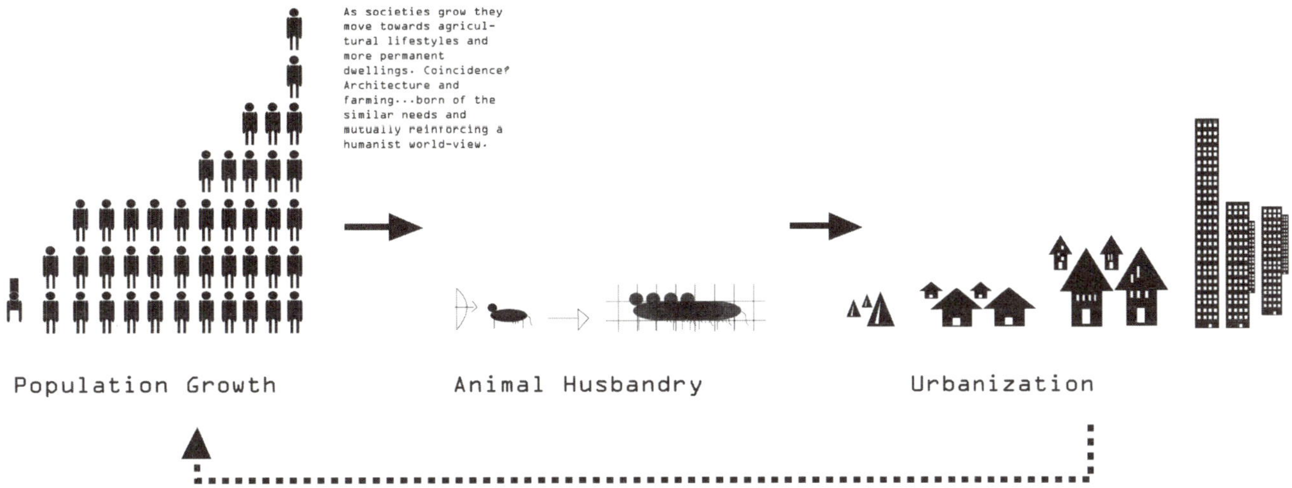

^ **As populations increase, civilizations move towards agriculture. Agriculture reinforces a stationary living style and the development of static architecture.**

We build the way we farm. We factory farm our food and we factory produce our homes. The way we build reflects the way we view ourselves in relationship to the beings that we farm and consume. We build walls, delineate one species' space from another's, give preference to some domesticated species (such as canine and feline) while rejecting the majority of others. We house ourselves in private homes down long cul-de-sacs and keep to ourselves. Currently, farming and living appear to mutually reinforce each other. How would a change in one practice affect a corresponding change in the other?

The idea of living well and eating well first appears in western text in 1826 from the hand of Anthelme Brillat-Savarin who wrote in the *Physiology du Gout, ou Meditations de Gastronomie Transccendante* "Dis-moi ce que tu manges, je te dirai ce que tu es." (Tell me what you eat and I will tell you what you are.) The phrase was later picked up by Ludwig Andreas Feurbach in 1863 who wrote "Der Mench ist was er ißt." (Man is what he eats.)[1] "You are what you eat" seems to have reached public consciousness throughout the 1940s, 50s, and 60s as the macrobiotic and organic foods movement began to gather popular exposure. Today, with the successful arrival of businesses such as Whole Foods Market, and increasing numbers of farmer's markets in major cities, it remains in the public consciousness. There is little doubt that macrobiotic or free-range foods improve our health, but what about our living habits?

In 2007, the United States slaughtered eight billion animals. Currently a mere four companies in the U.S. control the production of 81 percent of U.S. cows, 73 percent of sheep, 57 percent of pigs and 50 percent of chickens.[2] Almost 80 million pigs (out of a total 95 million) are killed each year on factory farms and according to the Worldwatch Institute, 74 percent of the world's poultry, 43 percent of beef , and 68 percent of eggs are produced through an automated, factory-farmed process.[3] In addition to these staggering death figures, the concentrated amount of animal waste generated by factory farming threatens local ecologies and pollutes groundwater, lakes, and rivers. However, despite these concerns and the growing movement in alternative foods, factory farming continues to be the dominant method of food production in this country and around the world.

Factory farming provides a clairvoyant portrayal of who we are. It not only reflects the ways we live, but underlines the way we live with others—a cultural position in relation to the beings we farm. Factory farming follows a Fordist model; efficient, mechanized, and driven by the bottom line. Similarly, our houses are driven by market

1. Information accessed at: http://www.phrases.org.uk/meanings/you%20are%20what%20you%20eat.html

2. Testimony by Leland Swenson, president of the U.S. National Farmers' Union, before the House Judiciary Committee, September 12, 2000.

3. Shen, Fern. "Md. Hog Farm Causing Quite a Stink," *The Washington Post*, May 23, 1999; and Plain, Ronald L. "Trends in U.S. Swine Industry," U.S. Meat Export Federation Conference, September 24, 1997, cited in Scully, Matthew. Dominion, St. Martin's Griffin, p. 29.

and real estate forces. Housing developments are also often mechanized, isolated, and rapidly reproduced. The relationship between our current methods of farming and living is not a chance occurrence.

The historical links between agriculture and architecture is long and as old as society. As societies grew they outstripped available herding and foraging resources. The local human population was then left with the choice to relocate or to control and manage the food resources around them. Farming is only a viable method of food production for stationary civilizations. In almost every early civilization, animal pens and human shelters arrive at the same time—agriculture and architecture coincide and progress in tandem.

The burgeoning field of tissue engineering, initially pioneered for the medical industry, is finding an audience in factory farming and has the potential to redirect the agriculture/architecture pair. Briefly, tissue engineering operates by seeding living cells harvested from a body into a biopolymer substrate. Over time, the cells grow to consume the substrate and create a single tissue that is the product of both the underlying substrate and the seeded cells.[4] Tissue engineering is simultaneously frightening and promising, and requires careful consideration. Tissue engineering offers a potential for reinvention; for redefining our roles vis-a-vis animals, and particularly the animals we consume. On the one hand, tissue engineering offers a science for growing and consuming meat without the environmental and economic downfalls of factory farming. It also promises victimless meat—meat without killing a single animal. On the other hand, it is another form, if not an extreme case, of factory farming. Henry Ford dreamed of a world where industrial products were made-to-order, where there was no surplus storage and the factory produced only what was immediately needed. Tissue engineering is Ford's dream applied to farming—meat on demand.

While some might take issue with its claim of victimlessness, tissue-engineered meat is not at the scale of current victimization of animals, but, regardless of its ethical position, the greatest promise of tissue-engineered meat is its unique position to other animals, its semi-livingness. The semi-living quality of tissue-engineered meat offers a new way of considering what we eat, how we eat, and the way we live among other animals. This task of reinvention, and reengagement is, as Jacques Derrida and Donna Harraway suggest, a question of how to enter into a conversation with the animal.[5] Tissue engineering, by virtue of its conflation of the living and nonliving, initiates this dialog. The task then given to us is to extend this conversation into our daily practices. To this end, all species-based distinctions should to be erased. Architecture should become reactive to animal and environmental actions—animals should be considered as design factors establishing a dialog between architecture and wildlife.

PolySpecies Park

PolySpecies Park is a proposal for the research and production of tissue-engineered meat in Smithfield, Virginia on the site of what is now the country's largest producer of pork products, Smithfield Foods. PolySpecies Park is equal parts farm, tissue-engineering factory, eco-resort, and research facility. The Park is both a producer of new types of engineered meat and a facility for studying dietary aspects of the new foods. It demonstrates the new dialogue possible between living and non-living systems. Each facility, while ostensibly rendering current factory farming obsolete, will need to keep a modest population of animals on site, not only for tissue engineering purposes (especially cell harvesting), but also for the study of meat texture and quality.

The Park is composed of roughly two dozen facilities of four types, each catering to different species of animal and type of engineered meat. These animals are organized according to their method of interaction with larger environmental strata—terrestrial animals (hoofed and herding animals); avian animals (things with wings); aquatic animals (anything that lives most comfortably in the water); and subterrestrial animals and microbes (animals in soil). Each facility utilizes a range of species-specific tools to lure, control, and study each specific animal. There is potential for cross-fertilization though production will be organized in groups. Each facility also contains the infrastructure needed for tourism, research, and on-site living quarters.

By tapping into the individual sensorial ranges—audible, olfactory, or haptic—of each species, the Park redefines all of the architecture, infrastructure, landscape and individual organisms as living players in a much larger system. Architecture becomes attractor, program container, and animal/architecture interface. Animals influence the siting and design of individual buildings. Each building helps to manage and attract a species of animal, while also being subject to larger migratory and environmental patterns. But, more than providing a production facility for new types of engineered meats, the Park offers the visitor an experience of a dynamic, non-heirarchical species environment.

Animals, humans included, are somewhat predictable. Every species has behavioral patterns, and each pattern has spatial implications. These are witnessed, for example, in fish schooling, bird flocking, mammal herding patterns, and animal territorial demarcation , among others. Each facility, therefore, enters into these species-specific patterns more closely.

To do this, it is necessary to understand the basic inputs and outputs of each species. The most easily influenced animal factors are food, sex, and shelter. The architecture, therefore, becomes one of sensory emission and soft control rather than hard-lined establishments. The position of each building, and the design of architecture is

4. In-vitro Cultured Meat Production, P.D. Edelman, M.SC., in *Tissue Engineering*, Volume 11, Number 5/6, 2005.

5. "My point is simple: once again we are in a know of species coshaping one another in layers of reciprocating complexity all the way down. Response and respect are possible only in those knots, with actual animals and people looking back at each other, sticky with all their muddled histories." Harraway, *When Species Meet*, 42. "To have a plural of animals heard in the singular." Derrida, "And say the animal responded," 39.

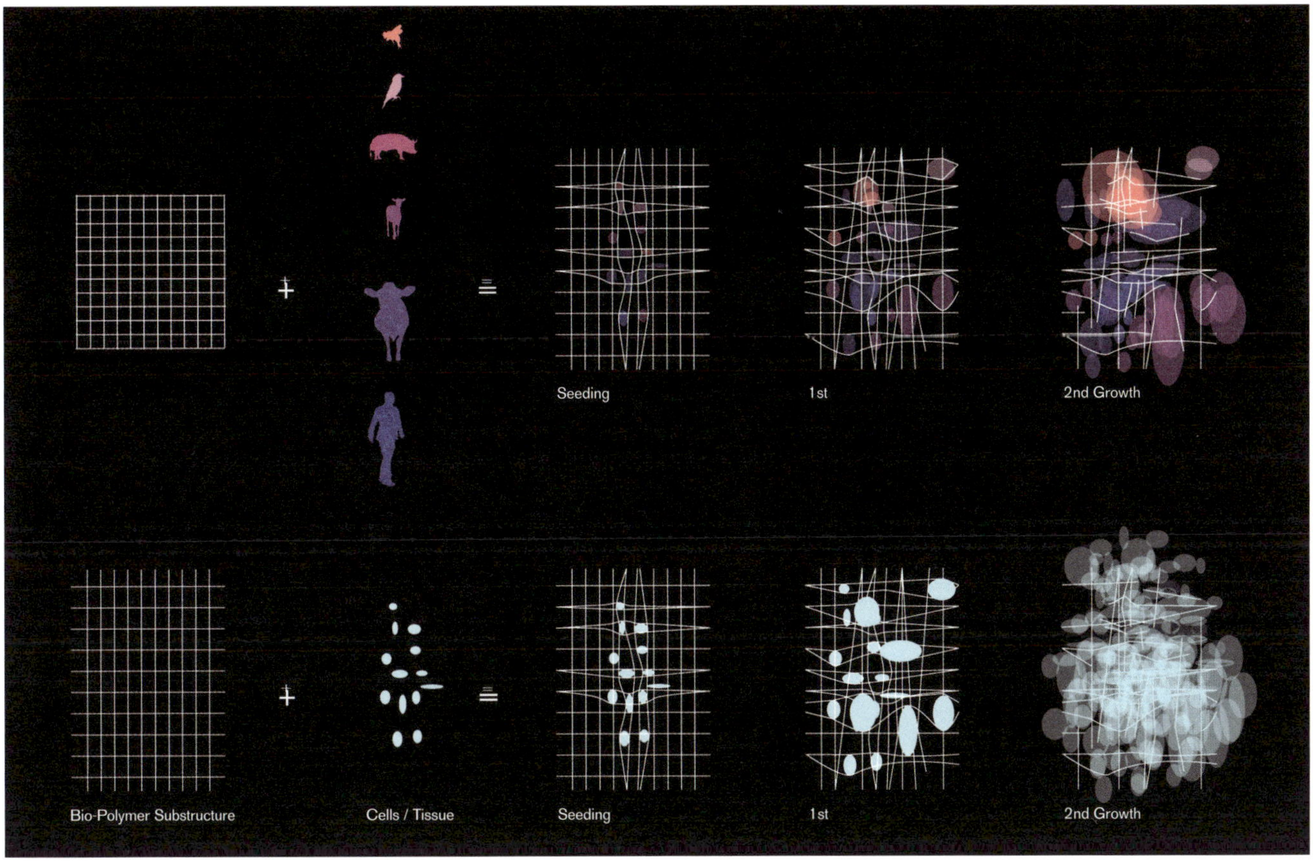

^ **Scaffold tissue engineering diagram (bottom) compared to an extrapolated and expanded argument for PolySpecies Park.**

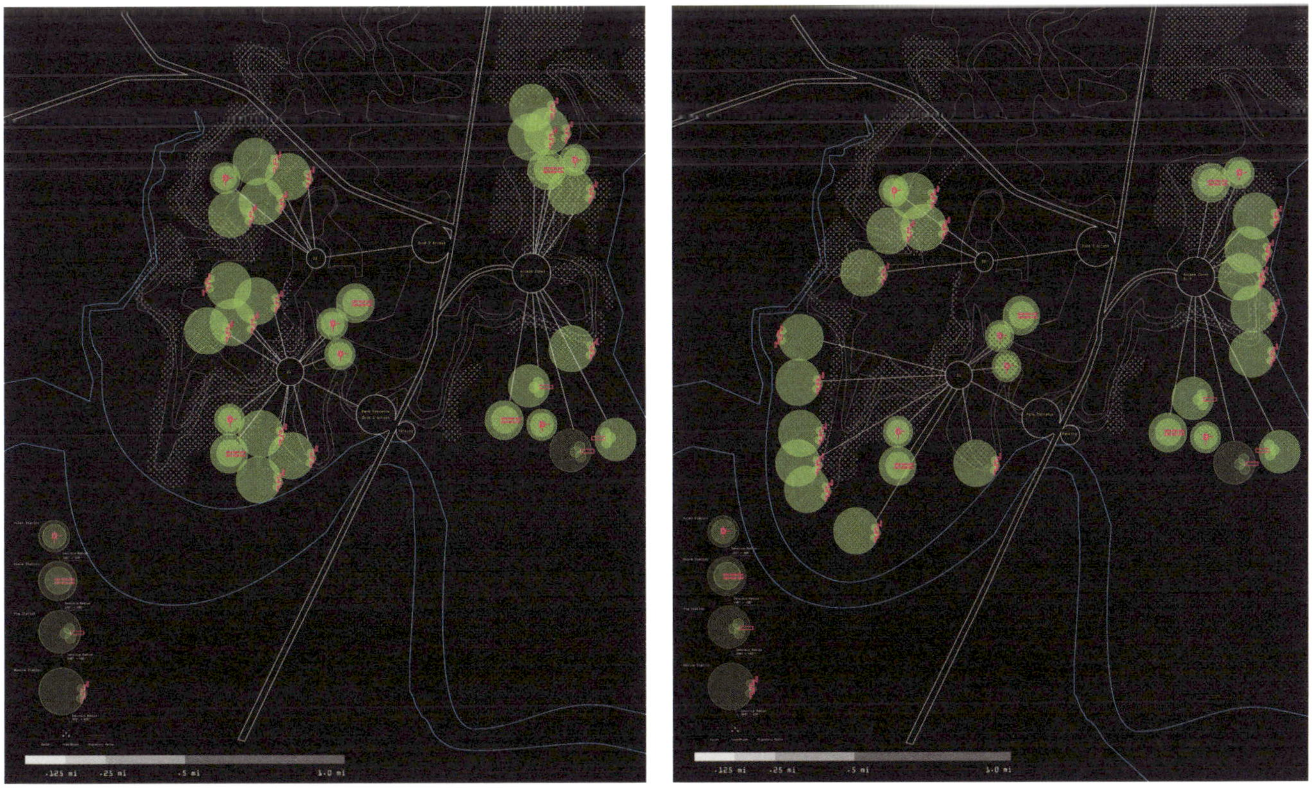

^ **Two site plans for PolySpecies Park reflect seasonal changes in facility siting based on average rainfall.**

^ **Four types of meat-engineering stations are organized throughout PolySpecies Park.**

attentive to the specific ethological needs and patterns of their target species.

Each tissue-engineering meat facility participates in a lager site system, allowing for crop rotation through seasonal migration. The site is diagrammed and zoned to accommodate a wide array of animals and their behaviors. Each circle represents an acre. As different species are taken into consideration, the grid becomes differentiated and zoned for more particular species. Just as the substrate in tissue engineering is manipulated to become more responsive to biological life, so too does the overall site plan respond to the various needs and particularities of agricultural livestock.

PolySpecies Park offers an example of animal-responsive design—for extending a polyspecies alternative into our human-centric environment. Harraway and Derrida have proposed to engage all species on an equal plane of respect and, as designers, it is now our challenge to enact a new environments and ways of living that reflects their missive. Hopefully, it will be a challenge we will meet and expand upon.

>>> **Edward Dodington is a Houston-based artist, designer and entrepreneur.**

> **Following:**

1. Terrestrial Station (page 34). The terrestrial factory is designed to participate in herding and animal migration patterns.

2. Aquatic Tissue Station (page 35). These facilities operate within the sonic and olfactory senses.

3. Sub Terrestrial Farming Station (page 36). This factory specializes in seismic and subsonic communication.

4. Aerial Tissue Production Facility (page 37). Here sonic and tactile qualities can be manipulated to effect avian behavior.

bracket—

Terrestrial Animals

The relative position of the mobile units to each other is dependent upon the type, path and number of terrestrial animals desired for meat production or study.

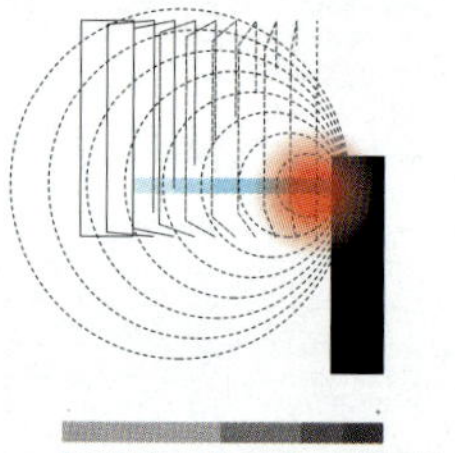

Animal Characteristics

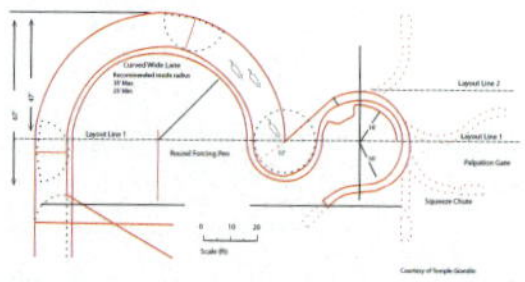

Design Elements

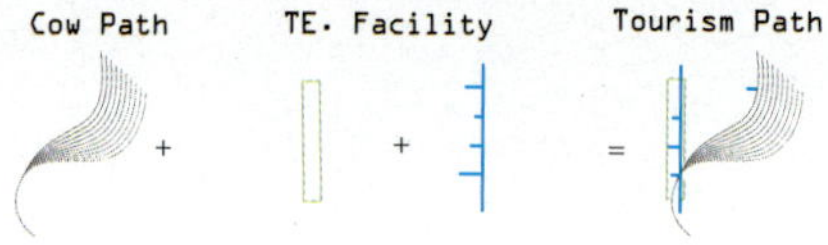

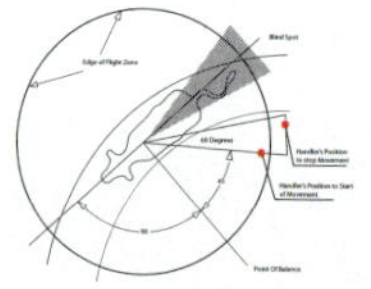

The System

In a biased field, intensities of sound smell and texture can be engineered and manipulated to influence terrestrial animal movement and behavior.

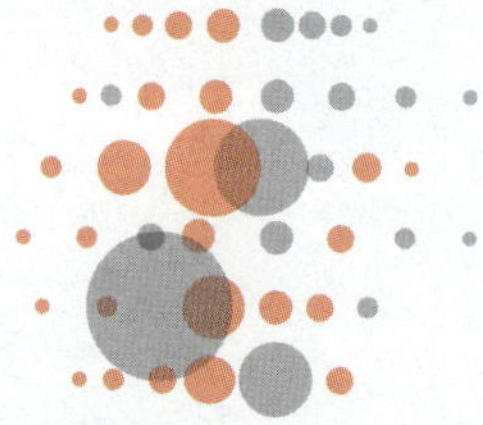

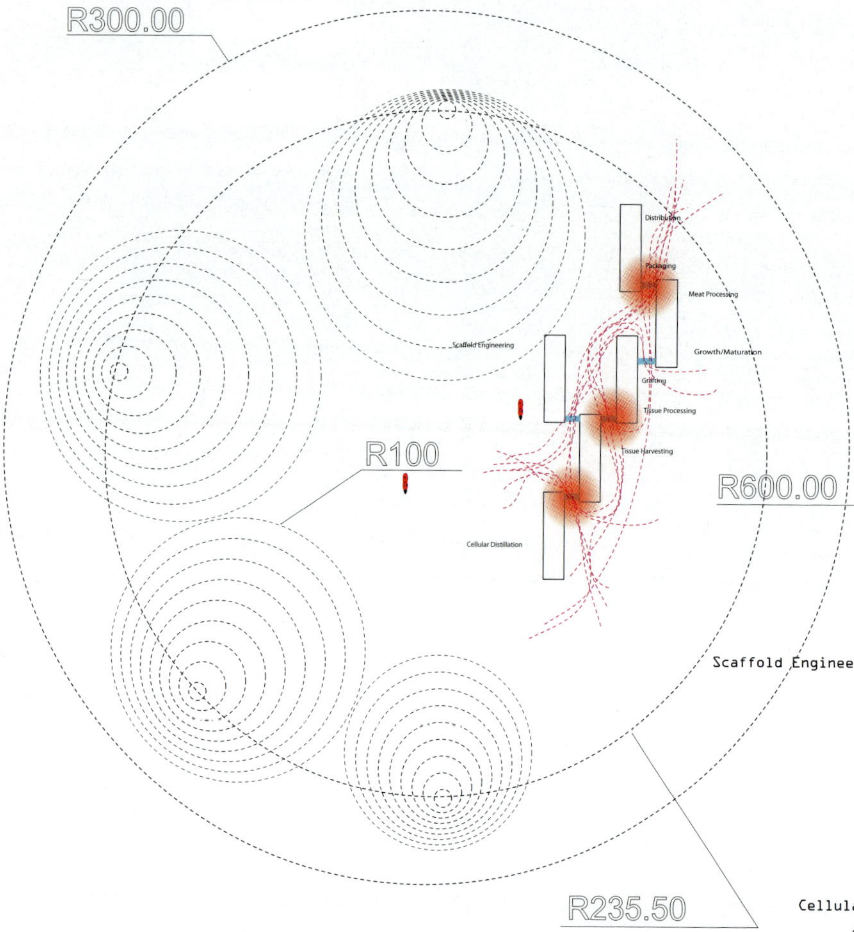

Plan

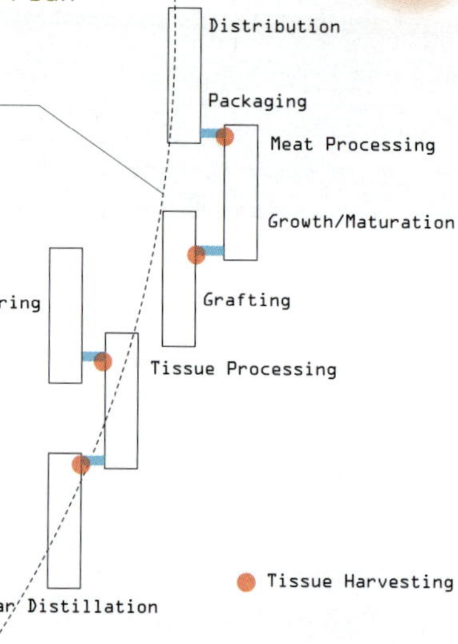

Potential perspective

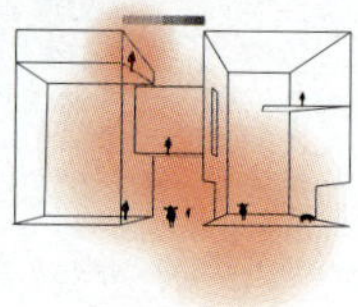

Since each factory is tuned to a specific animal they each can be arranged to accommodate a variety of terrestrial, herding animals.

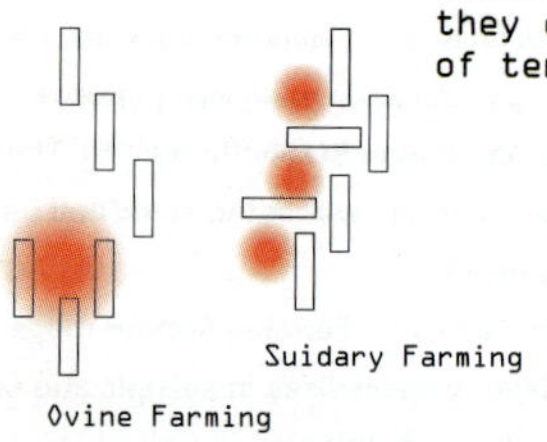

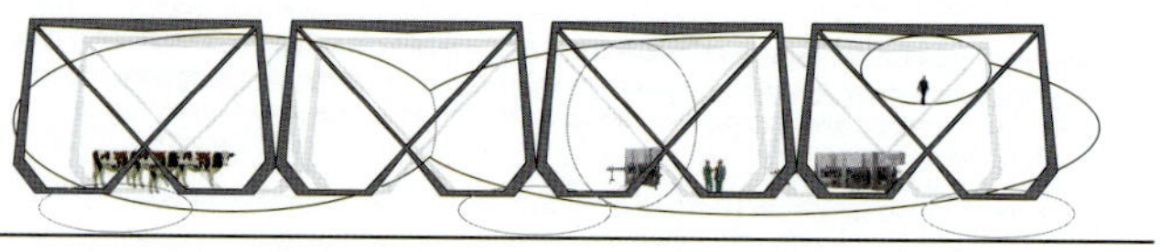

Aquatic Animals

The aquatic factory acts as a section through the various strata of the water, interacting with different species at different depths.

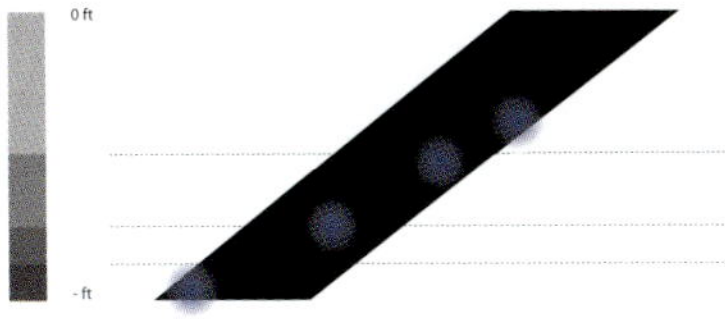

Animal Characteristics

Quantity/acre	50 - 200
Scent Radius	300 - 400'
Audible Range(hrz)	10 hz - 8 kHz
Audible Limit (db)	95
Visual Concerns	Sudden movements

Design Elements

Feeding and Herding Behavior

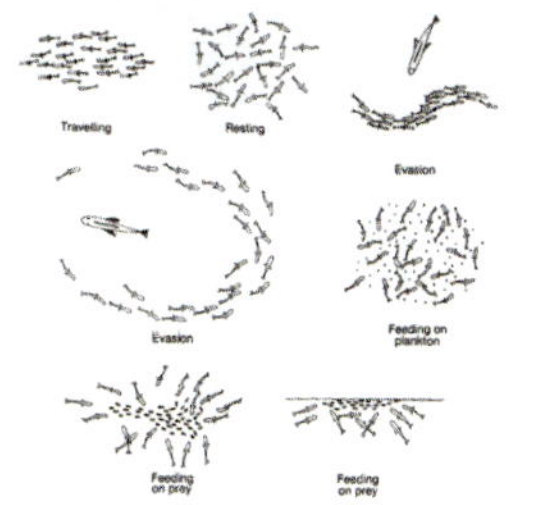

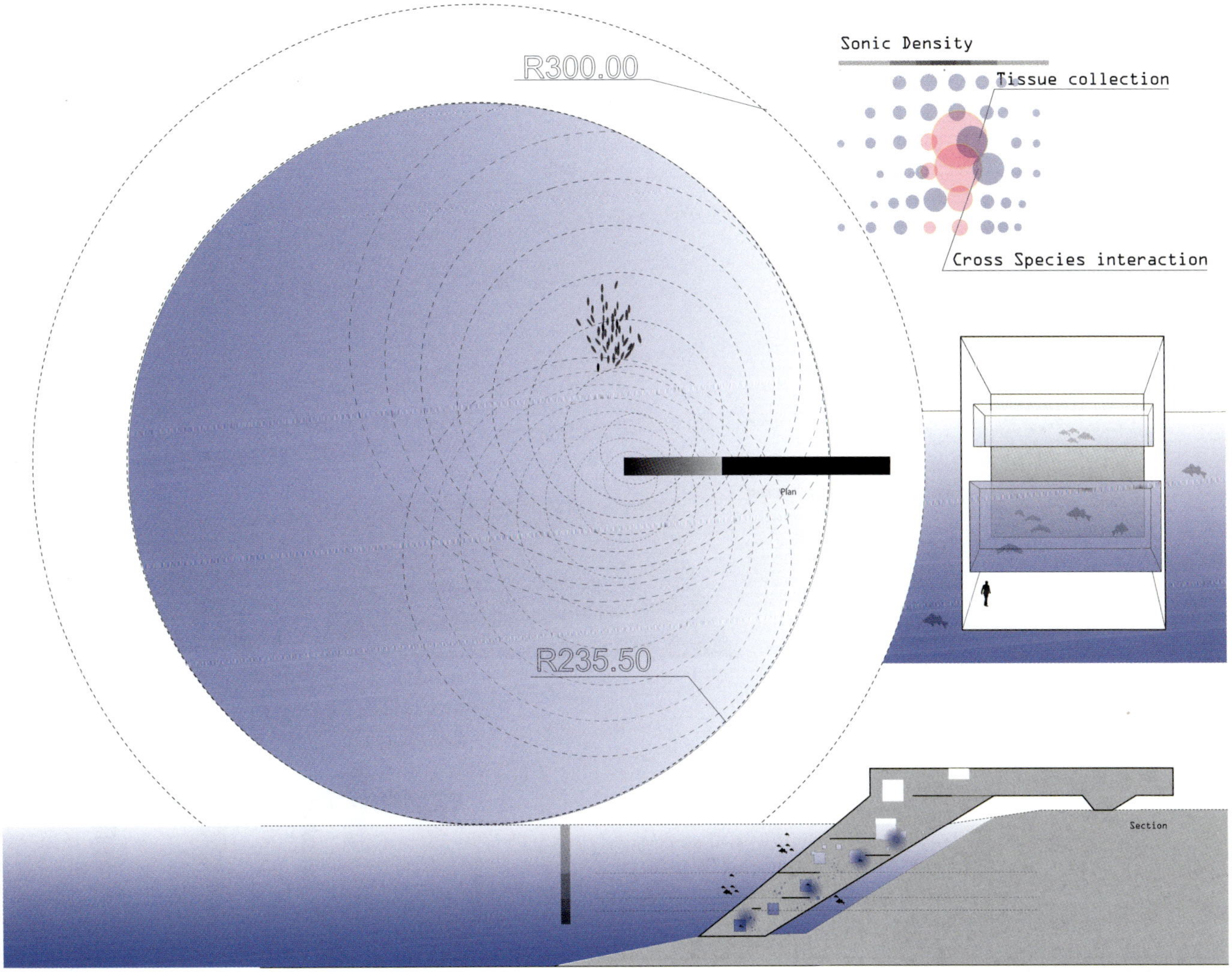

Chemoreception (Smell and Taste) is very well developed in the fishes, especially the sharks and eels which rely upon this to detect their prey. Fish have two nostrils on each side of their head, and there is no connection between the nostrils and the throat. The olfactory rosette is the organ that detects the chemicals. The size of the rosette is proportional to the fish's ability to smell. Some fish (such as sharks, rays, eels, and salmon) can detect chemical levels as low as 1 part per billion.

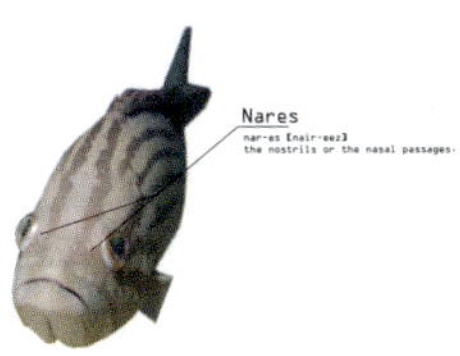

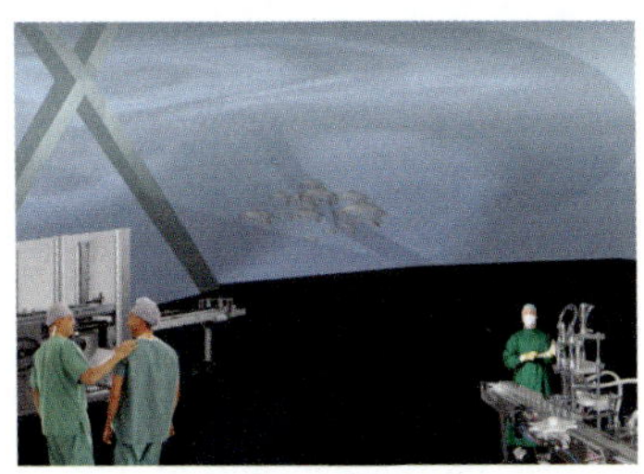

Subterrestrial

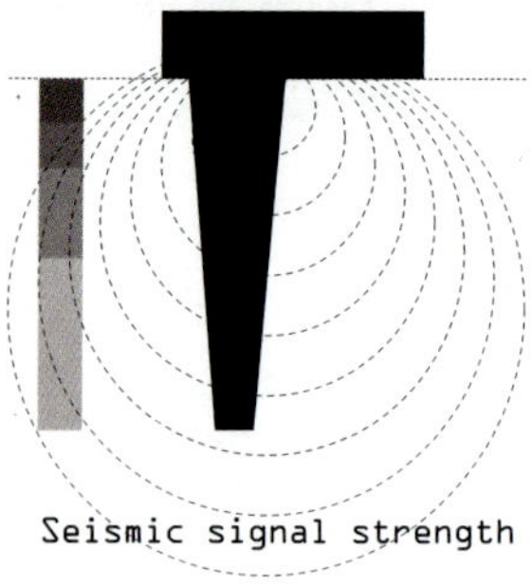

Animal Characteristics

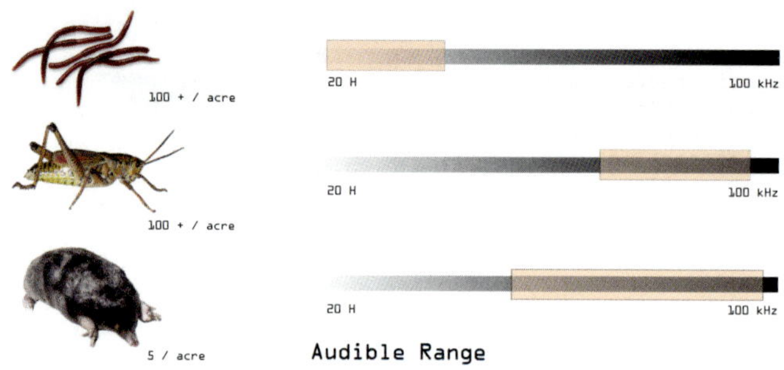

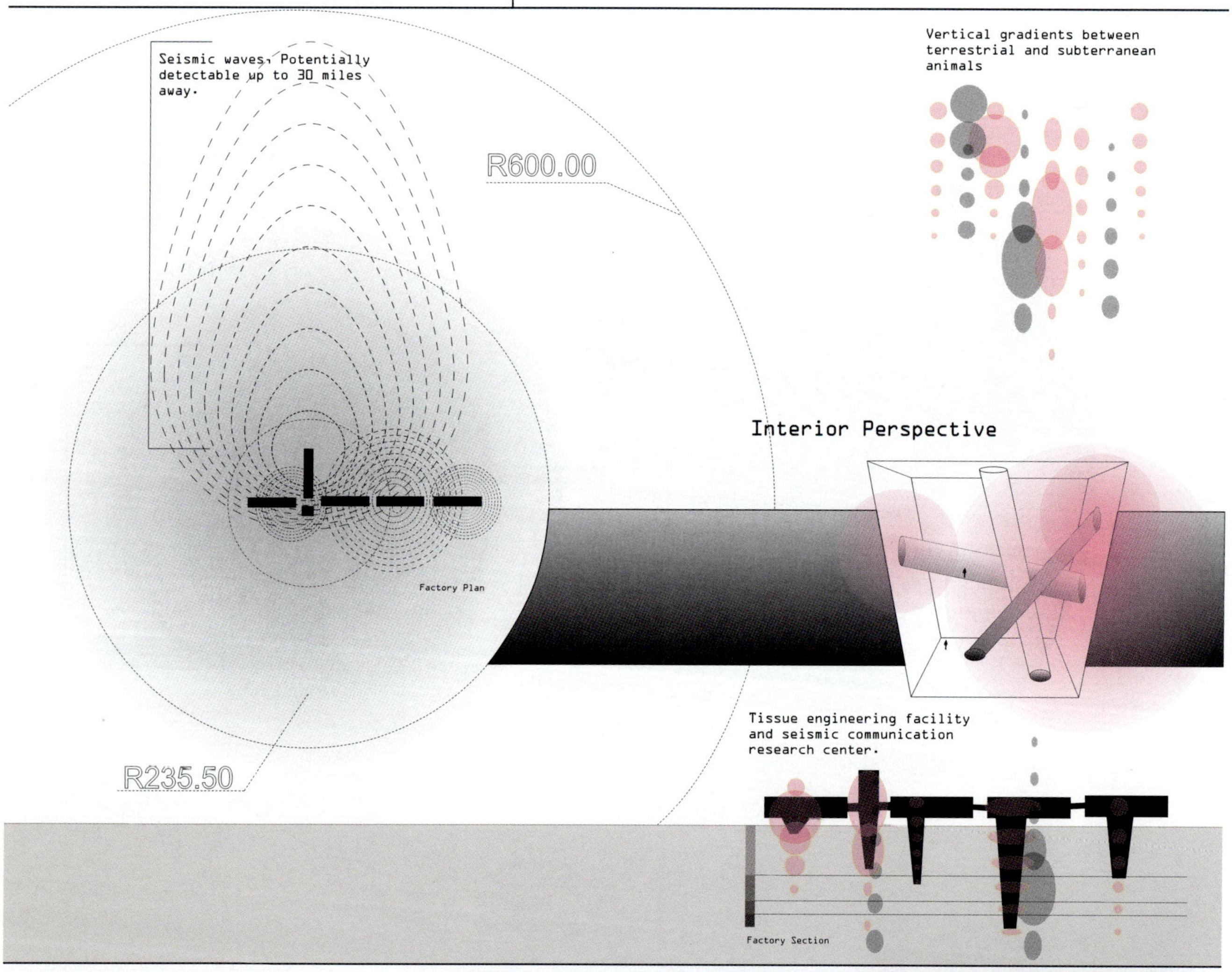

Sound in the ground

Seismic communication, low-frequency and patterned substrate-borne vibrations are normally generated by thumping. These vibrations travel long distances underground, and are important in the nonvisual communication of subterranean animals. This importance pertains both intraspecifically in adaptation and interspecifically in speciation. Neurophysiologic, behavioral, and anatomic findings in this study suggest that the mechanism of long-distance seismic communication is basically somatosensory and is independent of the auditory mechanism. Seismic communication thus appears to be a channel of communication important in the evolution of subterranean mammals that display major adaptation to life underground.

An estimated 200,000 species communicate through seismic vibrations

Aerial Animals

Sound intensities

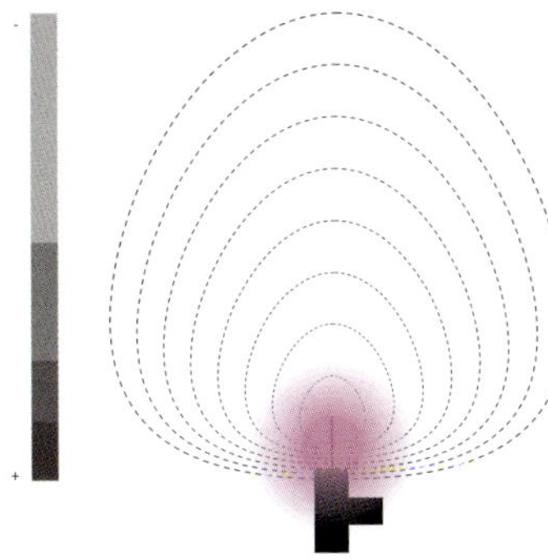

Animal Characteristics

Quantity/acre	50
Scent Radius	
Audible Range(hrz)	125-2,000
Audible Limit (db)	95
Concerns	Flight Zone Canopy

Design Elements

Point Path

Canopy

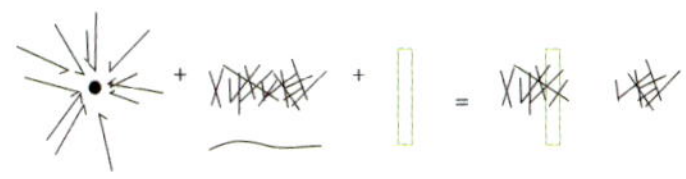

Point Path

Flight Zone

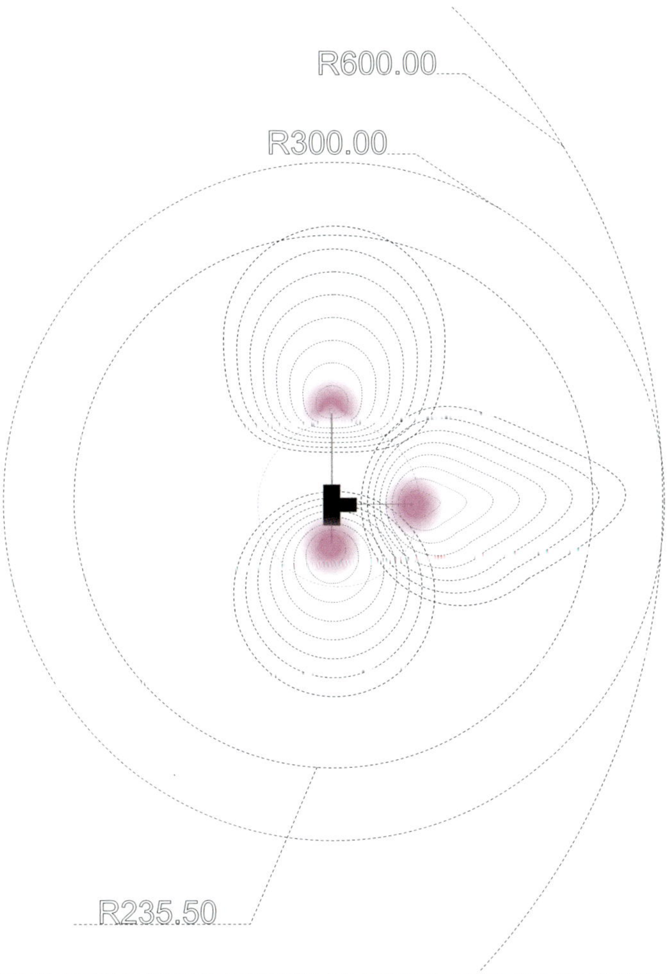

Interior perspective

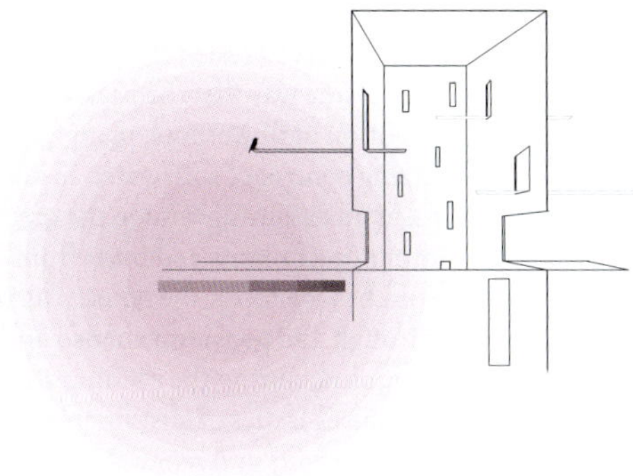

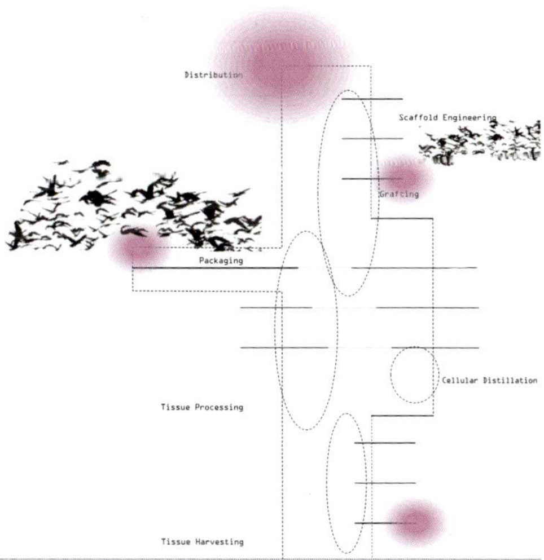

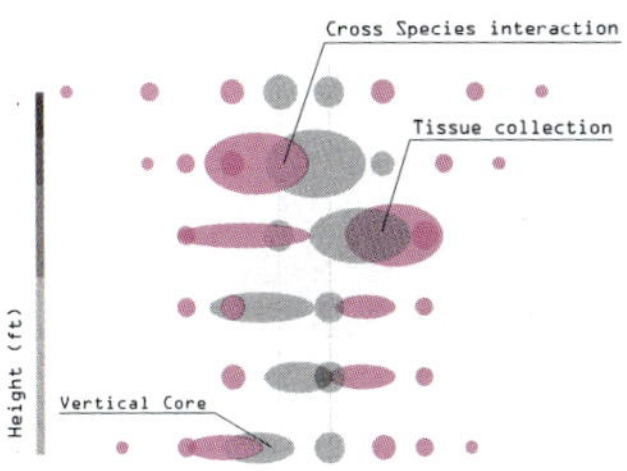

Hearing in many species of birds is comparable with that of mankind, having a greatest sensitivity between 2000 and 4000 hertz. This is partly why bird song is so useful in bird identification - it is easy for us to hear - and partly why we find bird song so pleasant. As a whole, known avian hearing ranges from a lower limit of below 100 hertz to over 29,000, though not all birds have this range. The common Mallard for instance has a range from 300 hz to 8000 hz.

ager-agri

ctrlz architectures (francesco cingolani & massimo lombardi)

"Ager-Agri" is a proposal for a house, about eighty square meters, situated on the top of a hill just south of the city of Ancona, in the Marche region of Italy. The site is located within land zoned for park. Parco del Conero is a nature sanctuary extending for 5,800 hectares of protected land, including areas of great beauty: the bay of Portonovo, the beach of the *due sorelle*, the northern viewpoint, *Pian Grande, pian dei Raggetti*. In addition, there are eighteen hiking routes among strawberry, ilex, and pine trees situated within the unique Mediterranean maquis. The area provides nesting grounds for several species of birds, some very rare, and a rich diversity of fauna. Many of the species within the Mediterranean maquis are protected and represent a significant portion of the Marche. The site's landowner is a young woman fond of nature and animals. She is strictly vegetarian and inspired by health and wellness. The project approaches the concept of 'living within nature' in a different way, linking the inherent qualities of the site with the client's lifestyle.

From the top of the hill, the views of the waterfront and the sea are remarkable, and this direction perfectly fits the best orientation at this latitude (an east-west axis, slightly shifted north), to get sunshine early in the morning and late in the afternoon. Instead of using the hill as a garden and basement, the house is elevated on four pilotis to occupy the site as little as possible. The pilotis serve as shelters for the small mammals and several species of birds living in this area, in the same way that the park shelters the house. The elevated strategy also invites a vertical distribution of flowerpots, plants, and vegetables boxes that are fixed to a steel structure. By optimizing control of soil, water, insulation, and wind, the house performs in a manner similar to a greenhouse. During winter, the greenhouse balconies are covered by plastic around the perimeter to protect plants and buds from frost. This strategy also helps to keep warm air inside, which reaches the top of the house. During summer days, the open green balconies allow wind to pass through and provide a shaded and colorful garden. This also encourages natural ventilation and cooling during the evening and night.

The project raises issues of sustainable development in architecture, though not just by using ecological solutions and renewable energy sources that are applicable everywhere, but by suggesting an entirely different way of living. "Ager-Agri" privileges minimum consumption and waste, but, most of all, exchange in a self-sufficient way that is symbiotic with nature.

>>> CTRLZ Architectures is a collective founded in Paris, formed by two [not so] very confused architects, Francesco Cingolani and Massimo Lombardi.

^ Located on top of a hill within a regional park, the site offers views towards the Italian coast.

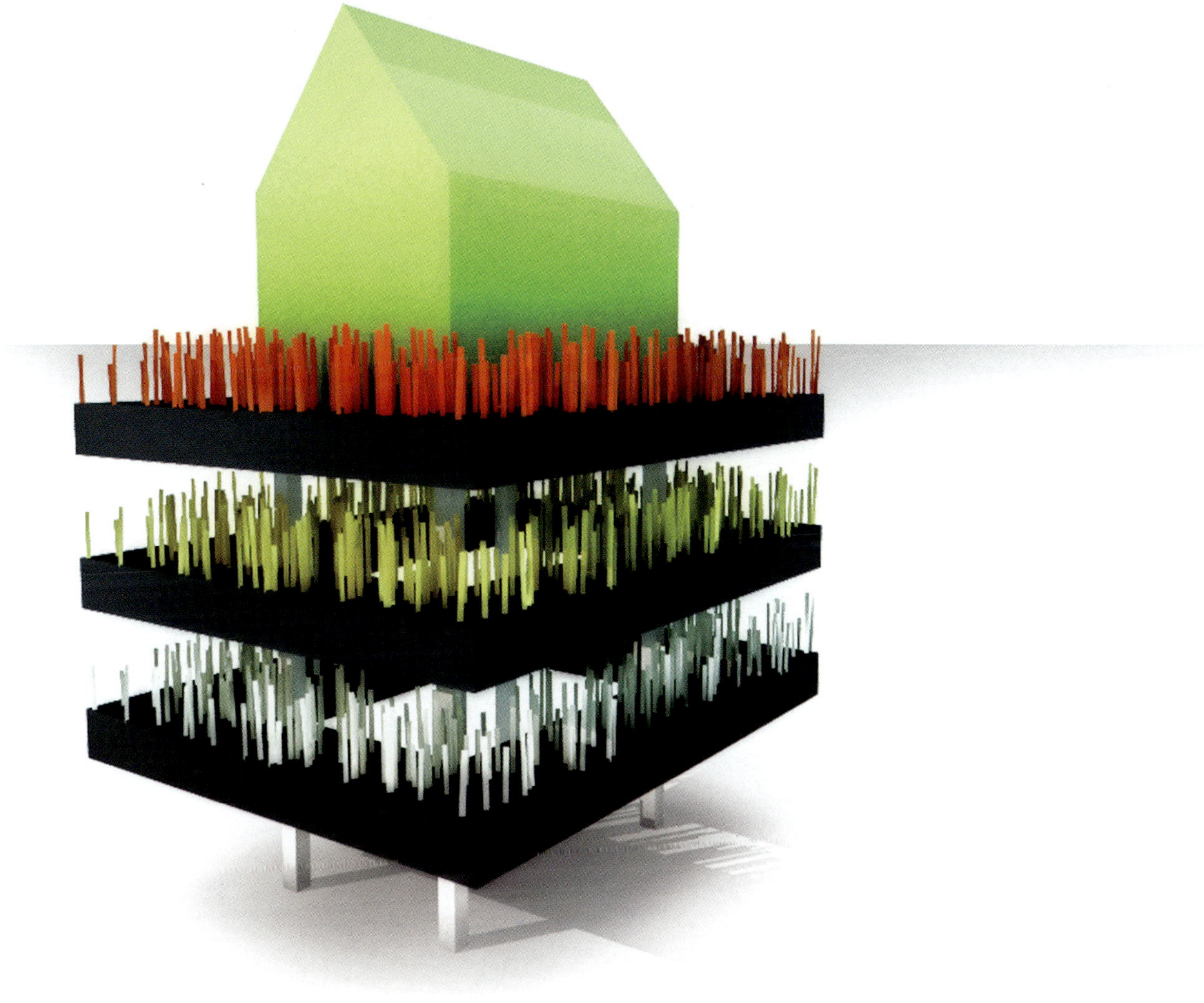

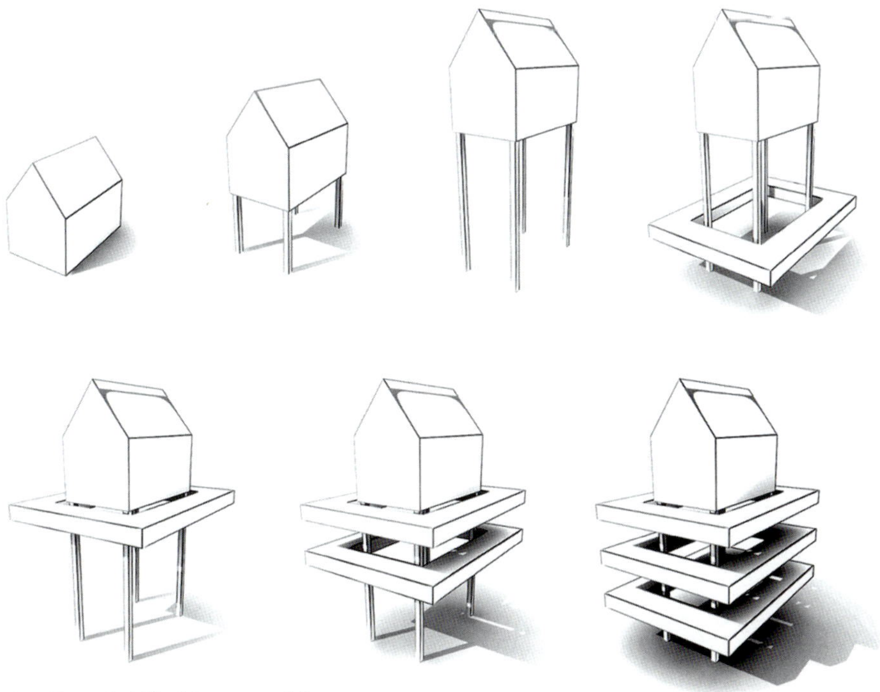

^ **The project lifts living space off the ground and provides balcony rings that serve as gardens.**

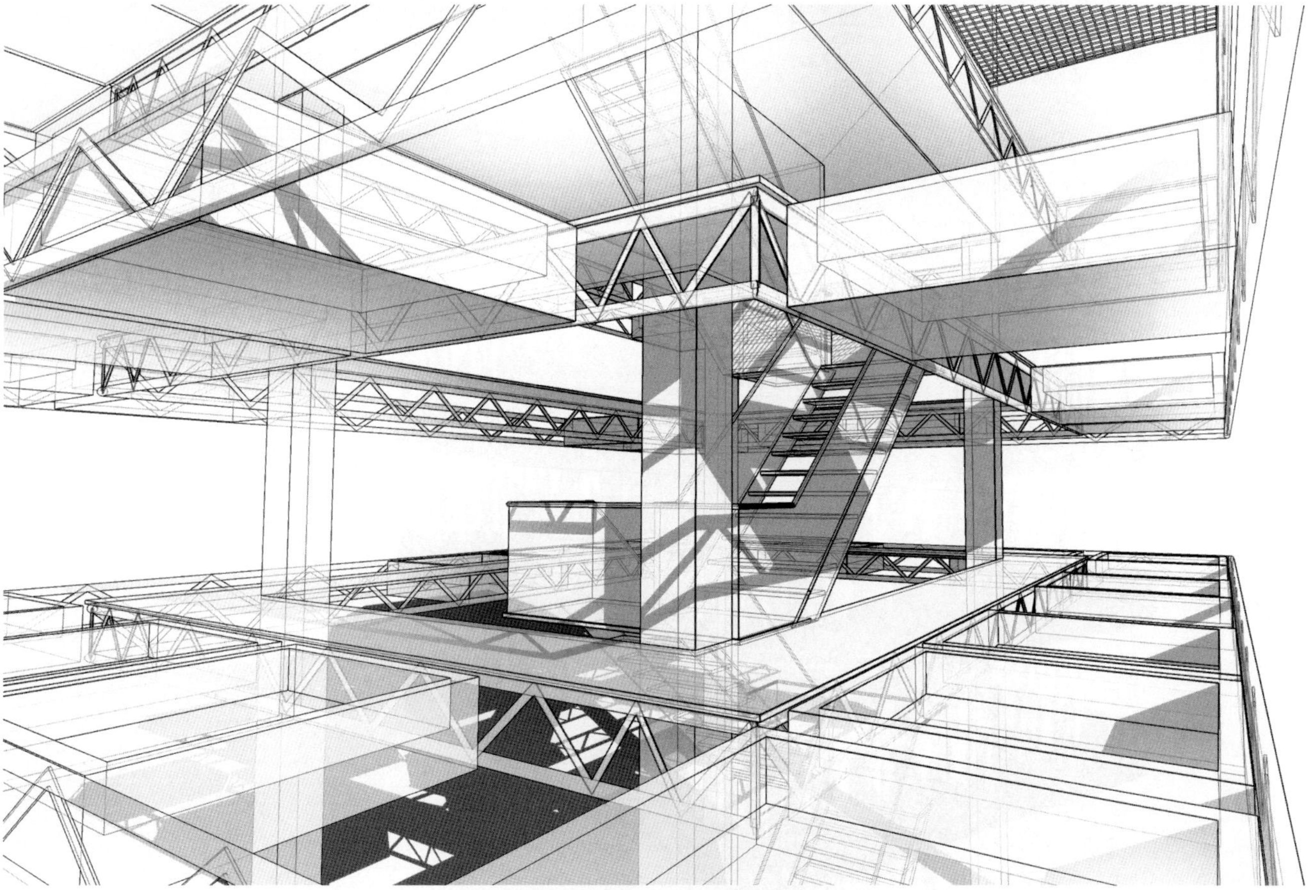

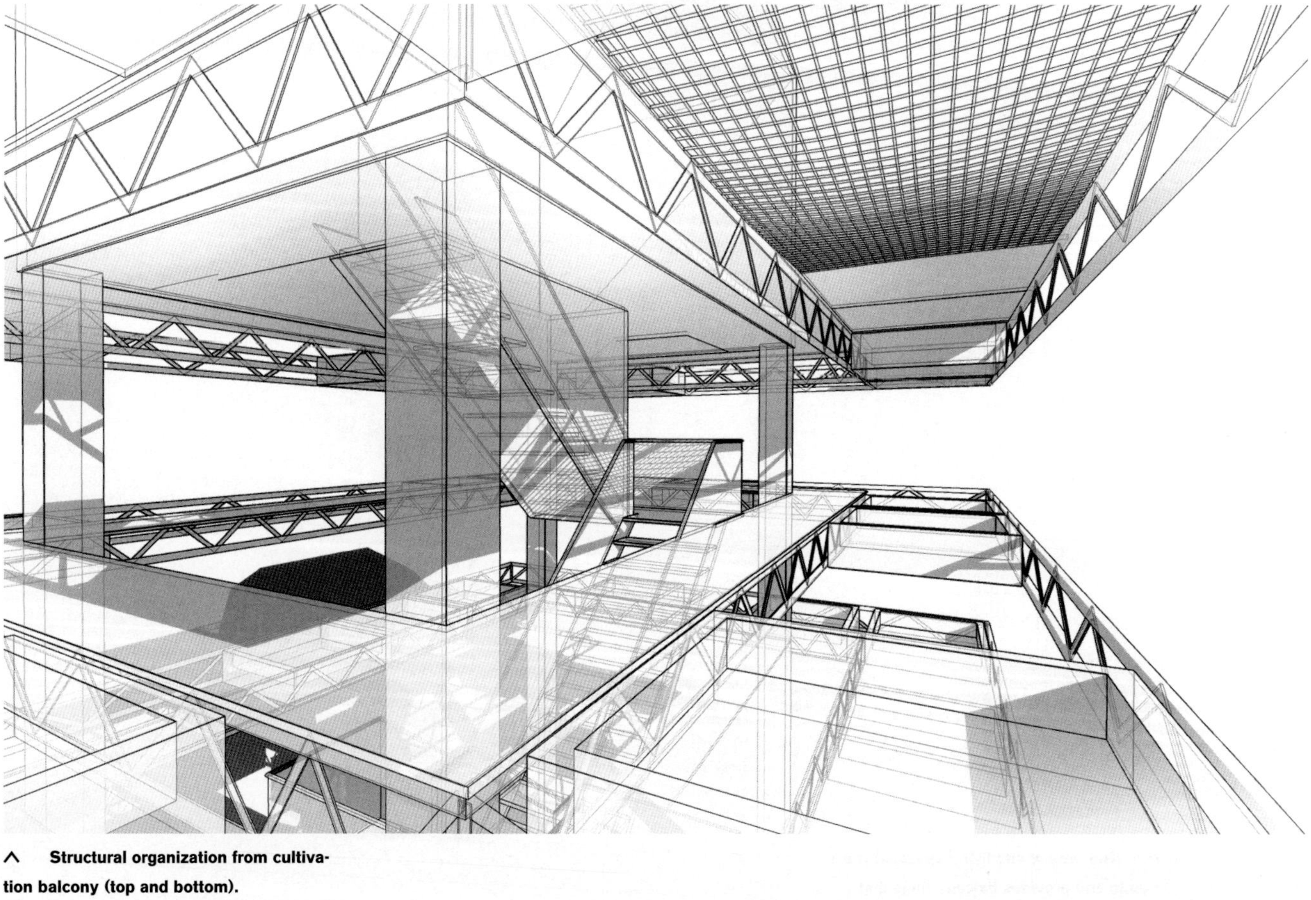

^ **Structural organization from cultivation balcony (top and bottom).**

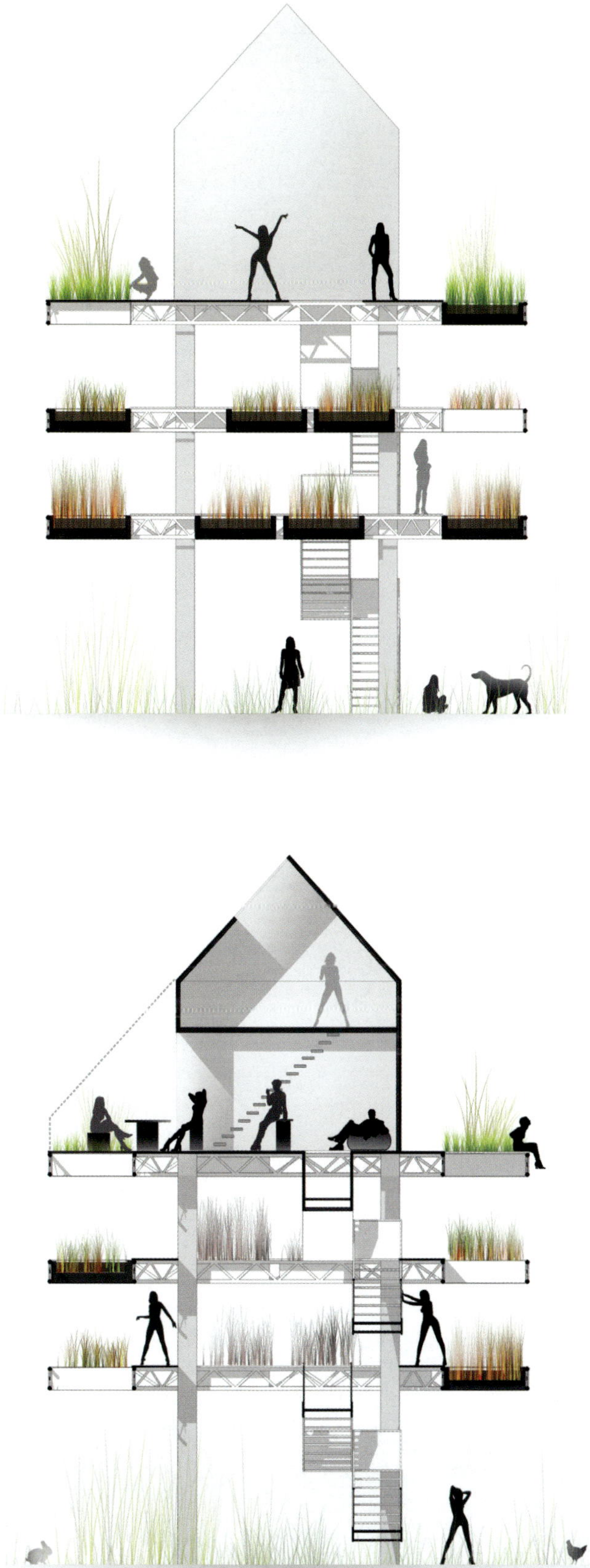

^ **House cross sections.**

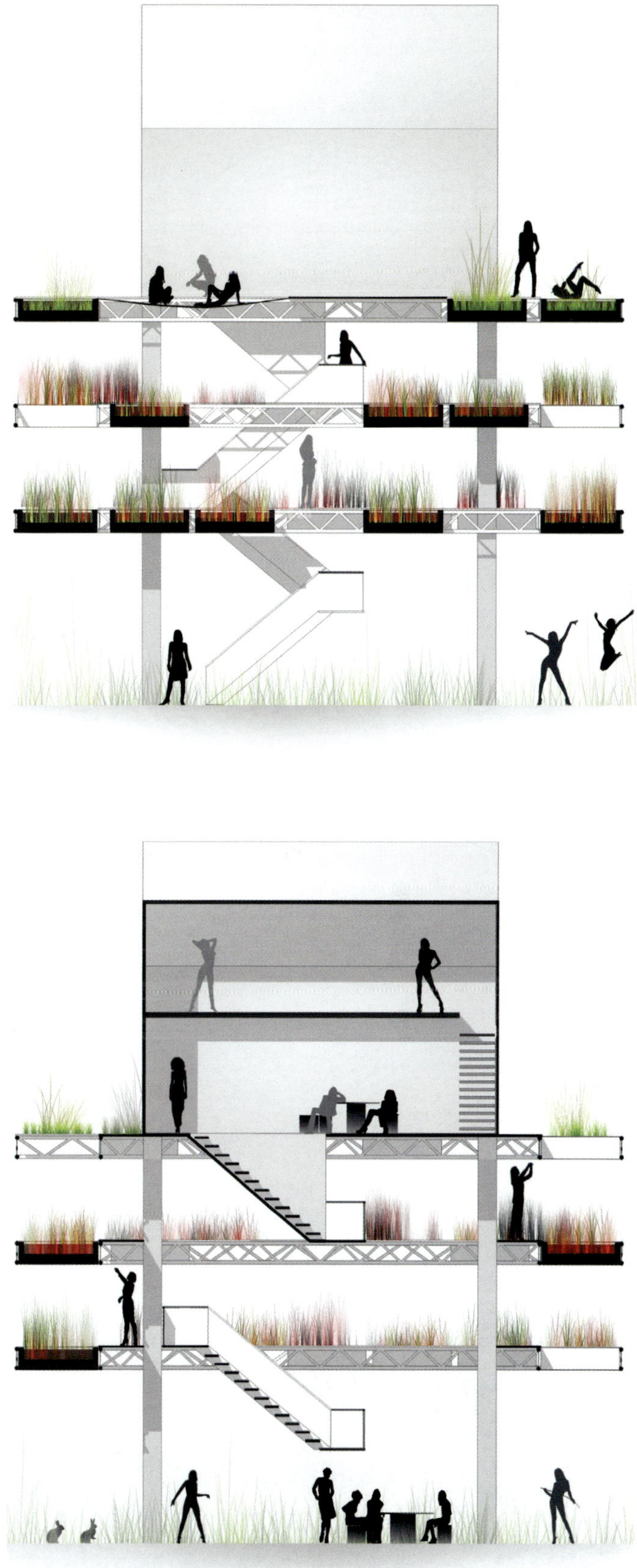

^ **House longitudinal sections.**

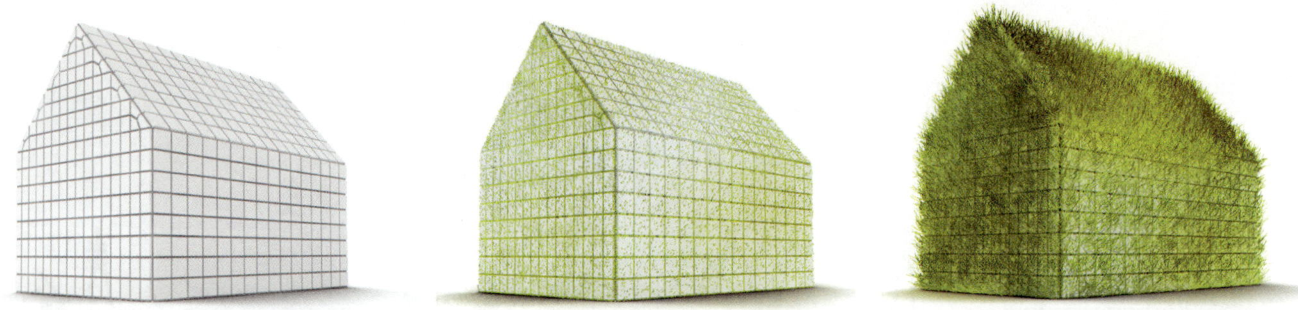

^ Exterior walls are clad with bricks embedded with flower and grass seeds. Growing over the course of the year, the walls' insulating capacity increases and bodily comfort optimized.

^ Autumn view with full growth of seeded bricks.

microcosmic aquaculture

bittertang (antonio torres & michael loverich)

^ **The Orb's Microcosm: Section prior to harvesting of fish. Captive fish flourish in a lush environment surrounded by newly established wild fish stock and plant life. This orb has been engineered to sustain its occupants by mixing feed into the gelatinous matrix which will slowly release as the gelatin dissolves.**

We imagine a future where the vast and deep expanses of the ocean will teem with overabundant floating gelatinous reefs. Humans will be nourished by these new floating reef-worlds that sustain large quantities of harvestable fish. Farming in this project is not viewed as a monoculture but as a new ecology in which wild and captive marine life are raised for the future enjoyment of divers and fisherman.

By encouraging the establishment of new ecosystems, contemporary farmers and the public can reap physical and recreational benefits from environmental stewardship. The ocean is a vast resource, only able to accommodate an intensity of life when the proper conditions arise. These conditions can be catalyzing, allowing new ecologies to flourish around the simplest substrate. In the ocean, it's not a matter of making space for life but making objects that attract life. By combining manufactured products with marine and plant material, Microcosmic Aquaculture establishes a new aesthetic associated with the qualitative benefits of returning back to nature and environmentalism. A series of living orbs provide the structure from which this aesthetic will flourish.

The premise of Microcosmic Aquaculture relies on the dissolvability and permeability of the orb's gelatinous wall section. The giant orbs have a pre-defined lifespan determined by the customizable recipe of the gelatinous mix. The mix composition establishes the point at

^ **Gelatinous Armada.**

which the walls will dissolve and the orb collapses. A longer lifespan can be accommodated by increasing the mixture's density. The outer surface nourishes, shelters, and provides substrate for various marine life while the interior is used as an unmanned and low maintenance fish hatchery or aquaculture pen. This allows the spheres to be both reef and farm. Food can be embedded and released at specified times in unison with the dissolution of the walls. Growth on the orb's surface can be customized in advance with specific additives at key locations to attract or detract beneficial native species.

The advantage of this system is that farmed fish are grown within a larger reef system that manages and cleanses the waste and accumulation of typical farms. Large colonies of wild fish and shellfish will be attracted to the orb's substrate, to be harvested by an orb's owner or made available to the recreational public.

Orbs will not only be productive and efficient farms, but the giant spherical reefs will become a living object to be explored by divers and recreational fisherman. The shape and gelatin composition produces various living conditions and reef typologies, thus no two are the same. Once set loose in the ocean, the orbs will (literally) take on a life of their own, following currents, attracting life wherever they go, and mingling with other spheres of various ages, sharing their ecosystems, and then shedding them as they dissolve. In the ocean, this growth can happen very quickly. Within months of being launched, the orbs can be covered with living material becoming evolving edible worlds.

^ **I went fishing and caught a farm!**

> **Following: Ominous Orbs.**

>>> **From Bittertang, Antonio Torres currently tastes new flavors and Michael Loverich explores frothiness in Bushwick, Brooklyn.**

microcosmic aquaculture

geotube: vertical salt deposit growth system

faulders studio

GEOtube is a proposal for a new 170-meter tall tower for the city of Dubai. With an open structure and an exposed membrane skin, the vertical planes of the GEOtube tower are continually misted with local saltwater through an external vascular water system. The result is a continuous uniform growth of salt crystal deposits upon its vast external surfaces. GEOtube provides an evolving architectural icon for Dubai, and serves as a specialized habitat for wildlife that thrives in this environment, as well as a harvestable skin of crystal salt.

Dubai is situated in one of the most unique natural environments on earth. Aside from the Red Sea, the world's highest salinity for oceanic seawater is found in the adjacent Persian Gulf. Typical salt content for the world's oceans is 35 parts per thousand, whereas the Persian Gulf is estimated at 40 parts per thousand. This is due to high evaporation rates in the region from high temperatures and low freshwater influx. The city's coastal plains, known as *sabkhas*, are geological formations of salt flats created by local temperature and humidity conditions combined with high-salinity water. These climate variations, with thermal contraction at night and expansion during the day, lead to polygonal surface cracking. Deposits of halite, gypsum, and aragonite form some of the major subsurface hydrocarbon reservoirs in the Middle East region.

Salt crystals saturate the air around them with healthy negative ions. Pollution, in contrast, produces large quantities of positive ions creating an unhealthy electrical imbalance in the air. The concentration of negative ions is naturally higher around waterfalls and the ocean; where water droplets are dispersed, an electrical charge is created. Extensive research has demonstrated the therapeutic values of salt caves and their positive influence in the treatment of respiratory diseases.

Saltwater is supplied to the GEOtube tower through a new 4.62 km buried pipeline. The incoming water will be distilled on site, to increase salinity levels, and filtered prior to distribution throughout the tower's surfaces. As high-salinity water is misted onto the skin's mesh, the water evaporates with atmospheric temperatures ranging from 24 degrees in winter to over 41 degrees celcius in the summer. Aiding in the evaporation are the prevailing winds during the day (*Shamal*), and the off-shore winds during the

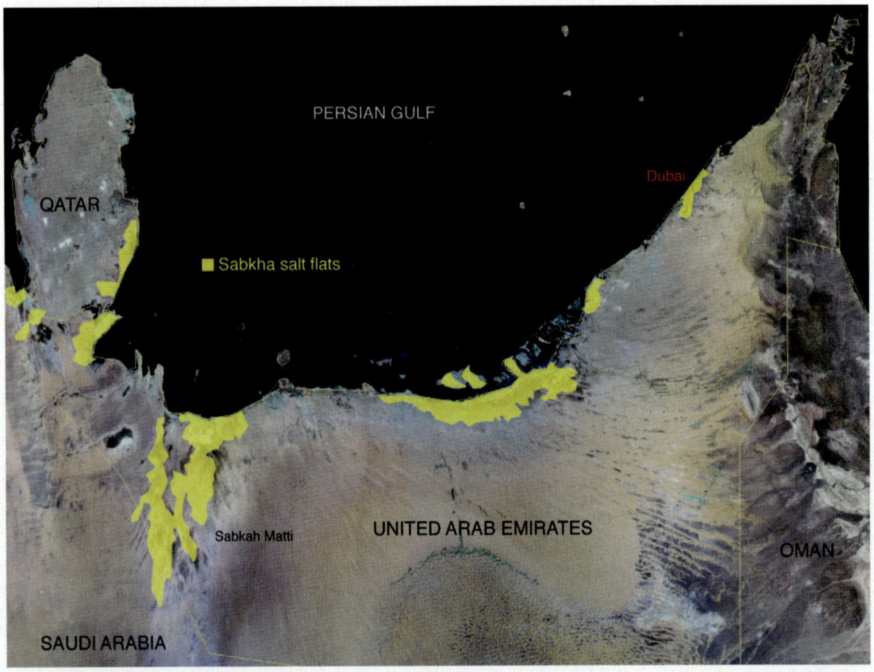

^ **Sabkha salt flats region.**

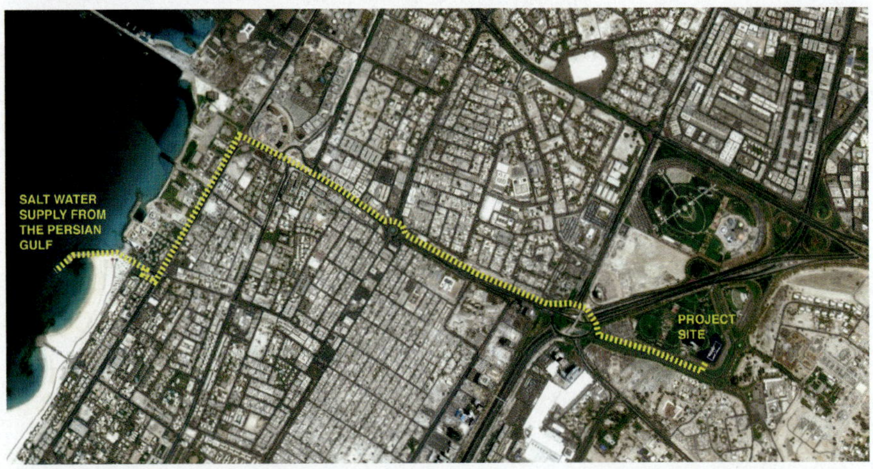

^ **GEOtube salt water supply map.**

^ **Open water salt harvesting farm.**

night. Regional rains are minimal, averaging approximately 80 mm annually. When rains do arrive, they constructively aid in cleansing and washing away lose particles from the salt-forming skin. Annual sandstorms (Al-Haffar, Barih Thorayya, and Al-Dabaran) provide a similar constructive role in scrubbing down rough edges and lose particles.

Integrating with these conditions, GEOtube is a new kind of tower. Gravity-sprayed with adjacent Persian Gulf waters, its building skin is entirely grown rather than constructed. Thus, it is in continual formation rather than fully completed, and is created locally rather than imported. As the water evaporates and salt mineral deposits aggregate over time, the tower's appearance transforms from a porous lattice to a white solid plane.

Design: Thom Faulders, Faulders Studio
Project Team: Jason Chang, Charles Lee, Devin Rutz, Scott Blew

>>> Thom Faulders is a San Francisco-based architect and professor at the California College of the Arts.

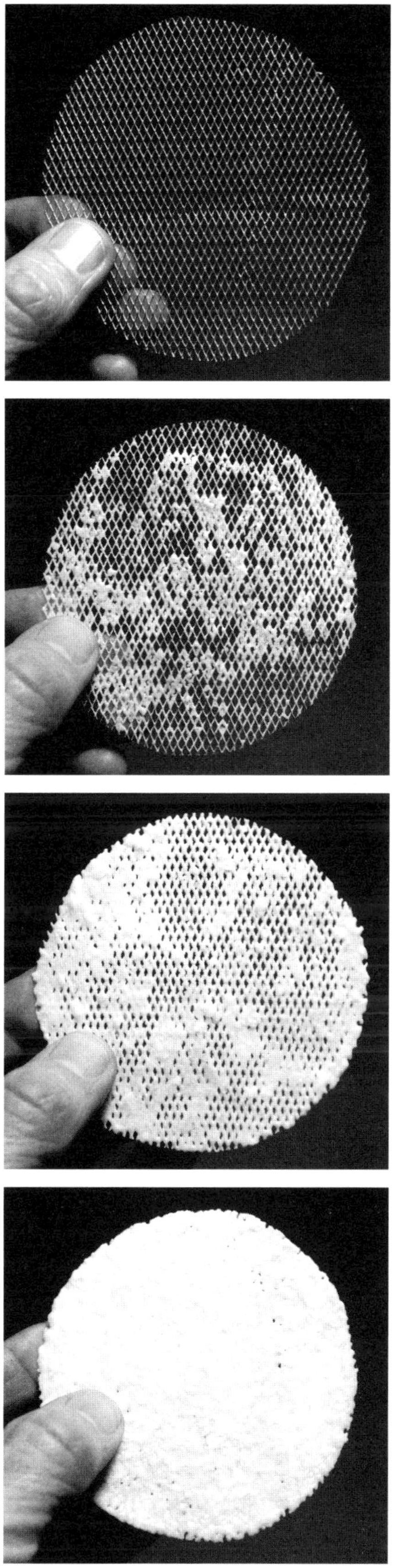

^ Salt deposit growth prototype: A mesh surface is washed in a saline water solution and allowed to air dry. Through repeated washing and evaporation, a solid salt skin emerges.

^ Evolution of GEOtube skin: As salt crystals deposit over time, the tower's skin evolves from a transparent mesh to a solid surface.

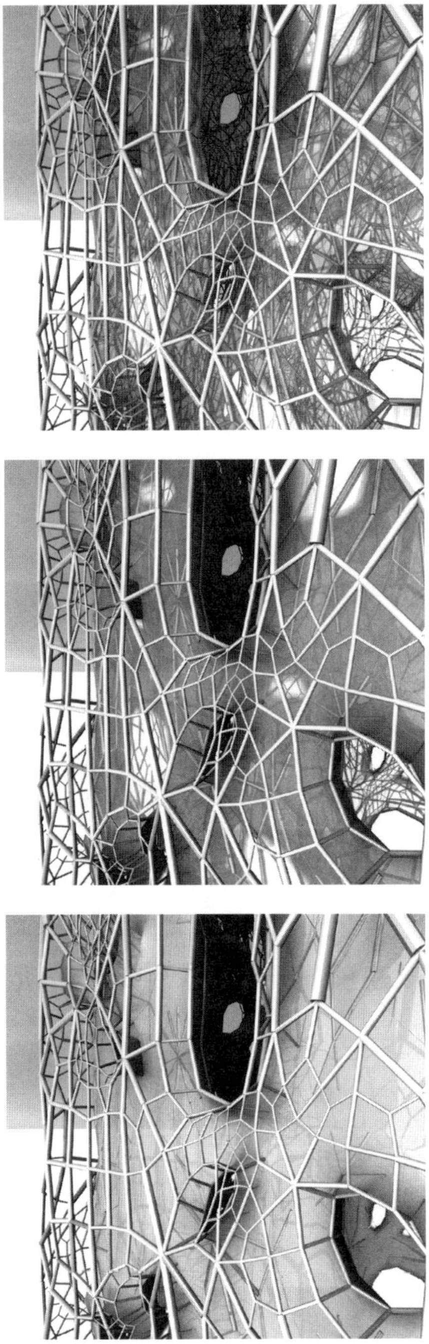

∧ **Detail view of skin growth.**

∧ **North elevation: looking towards top of tower. Salt membrane is layered behind open structural lattice.**

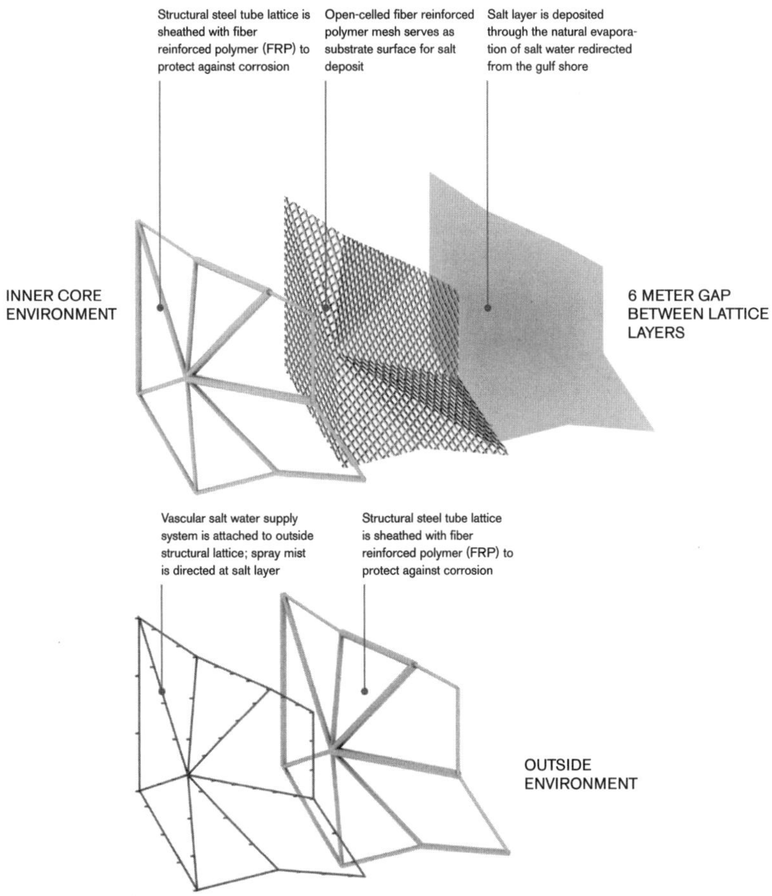

∧ **Membrane skin and structural lattice diagram.**

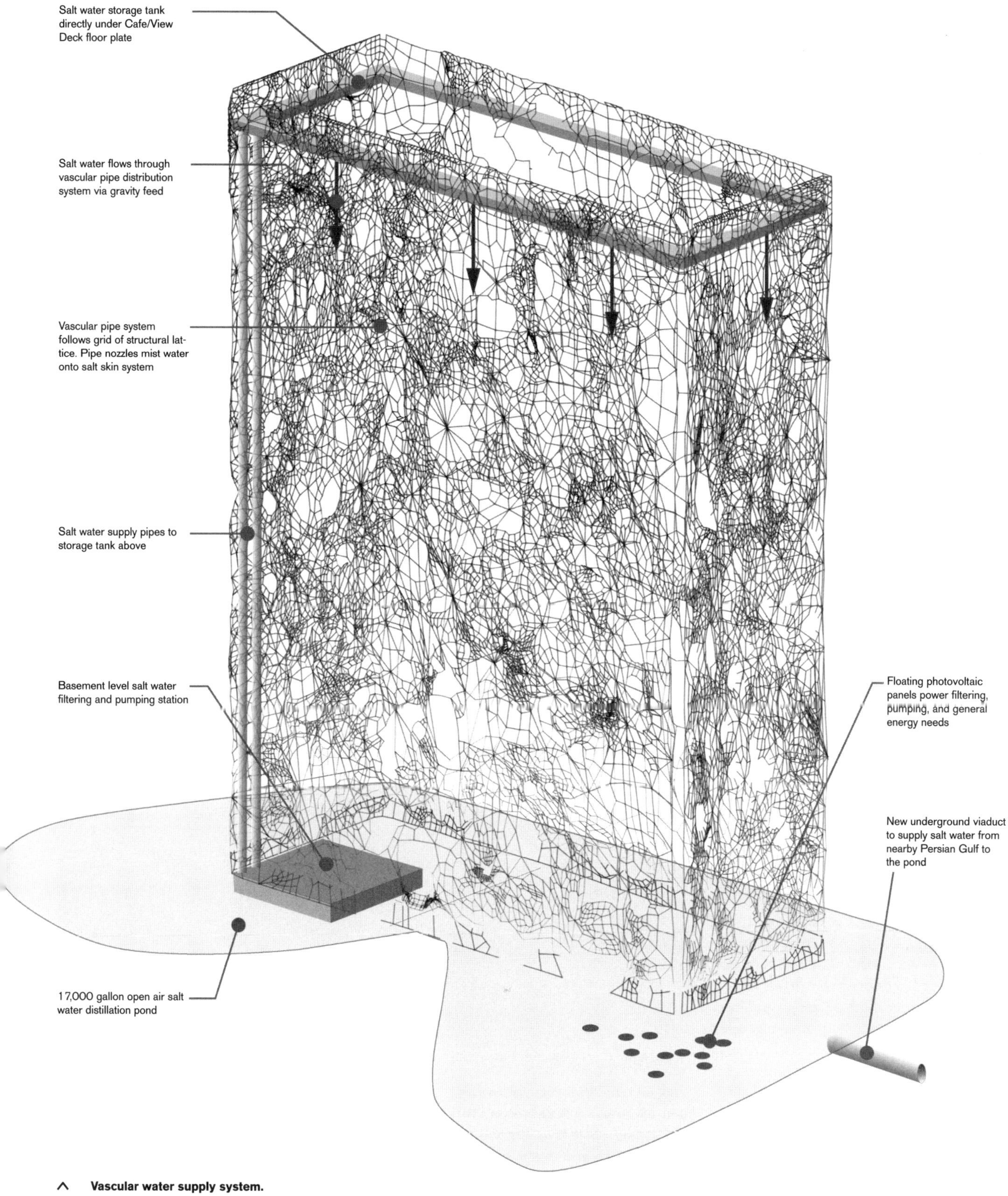

^ **Vascular water supply system.**

post-agricultural speculations

jeffrey l. day & fact

> Art Farm's physical presence is in its buildings and land. More elusive to describe is the ambiance—the subtle influence of the environment's impact on time and space. The sun and stars measure your time, not clock and calendar. Space is shaped by proximity to sound and silence. The sky, your eyes, your ears will fill with the sound and shapes of an incredible number of birds and bugs. And, like it or not, the weather will be your collaborator in all undertakings.
> —Ed Dadey, Art Farm Director

"Post-Agricultural Speculations" enacts the complex interrelationship of art, landscape, architecture, and history in the context of Art Farm, a rural art residency, exhibition site, and laboratory for explorations of creative inhabitation of the land. The ongoing project demonstrates alternatives to the predominant land uses found in the Great Plains and it asks several questions: Can art form the basis of a post-agricultural landscape? Can contemporary design practice engage the multiple histories of rural settlement? Can art engage history, land, and global culture with the same gesture? Can creativity be farmed? Art Farm, and the design proposal for its expansion, offers an alternative mode of resistance to the erasure of history found in the Great Plains. Opposed to nostalgic and sentimental attempts to preserve a lost narrative of inhabitation, Art Farm is a framework for new histories to be written on the land.

New Narratives

Art Farm is a non-profit art residency program situated on a forty-acre working farm in Marquette, Nebraska (located in south central Nebraska, approximately 80 miles west of Lincoln). It is a working farm of diverse agriculture, from hybrid corn to prairie grass. Art Farm and the surrounding area are part of a drainage basin forming the headwaters of the Big Blue River. The landscape varies from flat plains to gentle rolling hills to shallow wetland basins, and is mainly covered with cornfields. Two miles northwest of Art Farm is the Platte River where the flat landscape changes into clay bluffs. Established in 1993, the site currently includes a large, informal sculpture "Art Prairie," and approximately fifteen separate standing and salvaged buildings.

Each summer a diverse group of artists from around the world work in residence, occupying the salvaged structures, living and working on the land. The products of their labor are not the fungible commodities of the neighboring farmers but installations,

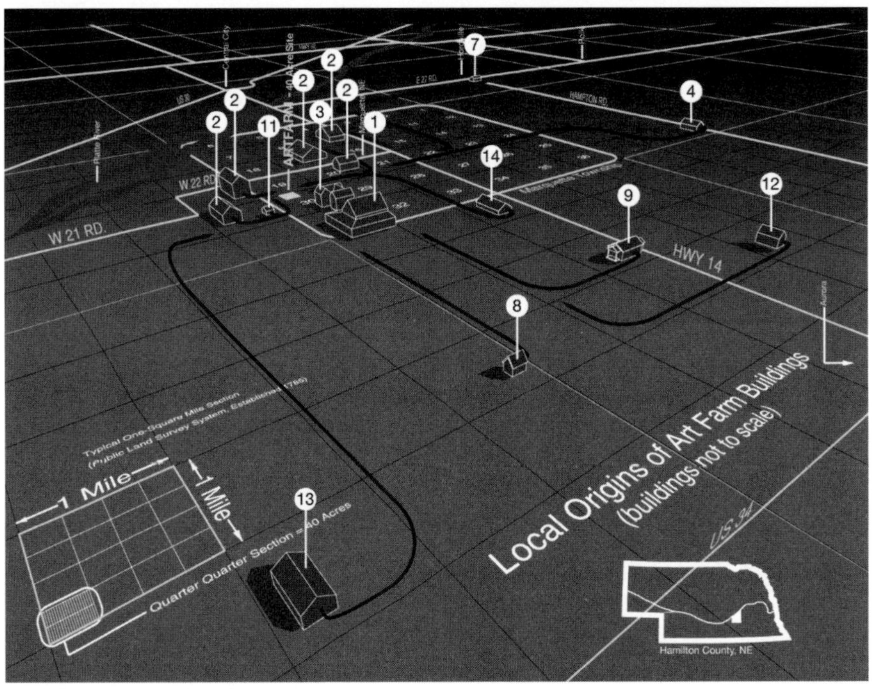

^^ Art Farm inhabits the Jeffersonian Grid. The institution is situated on a forty-acre working farm in Marquette, Nebraska. It is a working farm of diverse agriculture, from hybrid corn to prairie grass. Art Farm and the surrounding area are part of a drainage basin forming the headwaters of the Big Blue River. The landscape varies from flat plains, to gentle rolling hills, to shallow wetland basins and is mainly covered with cornfields.

^ All buildings at Art Farm are repurposed farm structures. Most of these have been salvaged from defunct family farms across the surrounding region and moved to Art Farm. 1. Red Barn Gallery; 2. Mutant House; 3. Victoria; 4. Lone Pine Studio; 5. Earth Room; 6. Campus Dormitory; 7. T.V. House; 8. Red Shed Video Lounge; 9. School House Studio; 10. Corn Crib Installation Space; 11. Toxic Dump Studio; 12. Granary Studio; 13. Tinned Wood Shed Studio; 14. Pig Pottery.

performance and speculations on the relationship between land and creativity. Although these artists may harbor similar motivations to those of the Land Artists of the 1960s and 1970s, who famously went outside to escape the strictures of the gallery system, Art Farm residents tend to be less focused on the physicality of the land—its topography and material qualities—than on the lifestyles embedded in the inhabitation and transformation of land. "Post-Agricultural Speculations" reinterprets the acts of inhabitation and stewardship of the rural landscape that underlay the history of the Great Plains and the American West. Thus, Art Farm reorients the artistic enterprise from a preoccupation with the visual understanding of landscape to focus on the social, cultural, and historical underpinnings of the contemporary rural environment.

An inherent contradiction exists between the modus operandi of the farmer and that of the artist—farmers by nature are not typically speculative. The farmer's livelihood is wholly dependent on predicting the yield of crops. To take the risks of speculation is to risk all. Consequently, farmers tend to follow guidelines and recommendations—plant when the almanac says so; obey the market; employ new methods only after proven by university extension officers. Artists, on the other hand, are encouraged to take risks. Few artists succeed by following the well-worn paths of their predecessors. Risk-taking is the livelihood of the artist. The overarching project of Art Farm, encompassing the work of artists-in-residence and the development of the site, is to reinterpret the nature of settlement and the productive capabilities of the land.

Both art and architecture at Art Farm tackle the multiple histories of objects (such as buildings) and landscapes, reinterpret them, give them new life, new histories, and new narratives. The future of Art Farm will be a reenactment, albeit in a microcosm, of the history of settlement and habitation in the Great Plains.

Organizing Art Farm

In 2003, Min | Day in partnership with FACT (Fabrication And Construction Team), a student-staffed design lab led by the author at the University of Nebraska, began a working relationship with Art Farm. At the outset, our primary task was to create an organizational infrastructure that can both accommodate and structure the development of Art Farm over the next ten years and beyond. The project developed with designs for several buildings, many which are renovations of salvaged structures moved to Art Farm over the last 30 years while others are entirely new.

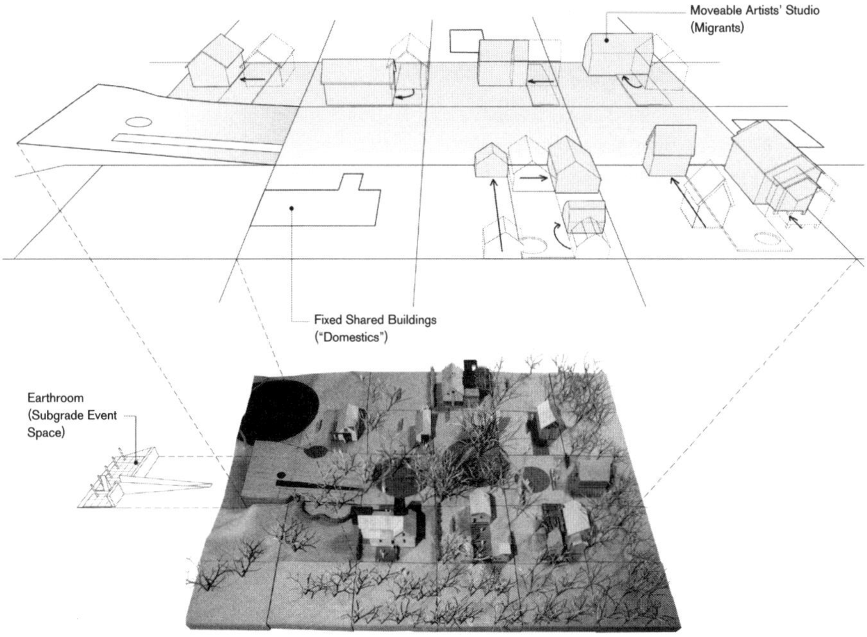

^ Artists' campus quadrant: The organizational plan categorizes individual buildings as either "domestics," structures fixed in place, and "migrants," structures that are free to roam within the bounds of a 66' x 66' "mini-section." Artists may relocate migrant studios as they wish.

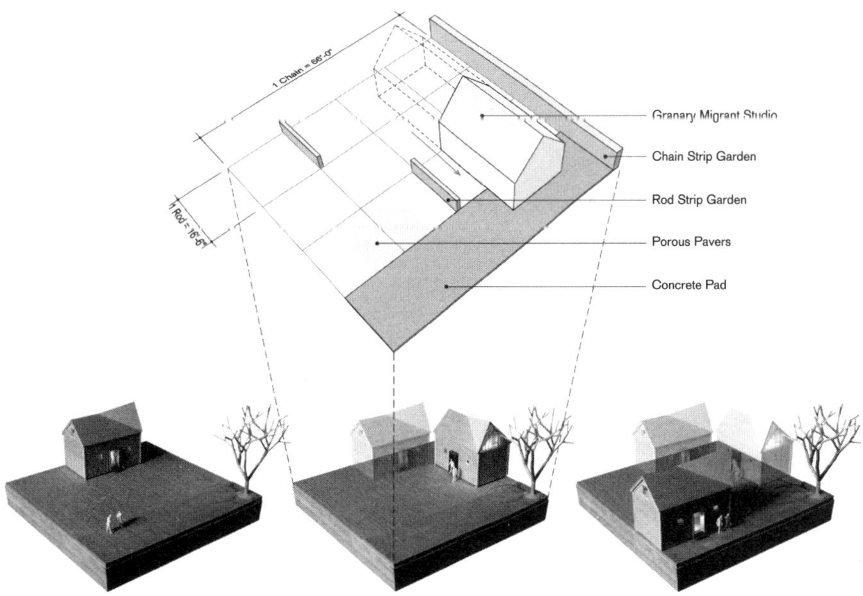

^ Diagram showing a typical artist studio territory (the Granary studio). Resident artists are assigned studio buildings on a 66' x 66' "mini-section" within which they may move the building in order to personalize workspace. Diagram with model photograph, Larry Gawel.

Although we may be fortunate to construct the new buildings and modifications to those that exist (such as the Red Barn Gallery), the very nature of Art Farm suggests that much of the work will be the product of the improvisations of multiple participants, in a similar way that the agricultural landscape of the Great Plains developed in the 19th and early 20th centuries.

To project a future plan for Art Farm, our primary design procedure is organizational; it is not concerned with geometric order or pleasing imagery. Our intention is to offer a set of guidelines that are flexible, yet controlled, so that Art Farm may develop over time without compromising a clear identity. The plan leaves room for improvisations and additions and assumes transformations with each of its passing occupants. The only comprehensive product of the planning process, a multi-layered site map, contains a series of overlapping systems of order, each of which loosely guides a different aspect of Art Farm's development. The primary systems are two grids—one composed of a 1/10th acre grid and the other derived from a scaled-down map of the regional sections from the Public Land Survey system. The subtle offsets between these grids provide an interference pattern that creates unique built or planted conditions within the grid. The lines of the grid will not ultimately be very visible. However, the guides function as property

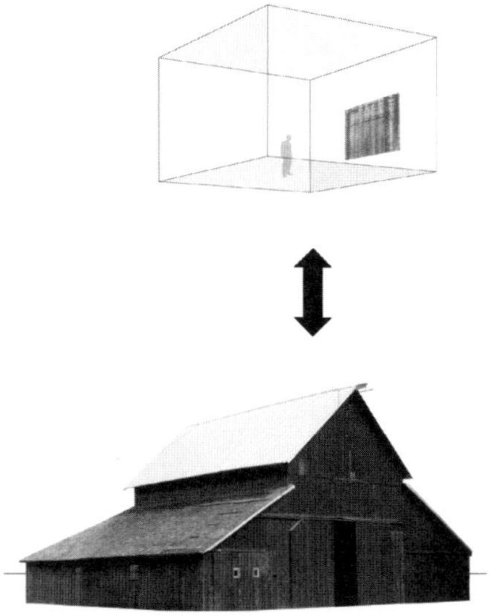

^ Beyond the White Cube. The Red Barn Gallery resists but does not eliminate neutrality of the white cube gallery typology. Instead, the flat white walls one expects give way to inflected surfaces that bend to accommodate old wooden structure and deflect light.

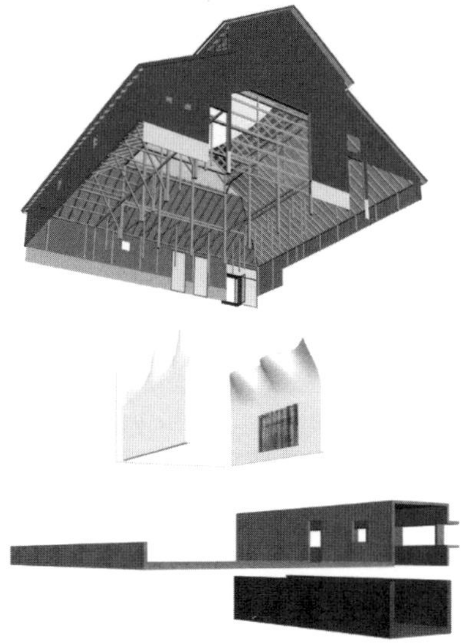

^ Exploded perspective of the Red Barn Gallery showing the salvaged barn, white cube gallery, concrete "carpet" and storm cellar.

^ Approach to the Red Barn Gallery with roadway and parking area defined by controlled plantings of native grasses.

lines. Within this system we have located the structures—each with its own 1/10 acre mini-section property—four separate path networks, a series of nodes for art installation and unique plantings. The ground of Art Farm is being restored as a native prairie, though with a distinctly organized pattern.

A semi-underground structure, the earthroom, is the only entirely new structure proposed for the site. This structure, an inversion of the ground plane, unites built form and landscape and provides a place of entertainment and refuge from the working environment of the campus (as the collection of artists' studios is known). It is constructed from stone gabions and rammed earth, and is covered with a sod roof supported on a structure of salvaged steel beams. The earthroom questions the presumptive location of buildings on the land. Instead, it makes space where there was none.

A key element of the project and the first to start construction is Victoria, a salvaged farmhouse purchased for $400 and moved to the site. Victoria will become the primary domicile for the resident artists who visit the Art Farm. To remodel Victoria for its new use on its new site, a new stair tower is added that will allow the house to be used in different ways with controlled access to its apartments. The tower will extend above the roof of the house to provide a viewing room of the expansive agricultural landscape. All new construction is distinct from the old house to establish clarity about new uses. All of these new elements utilize salvaged materials that are already stockpiled on the site: wood lath, wood siding, steel beams, aluminum and wood windows, and old agricultural machinery.

Beyond the White Cube

The Red Barn Gallery will be the entry point for the public and the interpretive center of the multiple ecologies of Art Farm. Currently, the residency program has no dedicated public space, nor is the site structured for public visits except during special events. As with Victoria, Art Farm purchased the Red Barn from the owner of an abandoned farmstead and moved it over land to Art Farm. Architecturally, the building represents the juxtaposition of the vernacular, place-bound physicality with the abstraction of universal 'White Cube' art gallery space. The Red Barn Gallery resists but does not eliminate the neutrality of the white cube gallery typology. Instead, the flat white walls one expects give way to inflected surfaces that bend to accommodate structure and deflect light. Filtered light from the polycarbonate roof is further diffused with a translucent light deflector (what we call "the surfboard") shaped specifically

^ Interior of the Granary, a typical "migrant" artist's studio in the Campus.

^ Victoria House modifications under construction, 2008. A key element of the project, and the first to start construction is Victoria, a salvaged farmhouse that will become the primary domicile for the resident artists who visit the Art Farm. Photograph, Ed Dadey.

^ Interior of the Red Shed Video Lounge. The Red Shed is representative of the shared attitude towards objects that emerges from diverse practices at Art Farm. The process of repurposing involves the deconstruction and preservation of the basic structure and the fabrication of a contrastingly smooth interior surface designed to support the Shed's new function as a gallery and worksite for New Media. Photograph, Larry Gawel.

by sun angle calculations. The east wall of the central gallery is eliminated in favor of an aluminum and glass garage door (in homage to New York's Chelsea gallery district) that opens to the restored prairie grass art park.

While the Red Barn Gallery is a focal point, all of Art Farm can be read as a recontextualization of the white cube. Here one experiences the white cube as a framing device for artwork that cultivates a relationship between overlapping contexts of universal contemporary art culture and the specific, historicized location of its making. The architectural interventions at Art Farm encourage a continual process of re-contextualization through which the relationship between art and everyday life is questioned. This inspired a saying that the goal of Art Farm is not to make life more artistic but to make art more "life-istic." Visitors to Art Farm confront artwork in the context of its making rather than in the isolation of the neutral gallery. Artists operate with a series of spaces from landscape to built structures in which the distinctions between studio, gallery, art park and raw materials are constantly being renegotiated.

Post-Agriculture

The project evolved beyond its original scope into an exploration of building and landscape within the post-agricultural condition. With the dissolution of the family farm and the shrinking populations of many rural towns, the Great Plains landscape is increasingly uninhabited. The changing demographics of the region are accompanied by transformations in land use patterns and building habits as one would expect, but there often remain vestiges of prior development patterns that are slow to disappear. The organizational system of the Jeffersonian grid is an obvious example but the disintegrating farmsteads and barns are also significant. Art Farm demonstrates the ways in which art and design can opportunistically inhabit these sites with new and provocative uses. The project represents the possibilities of intertwining creative disciplines within an expansive, productive context.

>>> **Jeffrey L. Day, AIA is Associate Professor of Architecture at the University of Nebraska-Lincoln and a principal of Min | Day, an architecture & design practice based in San Francisco and Omaha. Jeff also runs FACT (Fabrication And Construction Team) an academic/professional collaborative design lab.**

^ Installation at Art Farm: "Ring of Fire," 2003, J.D. Messick.

^ Installation at Art Farm: "Upland Game Fowl," 1998, Dawn Detarando.

^ Installation at Art Farm: "Shadow House," 2004, Beili Liu. Photograph, Ed Dadey.

harvesting space

nathalie de vries

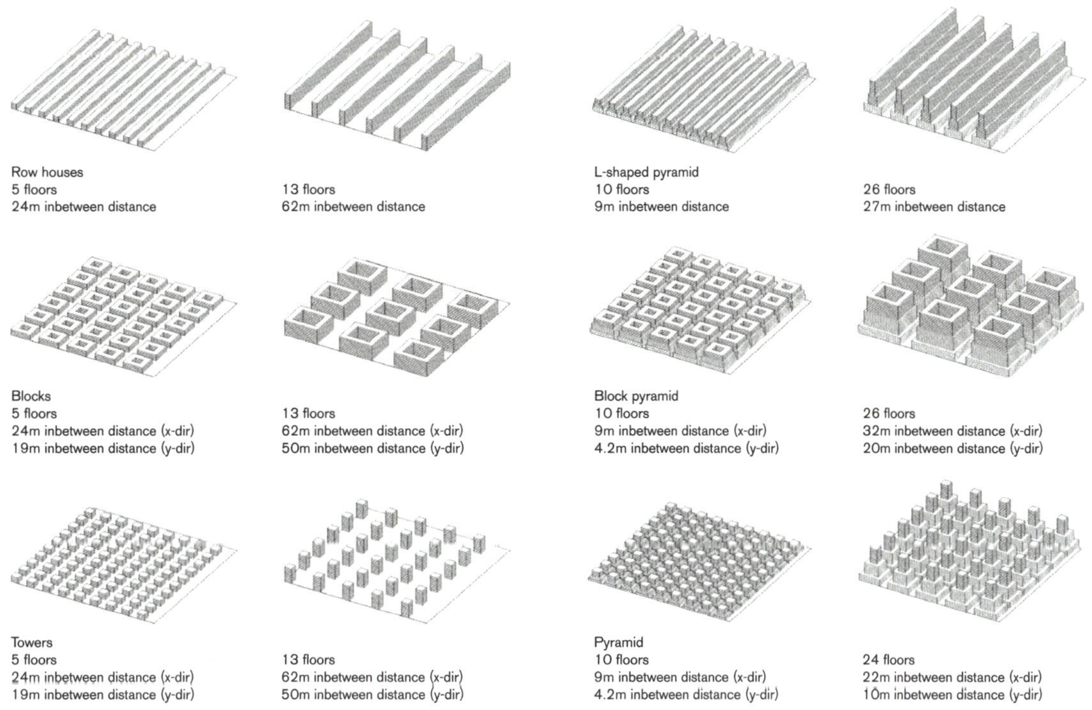

^ Function Mixing. Effective zoning and function mixing can save space, and efficient stacking of functions allows for a better three-dimensional use of space.

In recent decades, we have become more and more aware of the fact that our way of living has a substantial ecological footprint. At some point in the 19th century we lost our "CO_2-innocence," as the German philosopher Peter Sloterdijk recently stated. We are using more space than the earth is able to provide. We know now that everything we do and produce requires space, including compensation for losses. We know we can calculate how many "earths" we would need based on our current lifestyle. Since we consume too much anyway, we have to decide either to reduce the consumption, or use the earth's surface more efficiently. The latter could be considered as farming. Instead of only consuming "less" we can still have some of "more."

In current times, nobody argues anymore the necessity to be more efficient and productive in for example, land use. But it is also no coincidence that in Europe the idea of urban farming has come up at the same time as the notion of shrinking cities. In North America, however, urban farming is rooted in the practice of producing your own food again, against the backdrop of poverty and an increasingly distanced relationship between people and food production.

In recent years, MVRDV has dealt with the other interpretation of farming, which is also discussed in this issue of *Bracket*. The Dutch are accustomed to having to create land, or polders, and to being highly productive per square meter in terms of land use or agriculture. In recent years, we have tried to be as efficient as possible with land use in urban planning and to always search for ways to make at least two uses of project sites, or try to find a way to compensate for any loss of open space. Effective zoning and function-mixing can save space, and efficient stacking of functions allows for a better three-dimensional use of space. We still have not reached the limits of understanding how mixing functions can lead to integrative designs, where buildings produce cycles within themselves as with a heat exchanger or a heat pump.

Investigations into the double usage of land are also effective. For example, cars continue to occupy enormous amounts of space. Allowing cars to park in the street should be seen as outrageous as the fuel consumption of SUVs. It should be considered an obscene waste of space. But much of this has already begun to change. For example, cities promoting green roofs are now as common as cities promoting tourism.

Recently, more and more neighbourhoods are being designed to produce their own energy by integrating renewable energy systems such as solar panels, solar boilers, and wind turbines that are placed on the rooftops or in the public realm. Also water remediation and purification can be integrated, which often comes with a considerable cost of usable space. But, the integration can also have a powerful effect on the inhabitants, who are then made aware of what it takes to produce clean water. These neighbourhoods become sometimes so effective that they can even start their own energy company to sell off the surplus energy they have produced.

It would be interesting to see if we can broaden the roles of architects and urban designers from creating culturally and functionally good designs to making productive designs as well. Some of the previously mentioned qualities could be incorporated into briefs and, as necessary, into budgets. But it gets even more interesting in cases where the desired productivity is considered parallel with the design, as in agricultural aspects, life of plants and animals, floating and flying cities, or by using energy produced by humans through body heat, movements, and activity. The goal is to develop clever contemporary designs, in other words, a 21st-century 'Frankfurter Küche.'

The most exciting designs are now being produced when so called by-products can be harvested—products or designs that are coincidentally related to the project, such as the possibility of generating energy on the same surface as another activity. Many of the projects and proposals in *Bracket*'s "On Farming" demonstrate this particular fascinating possibility of harvesting as it relates to architecture and urban design—the beautiful inadvertent combinations of productive designs.

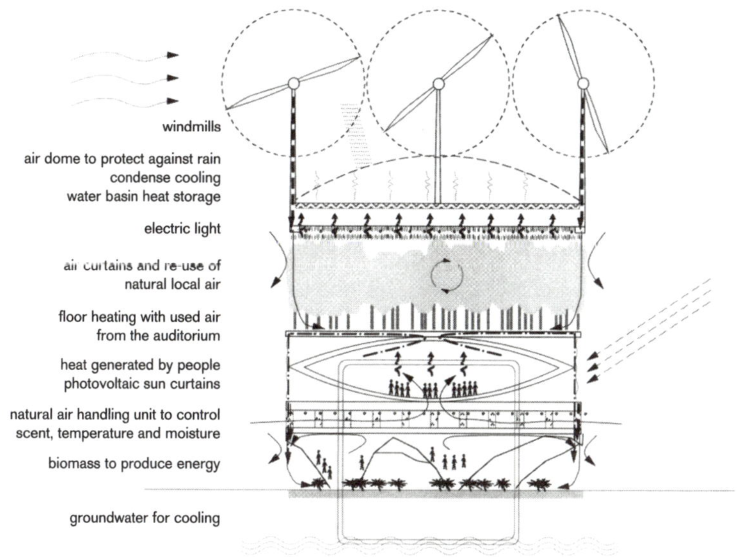

^ **Expo 2000 Pavilion for the Netherlands, 2000. Energy use.**

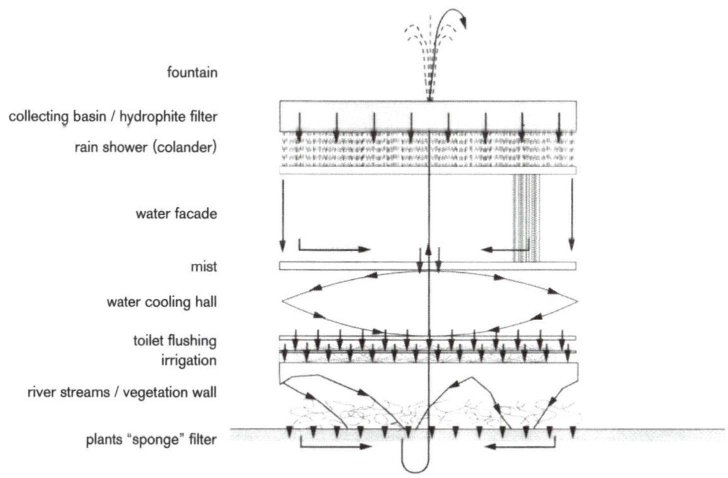

^ **Expo 2000 Pavilion for the Netherlands, 2000. Water circulation.**

Expo 2000 Pavilion for the Netherlands

Can increasing population density coexist with an increase in the quality of life? What should be the conditions and what role will nature play? This kind of effort can be the Netherlands' specific contribution to the world fair, which seems to be devoted to a nostalgic glimpse of ecology.

The Dutch entry shows a mix of technology and nature, emphasizing nature's producability and artificiality. It provides multi-level public space as an extension to existing public spaces—a monumentalized multi-level park. This act of stacking saves space, energy, time, water, and infrastructure.

^ **The Pig Towers allow for each pig to have access to outside space and fresh air by means of balconies.**

Pig City

In 2000, pork was the most consumed form of meat at 80 billion kg/year. With a production of 16.5 million tons of pork, The Netherlands is the chief exporter of pork within the European Union. In 1999, 15.2 million pigs and 15.5 million humans officially inhabited The Netherlands. One pig needs an area of 664 m^2, including current food processing. In the case of organic farming, pigs are fed with 100% grain, leading to a required 130% more field surface due to the reduced grain production. This would cause a demand of 1726 m^2 per pig, including the organic food processing. This would mean that there would be only 774 m^2 per person left for other activities. In other words, 75% of the Netherlands would be dedicated to pigs.

Can we combine organic farming with a concentration of production-activities so that there will be space for other activities? Is it possible to collect all the pig production needs within concentrated farms, therefore avoiding unnecessary transportation and distribution, and subsequently reducing the spread of diseases?

^ Vertical pig farms along Slufter Beach in the port of Rotterdam.

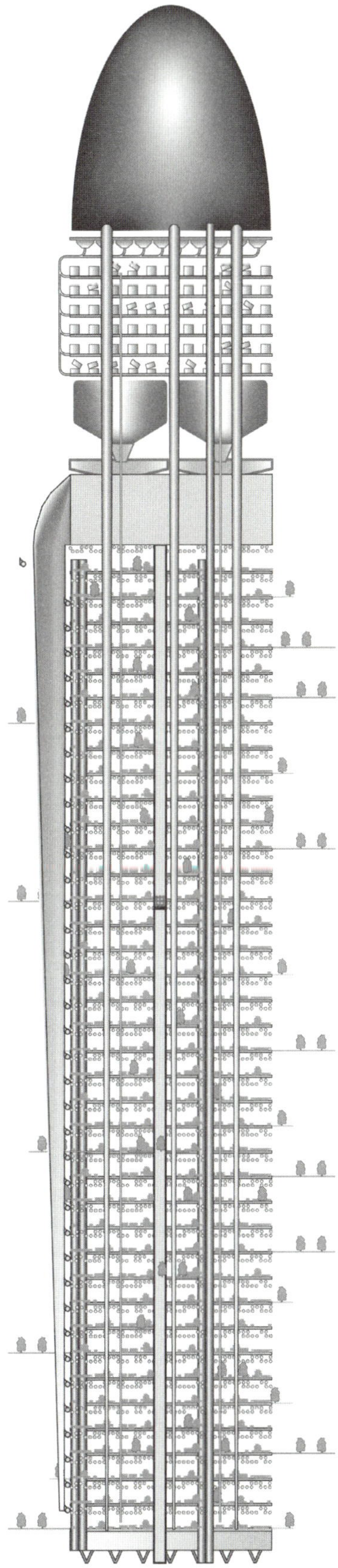

^ Pig Tower cross section.

^ While the building occupies only 10% of the entire site, the remaining open space becomes an 'eco park.' Photovoltaic cells are distributed across the park, generating energy for the 3000 apartments and reducing CO_2 emissions by 6,000 tons per year.

Eco Ciudad Montecorvo Logroño

A medium sized city of approximately 130,000 inhabitants, Logroño is located in the wine region of La Rioja in the north of Spain. This proposal for an Eco City consists of approximately 3,000 social houses and complementary program—such as schools, social buildings, sports facilities—all developed in a sustainable way. By producing all the energy needed on site, the new neighborhood achieves a CO_2 neutral footprint.

The 56 hectare site, just north of Logroño on the two small hills of Montecorvo and la Fonsalada, has considerable potential to create a new neighborhood. Besides the fact that the hills provide a beautiful view over the city, the slopes are orientated to the south, allowing for the convenient generation of solar energy. A tapestry of photovoltaic cells clad the mountain, as if the hills are covered in gold. On top of the hills, windmills are generating part of the energy needed for the housing units and at the same time serve as landmarks. 100% of the energy demand is generated on site by a combination of solar and wind energy. The development is designed in a compact way to minimize undesirable impact on the landscape. Only 10% of the site is occupied by buildings. The remaining space becomes a magnificent eco-park, a mix of landscape and energy production.

^ **Wind turbines located on top of the hill help generate energy for the 3,000 apartments.**

^ **The linear compact urban development meanders through the landscape. By making use of the height differences, every apartment has a beautiful view over the city.**

grafts/ hybrids

Moira Wilson

Jason Sowell

Austin Nicholas Tragni

Neeraj Bhatia, Marissa Cheng, Elizabeth Nguyen, Liu Peng, and Yang Jiang

Ellen Burke

Ryan Lingard

aquaculture seascape park

moira wilson

The Aquaculture Seascape Park is a dynamic, productive park designed to generate an inclusive culture of sustainable aquatic farming. Situated along the coast of Vancouver Island, Canada, the park is a hybrid of land and sea that changes with the rise and fall of tides. Living deep-water gardens and tide-pool gardens cultivate kelp, fish, mussels, sea anemones, urchins, and sea stars within polyculture ecosystems. A responsive anchor system allows seawater to flood the three levels of the park twice a day, setting up optimum growing conditions for various species. The tide pools function as both nurseries and gardens. Boundaries between land and sea shift from moment to moment, creating a spatial-temporal tourist experience. At each flood level, the visitor's experience of the park is unique. In imitation of intertidal ecology, organisms such as plankton, seaweeds, and sea stars are grown in horizontal bands at splash, high, mean, and low tidal zones. Physical gradients and biological interactions (filters, modifiers, and indicators) are designed as part of the park's living system to optimize production cycles and biological processes. Biofilms, algae, and molluscs that naturally accumulate onto surfaces of the park (bio-fouling) are incorporated into the production cycle dependent upon water flow, selection of surfaces, and the selection of species. Solid waste and water are recycled through the system using a biological treatment process.

Modular units provide the flexibility to configure, add, or subtract to the park for farming or development in various combinations, ratios of species, and degrees of publicness. As a mobile system, it also serves as infrastructure for coastal conservation

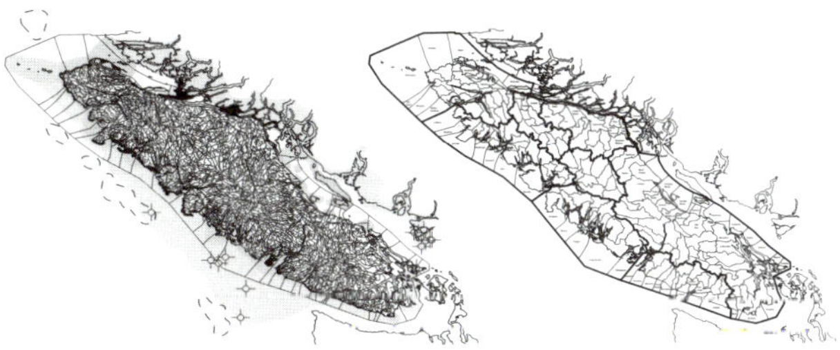

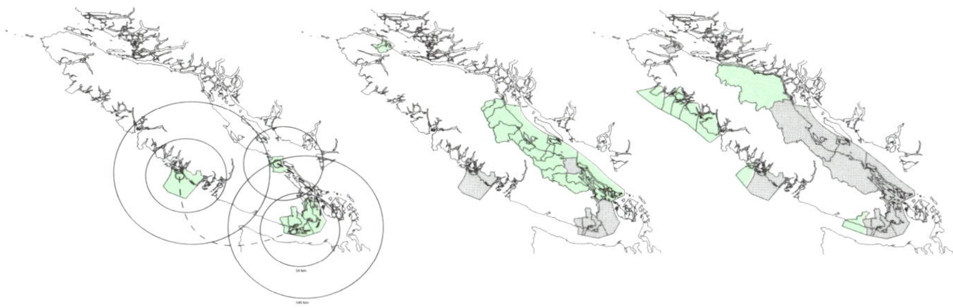

^ **The mobile park changes locations to stimulate tourism, research, restoration work, and to facilitate mapping of watersheds and coastline territory. Ocean ecologies and species can be monitored to track changes over time.**

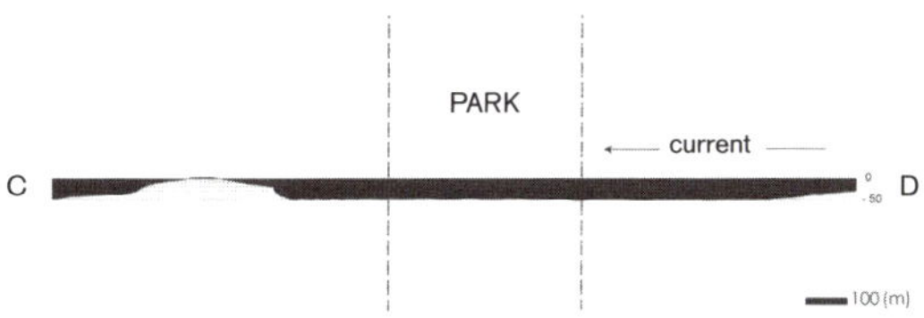

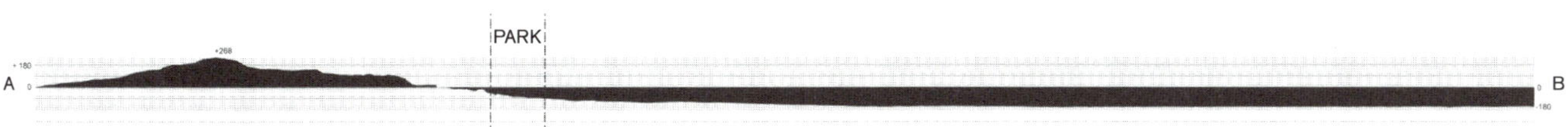

^ **A temporary location for the Aquaculture Seascape becomes an extension of the terrestrial public park system to demonstrate environmentally, socially and economically sustainable aquatic farming.**

^ **Mussel farming.**

Biofiltration
Horizonal and Vertical Surface (m)

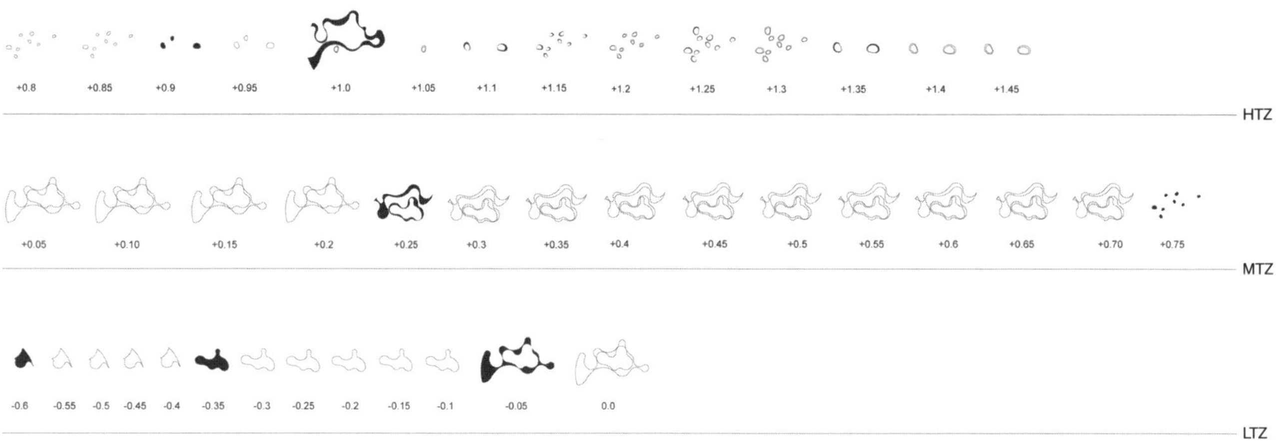

Intertidal Zones / Gradients

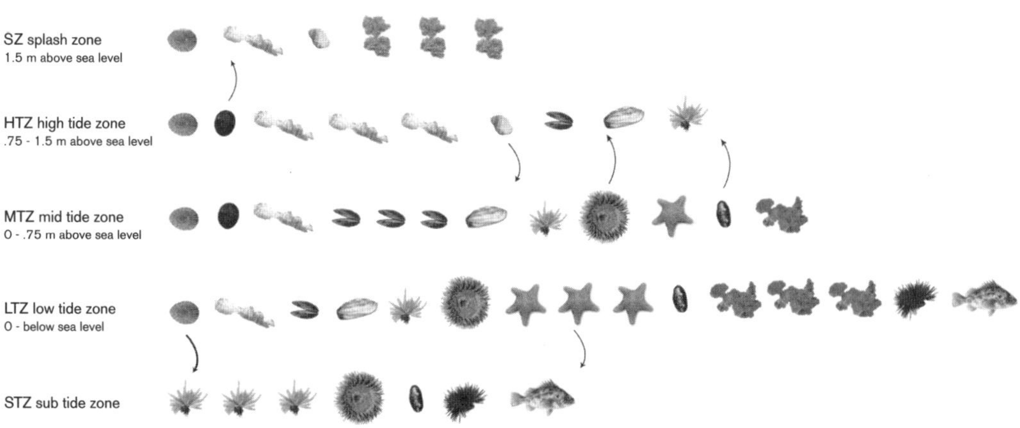

^ **Species are cultivated in vertical zones that inform the topography and food web of the park.**

groups, such as the detailed mapping and monitoring of coastal oceanic territory and seabed restoration. Eco-tourism can also be 'seeded' at specific nodes along the Island's coastline as temporary or seasonal events.

The park provides a physical interface for communication and collaboration among private industry, public interest groups, community members, researchers, government, and aquatic food production agencies. It offers a transparent, inclusive infrastructure to address sustainability issues of aquaculture activities, such as the degradation of habitats, waste production levels, transmission of pathogens between cultured and wild stocks, food and product quality, support to small farmers, and access rights and limitations to various types of activities by local communities. Thus, the park becomes a forum for education, information, research, and training.

The modular organization of space enables economies of scale and encourages both individual and collaborative work or learning. The design of the park, informed by traditional agrarian culture, is immersed in the rhythm of tides, the moon and sun, seasons, as well as wind and weather patterns.

Species Functions

Fish: Provide nutrients for shellfish, kelp and benthic community; absorb oxygen from mussels and kelp.

Mussels and Oysters: Filter water by absorbing organic particles, produce dissolved nitrogen to fertilize kelp, CO_2 released by mussel respiration is absorbed by kelp to metabolize photosynthates, feed on enhanced plankton created by kelp production.

Kelp: Absorb nitrogen (nitrate-N), phosphorous and CO_2 from fish, produce and release O_2 into water for mussels and fish, the increased plankton growth (created by slowed water circulation) is fed on by mussels and fish and sustains species of smaller invertebrates, which in turn provide feed for fish and other organisms, such as sea urchins, abalone, sea cucumbers. Kelp forests provide shelter and protection for fish fry, young fish, and small-sized species of fish.

Moira Wilson is a Toronto-based designer interested in green infrastructure projects and process-based design. She holds a Master of Landscape Architecture from the University of Toronto.

^ In public areas, tide pools are gardens and, in private areas, the tide pools are nurseries for commercial farmers or scientific researchers.

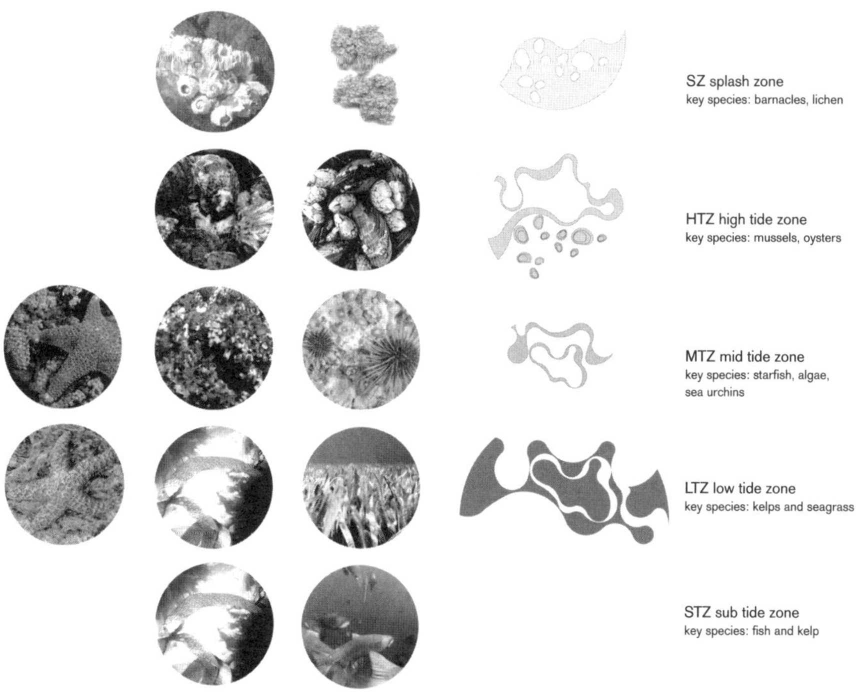

^ The key species identified for each vertical zone in the park has important impacts on the function, structure and/or diversity of a community.

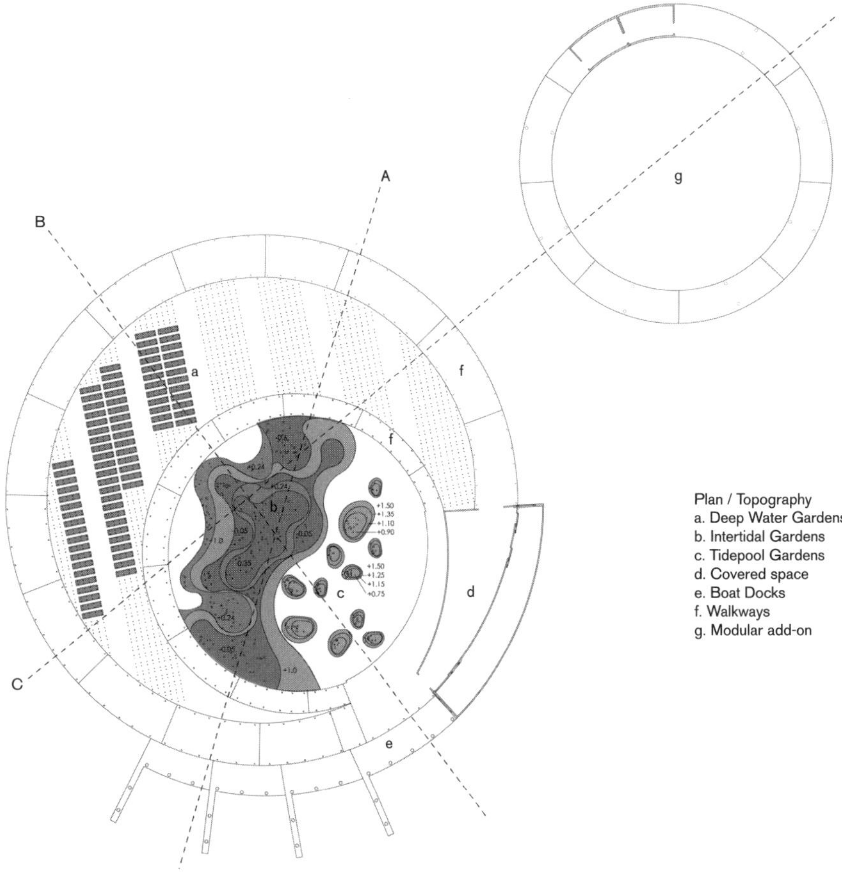

Plan / Topography
a. Deep Water Gardens
b. Intertidal Gardens
c. Tidepool Gardens
d. Covered space
e. Boat Docks
f. Walkways
g. Modular add-on

^ **Plan and intertidal topography.**

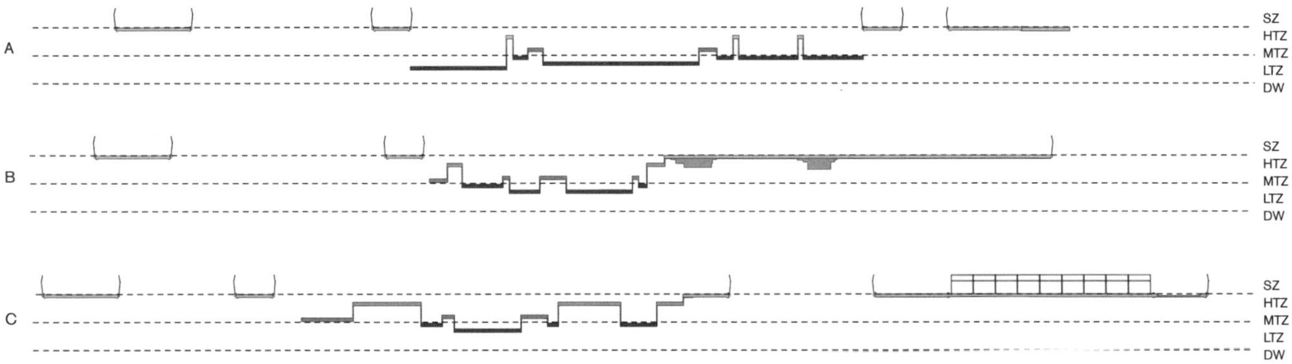

Submergence_Intertidal Topography

SZ	Splash Zone 1.5m+	0% Submerged	Seldom flooded, moisture from crashing waves and salty mist
HTZ	High Tide Zone .75 to 1.5m	0% Submerged	Flooded during high tide only
MTZ	Mid Tide Zone 0 to .75m	50% Submerged	Submerged and flooded for approximately equal periods of time per tide cycle
LTZ	Low Tide Zone 0 to -1.5m	90% Submerged	Only exposed at the point of low tide and for a longer period of time during extremely low tides

Section
a. Deep Water Gardens
b. Intertidal Gardens
c. Tidepool Gardens
d. Covered space
e. Boat Docks
f. Walkways

1m 10m

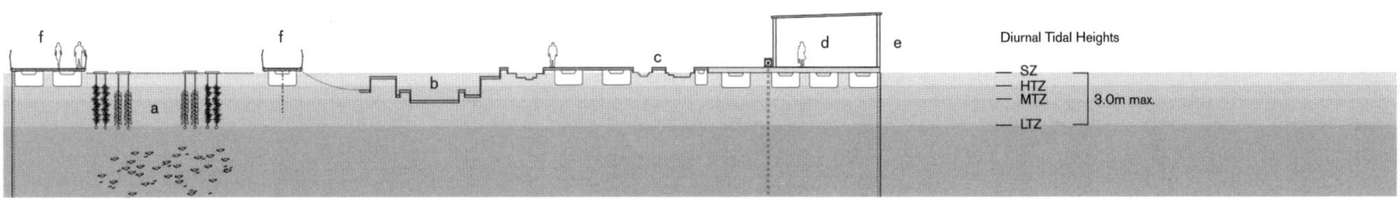

^ **Sections and intertidal topography.**

^ Kelp farming.

cash crops, energy landscapes

jason sowell

As a leading producer and consumer of electricity, Texas is poised to initiate some of the largest energy projects in the world. Overlaid onto traditional agricultural land, these new energy landscapes conflate crop production, livestock grazing, water harvesting, and wind farming. Subsequently, the land is prepared not only as an energy source, but also for energy processing and storage. The transition toward a renewable energy supply, however, occurs simultaneously to the continued extraction of nonrenewable resources, such that the production of food and fuel vie for groundwater stores. Inextricably linked to contemporary shifts in energy costs, development rights, or recovery technologies, the planning of wind farms, gas fields, and water harvests forecast the competitive nature of the state's infrastructural landscape.

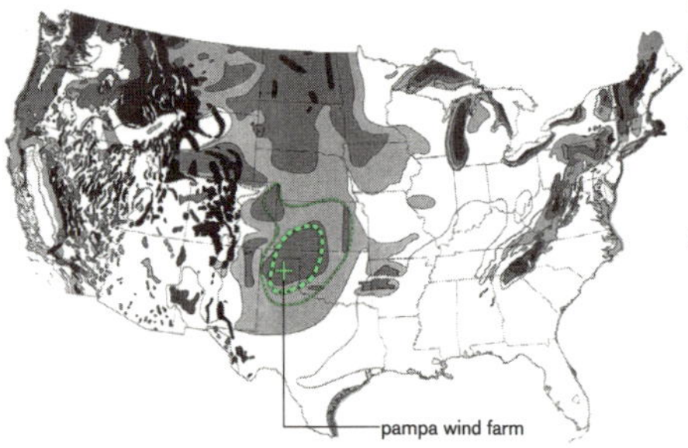

^ **Wind power class, annual average: Summer power class is shown in green National Renewable Energy Laboratory/ USDOE.**

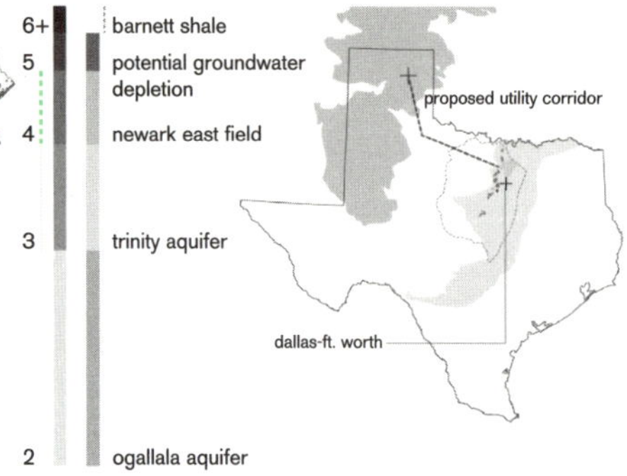

^ **Subsurface resources, Texas.**

Wind Farms

On July 8th, 2008, utilizing a combination of print media, television, radio, and internet advertising, Texas oil magnate and billionaire T. Boone Pickens released *The Pickens Plan*, a three-step, ten-year energy platform intended to facilitate the United States' transition toward energy independence. Led by private industry, "step one" would implement the world's largest wind farm; stretching across the Great Plains from Texas to Canada, harnessing wind speeds averaging nine meters per second as a renewable resource capable of supplying 22 percent of the United States' energy needs. "Step two" would construct the necessary electricity transmission infrastructure connecting the wind farm to metropolitan centers in the nation's Midwest, South, and Western regions. "Step three" would redirect natural gas resources currently utilized for electricity production to transportation needs. As a replacement for oil-based fuels, this last step would require the construction of a natural gas fuel station distribution network, and with that the re-engineering of the nation's automotive fleet.[1] At the time of the *Plan's* release, crude prices per barrel had hit an all-time high, with the week's trading average closing at $133.32.[2] For that same period, the equivalent volume of natural gas averaged $63.72.[3,4]

T. Boone Pickens' *Plan* provides a conceptual framework by which the United States could upgrade, if not significantly redevelop, a major sector of its energy infrastructure. However, the substitutions and offsets the *Plan* proffers are grounded in Pickens' ownership of extensive land holdings and mineral rights, which grant Pickens the ability to harvest large volumes of wind, natural gas, and groundwater; build power and pipe lines without the need to purchase additional utility rights of way; and transport energy and water to metropolitan markets facing future shortages, particularly water.[5] As a precedent for similar resource extraction enterprises, the regional economic and environmental benefits he has identified convey only a portion of the effects such change would impose on the Texas landscape. This raises the question of how the more

1. T. Boone Pickens, ***The Pickens Plan***, http://media.pickensplan.com/presskit/2008/pickensplan_press_unveils.pdf [accessed January 27, 2009]. The Plan is currently on indefinite hold due to a number of economic and legislative factors that have transpired since its release. Aside from meeting stiff opposition from the state's citizens, particularly those whose land would have to be leased in order to complete the utility rights of way, Pickens' BP Capital was weakened by the economic downturn, thereby reducing access to needed capital. In addition, the 10th Court of Appeals ruling granted Tarrant County access to water sources north of the Red River in Oklahoma. These sources would keep Lake Bridgeport at constant levels, thereby effectively eliminating an outlet for groundwater Pickens intended to harvest from the Ogallala Aquifer. Were Pickens' Mesa Power able to have secured the outlet, he would have been able to combine an electrical project with a water district, allowing the project to compel sale of land through eminent domain.

2. United States Energy Information Administration [EIA], ***Weekly United States Spot Price FOB Weighted by Estimated Import Volume [Dollars per Barrel]***, http://tonto.eia.doe.gov/dnav/pet/hist/wtotusaw.htm [accessed January 27, 2009].

3. ***U.S. Natural Gas Wellhead Price [Dollars per Thousand Cubic Feet]***, http://tonto.eia.doe.gov/dnav/ng/hist/n9190us3m.htm [accessed January 27, 2009]. Natural Gas averaged $10.62 per thousand cubic feet in July 2008. EIA conversion factors, per BTU output, sets the equivalent volume of natural gas as approximately 6000 cubic feet per barrel of oil. Multiplying the gas price by 6 allows for a rough one to one comparison in U.S. dollars. Refer to EIA Conversion Factors, http://www.eia.doe.gov/oiaf/archive/aeo01/pdf/apph.pdf.

4. Jordan Weissmann, "On the Road Again: Pump Prices Revive Appeal of natural Gas on Capitol," ***The Washington Post***, August 26, 2008. http://www.washingtonpost.com/wp-dyn/content/article/2008/08/25/AR2008082502546.html [accessed January 27, 2009]. The cost of natural gas for vehicles is roughly two-thirds of gasoline. Of 176,000 fueling stations in the United States, approximately 2,000 supply natural gas. It should be noted that natural gas powered vehicles emit 20 percent less gases than their gasoline counterparts.

5. S.C. Gwynne, "H_2NO: The Last Drop." ***Texas Monthly***, February 2008, 169-71, 261-2, 268-9. Projections from the Texas Water Development Board for Region C, the 11 counties of the Dallas-Ft. Worth Metroplex, anticipates a potential water shortfall of 1.93 million acre feet by 2060.

^ Desert Sky Wind Farm. Iraan, Texas: With a 160.5 megawatt (Mw) capacity, the wind farm contains 107 turbines spread over a 15 square mile area. At 20 times the size, T. Boone Pickens' Pampa-based wind farm would generate 4,000 Mw from 2,000 turbines placed over a 312.5 square mile area. In context, at present, the state contains 40 farms producing 7,116 Mw, with each turbine accounting for $3,000-5,000 in royalty fees per year.

negative impacts the natural gas industry imparts—particularly on groundwater resources—might be mitigated.

Gas Fields

Development of Pickens' 4,000 megawatt, Pampa-based wind farm—only the first component of a regional infrastructure—parallels the increasing extraction of light sweet crude and natural gas.[6] The state of Texas, as the leading supplier of natural gas in the United States, accounted for nearly 30 percent of the nation's production at 5.17 trillion cubic feet in 2006 alone.[7] During that period, the nation's second largest natural gas field, the Newark East Field, produced merely 716 billion cubic feet.[8] Part of the Barnett Shale, the core of the Newark East Field is located primarily in Wise and Denton Counties, just north of the Dallas/Fort Worth Metroplex. Although discovered in 1981, extraction from the field did not become economically feasible until gas prices at the wellhead remained at or above $4 per thousand cubic feet, a figure that occurred first in 2003, and has been sustained since 2005.[9] Currently, the state's largest private natural gas distributer, Devon Energy, operates 7,000 wells, and routinely extracts up to 30 percent of the Newark East Field's yearly production.[10]

The gas field rests below a region that has witnessed the sequential farming of extensive surface and subsurface geologic resources. Founded in 1856, Wise County is home to the nation's third largest aggregate mine, as well as the seventh largest natural gas processing facility. Although the county transitioned during the 1930s from an agrarian—cattle, cotton, corn, wheat—to a largely mineral- and energy-based economy—clay, crushed stone, coal, oil, natural gas—the county contains 2,696 active farms covering 83 percent of the county area.[11, 12]

Given the amount of land that remains in agricultural production, much of the gas extraction and transmission easements are leased from private landowners. The lease agreements typically grant the landowner 25 percent of the profits from each well. Relatively unobtrusive, both visually and physically, pumping stations and pipelines simply coexist side by side with cattle and crops. However, despite the ability to operate each type of farm simultaneously, the extraction process is not environmentally benign. For example, the extreme amounts of groundwater required for drilling threaten livestock and crop production.

6. Pickens' ability to develop the wind farm was predicated on his gas holdings, if not the rising energy costs prevalent at the time that he announced his plan. The decrease in gas and petroleum prices, in addition to the economic downturn, have contributed to the project's demise. At this stage, Pickens anticipates the construction of a number of much smaller wind farms, each with the potential to be expanded later as funding and investment become available.

7. EIA, *National Gas Annual Supply and Disposition by State*, http://tonto.eia.doe.gov/dnav/ng/ng_sum_snd_a_EPG0_FPD_Mmcf_a.htm [accessed January 27, 2009].

8. Ibid, "Appendix B: Top 100 oil and Gas Fields." *U.S. Crude Oil, Natural Gas, and Natural Gas Liquids Reserves 2006 Annual Report.*

9. Ibid, *U.S. Natural Gas Wellhead Price.*

10. *Devon Energy 2007 Annual Report*, http://media.corporate-ir.net/media_files/irol/67/67097/reports/AR2007/index.htm [accessed January 27, 2009].

11. The third largest aggregate mine, in this case, refers to area, not material volume. At roughly 5,000 acres, the combination of several quarries owned by four producers [Hanson/ Heidelberg Cement, Martin Marietta, TXI, and Latimore Materials Inc.] forms one of the largest mining complexes in the state, if not the nation. The statistic stems from several conversations with a number of plant managers and engineering staff from Hanson Aggregates. The *USGS Mineral Industry Survey 2004* does not provide individual breakdowns of either volume or quarry area for the leading crushed stone producers [it does list Texas as the leading producer of construction aggregates, particularly crushed stone, in the U.S.]. The ranking of the natural gas field is based on total volume extracted in trillion cubic feet. Refer to DOE/ EIA-0216 (2007) *Advance Summary: U.S. Crude Oil, Natural Gas, and Natural Gas Liquids Reserves 2006 Annual Report.* Economic history and transitions can be found in Gregg Rosale's *Wise County History: A Link with the Past* [Nortex Press, 1975].

12. Texas State Historical Association [TSHA], *Wise County.* http://www.tshaonline.org/handbook/online/articles/WW/hcw14.html [accessed January 27, 2009]. 48 percent of the agricultural land was devoted to pasture, with livestock sales totaling 25.7 million; 44 percent was devoted to crops and earned 7.6 million.

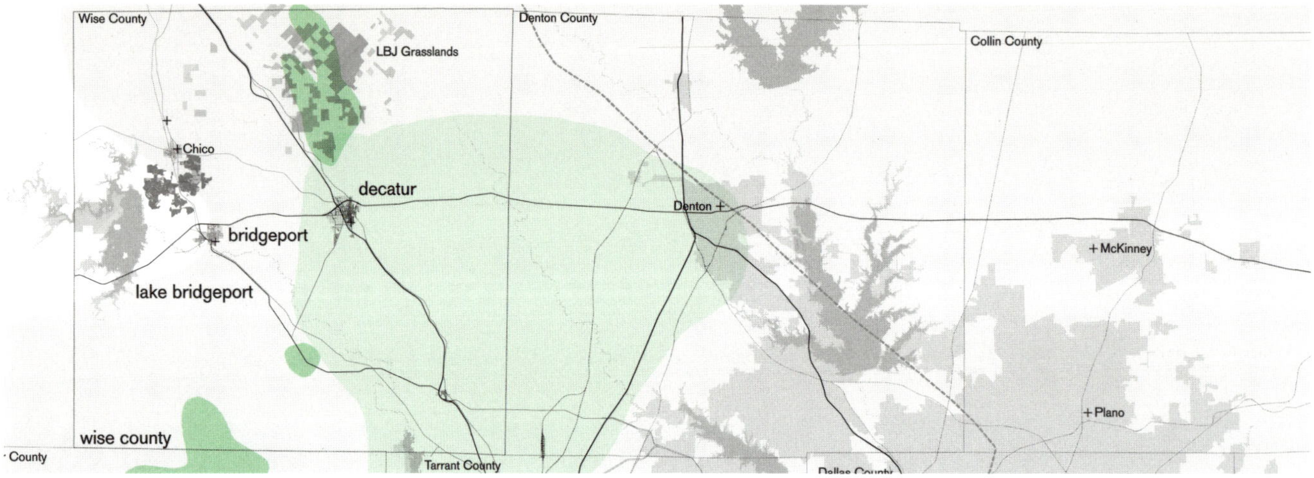

^ **Gas extraction, water depletion. Newark East Field: The use of groundwater—a relatively free resource as per Texas water rights—in the hydraulic fracturing process adversely affects rural users dependent on aquifer supplies for their water needs. Although the fracturing process uses less than 3 percent of groundwater stores, increased exploration could lead to a number of depletion zones throughout rural areas.**

Water Harvests

Faced with future water shortages, municipalities across the state have purchased large tracts of farm land in order to gain access to groundwater stores. The purchases were enacted in conjunction with changes in land use, whereby cropland, which utilizes more water through irrigation, shifts to pasture, which uses less water accommodating livestock. Provided the purchased farms overlay rechargeable aquifer sources, water districts maintain a renewable resource through sustainable water harvesting practices.[13] Seeking to cut water losses in industrial processes, which can account for ten percent of all water consumed, a number of recovery programs have been established by private industry, in particular those tied to gas extraction.

The Barnett Shale, an unconventional natural gas play with an anticipated reserve of 27 trillion cubic feet, has arisen in the past eight years as one of the nation's leading natural gas fields due to the development of hydraulic fracturing techniques.[14] With horizontal well depths reaching almost 5,000 feet, the initial drilling process utilizes as much as five million gallons of water per wellhead in order to access the trapped gas. And refracturing over the life of the well to maintain gas flow requires up to 20 million gallons.[15] Much of this water is tapped from groundwater sources, and accounts for up to three percent of the Trinity Aquifer's groundwater budget.[16] With 10,000 wells currently in place, water drawdown in sections of the aquifer could impact wells utilized for agriculture, namely watering cattle and irrigating crops. The drawdown, while a small percentage of total aquifer reserves, could pull water levels below depths that agricultural wells are able to reach.

During the fracturing process, 80 percent of the fracture water returns to the surface. Contaminated by both petroleum resources and suspended solids, the water's increased salinity—greater than seawater—requires disposal in either injection wells or evaporation ponds. The former has led to leaks into groundwater sources, while the latter requires landfilling of leftover sludge. In either case, up to four million gallons, or 12.2 acre feet, are removed from the aquifer with each wellhead costing almost $2,000. Over the life of the well, the potential tax on the aquifer budget could approach 16 million gallons, or 48.4 acre feet. The Texas Water Development Board estimates that with anticipated expansion in wellheads, total water demand in drilling operations could increase to 25,000 acre feet per year.[17]

Desalination efforts undertaken by Devon Energy recover 85 percent of produced water.[18] The distilled water from this process is harvested for future fracturing or injected back into the aquifer. For smaller drilling operations without mobile desalination units, remediation of produced water and its reintroduction to the aquifer or surface systems could also be accomplished through a series of storage ponds stocked with halophilic bacteria. These microorganisms are capable of living in extreme saline environments, and have been used in oil recovery and sodium carbonate production.

13. Fred Pearce, *When the Rivers Run Dry: Water—The Defining Crisis of the Twenty-first Century* [Beacon Press: Boston, 2006]. The transition from crops to livestock, ironically, fails to acknowledge the larger water budget, given that livestock feed requires large water inputs. The purchase of the farms is also not without contention, as evidenced by El Paso's purchase of 22,000 acres in nearby Dell City. For more information, see Thomas Fullerton's "Water Transfers in El Paso County, Texas," *Water Policy 8* [2006]: 255-68.

14. USGS National Assessment of Oil and Gas Fact Sheet, *Assessment of Undiscovered Oil and Gas Resources of the Bend Arch-Fort Worth Basin Province of North-Central Texas and Southwestern Oklahoma*, 2003, http://pubs.usgs.gov/fs/2004/3022/fs-2004-3022.html [accessed January 27, 2009].

15. Texas Water Development Board, *Northern Trinity/ Woodbine GAM Assessment of Groundwater Use in the Northern Trinity Aquifer Due to Urban Growth and Barnett Shale Development.* TWDB No. 0604830613. January 2007. Appendix 2. Total water use in hydraulic fracturing was estimated based on assumptions stated by the TWDB report. Refer also to M.K. Fisher, et al. "Optimizing Horizontal Completion Techniques in the Barnett Shale Using Microseismic Fracture Mapping." Society of Petroleum Engineers Annual Technical Conference, 2004.

16. Ibid, Executive Summary, 2.

17. EIA, TWDB No. 0604830613 Appendix 2.

18. Ibid, Appendix 2. Devon Energy was the first to initiate water recycling in the Barnett Shale, and remains the leader in water reclamation with 9 mobile desalination units. Since establishing the program in 2005, they have recovered 5 million gallons, or 15.3 acre feet, of water. See Devon Energy's *2007 Annual Report*, 15.

^ Drilling operations. Wise County, Texas: Increased energy prices since 2005 have led to a corresponding increase in the number of wells drilled. The fracturing process utilizes up to 5 million gallons of produced water, 85 percent of which returns to the surface during drilling. The water, which has absorbed subsurface salts and petroleum residues, must be disposed in either injection wells or in evaporative ponds.

V Mineral rights. Bridgeport, Texas: Much of the gas extraction and transmission easements are leased from private landowners, who are then granted a share of the well's profit. It should be noted that in Texas, mineral rights are separated from, and take precedent over, surface rights; those leasing the mineral rights can effectively dictate what activity can occur above. In Wise County, extraction wells account for 13,500 acre feet per year of fracture water, with each well providing lease profits of 25 percent.

VV Economic diversity. Bridgeport, Texas: 7,000 natural gas wells (at left) and 5,084 oil wells (pumpjack in background) are dispersed amongst Wise County's 2,696 active farms. Seemingly innocuous, the gas industry's water use competes with irrigation and livestock needs.

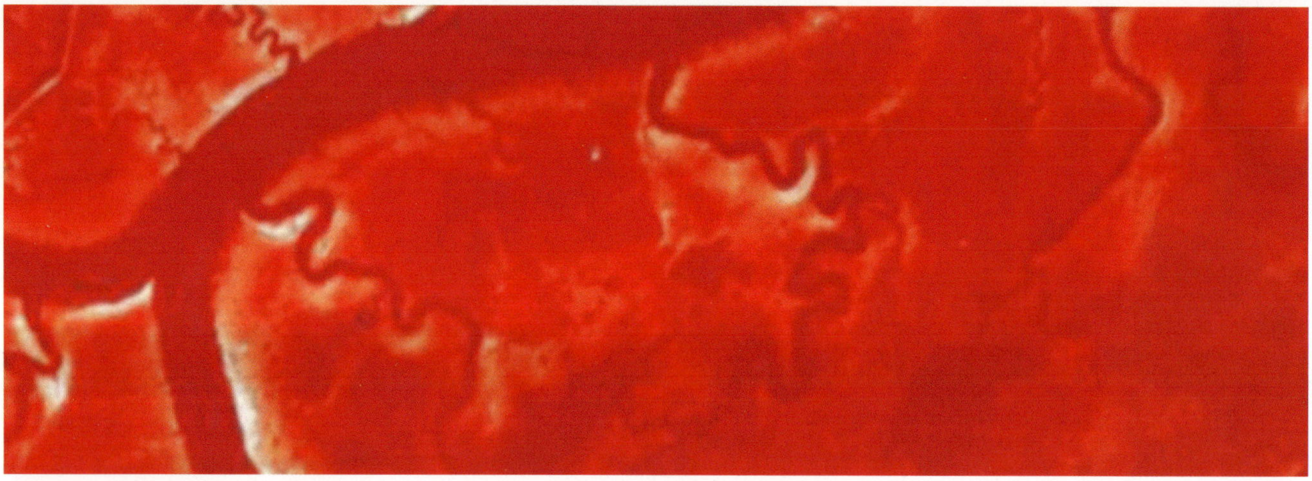

^ **Water treatment: Desalination can recycle 9,825 acre feet of produced water per year; the technology has proven economically viable, given that drilling firms spend $128 million per year in disposal fees, or roughly $1-2 per 1,000 gallons. As an alternative to mechanical desalination, halophilic bacteria can be utilized in the remediation of produced water. Tolerant of high salinity levels, the microorganisms metabolize petroleum contaminants, converting pollutants into an ATP energy source. Organized in a series of ponds constructed in abandoned aggregate quarries, the produced water can be trucked in, cleansed, and released into constructed wetlands prior to its reintroduction to surface water bodies or ground water sources. 2007 Google, 2009 TerraMetrics: San Francisco, Cargill Industries Salt Flats.**

The genus *Haloferax*, in particular, possesses the capability to metabolize petroleum-based effluents contained in the produced water.[19]

Agricultural Cogeneration

As the proposals above underscore, the productive nature of the Texas landscape involves the growing of food or the extraction of resources. These activities are frequently linked, though as seemingly singular enterprises, such that land area and respective water needs determine which type of operation is pursued. With state population projections anticipating increases from 22 to 67 percent, consumption of energy, food, and water will likely further stress available resources.[20] The scales of these industries and their related infrastructure thereby require a synthetic oversight, such that their implementation, cultivation, and renewal seamlessly utilize the wastes from one process as inputs for another. It is only through such a perspective that the state's cultural landscape will couple its economic ambition with its sustainable aspiration.

Research assistants: Michael Averitt, Ryan Buckley, Caroline Castello, Karen Pittman, and Sandra Veras. Sponsors: Dallas Urban Lab, Bridgeport Economic Development Corporation, and Jerome Frank Investments.

>>> **Jason Sowell is an Assistant Professor of Landscape Architecture at The University of Texas at Austin School of Architecture.**

19. A. Ventosa and J.J. Nieto, "Biotechnological applications and potentialities of halophilic microorganisms." *World Journal of Microbiology and Biotechnology*, vol.11, no.1 (January, 1995): 85-94.

20. Steve H. Murdock et al, *A Summary of The Texas Challenge in the Twenty-First Century: Implications of Population Change for the Future of Texas*, http://txsdc.utsa.edu/download/pdf/TxChall2002Summary.pdf [accessed May 05, 2009].

farming [park]: rail, roadway, and urban form today

austin nicholas tragni

All too often we see aerable land supplanted with parking and simultaneously the reclamation of abandoned parking lots into viable land. Long Island City Farming Park is a park-and-ride facility and urban agricultural farm that combines these two typologies to co-exist on one site, bringing the process of food production and consumption in contact with a major multi-modal transfer point between the car and New York City's existing public transportation network. The project will provide an alternative option for those accessing the city by car. It also challenges the conventional function of a park-and-ride facility by providing a greater good for users and the surrounding neighborhoods; connecting Long Island City and Sunnyside, Queens through a much-needed public green space.

Although designed for New York City, the park can serve as a prototype for addressing urban vehicular growth and food security in cities worldwide. The selected site in NYC is typical of what can be found surrounding other large urban areas whose existing public transportation infrastructure has an accompanying rail yard and a potential farming site. The project developed as a response to some of the impacts of urban growth, specifically addressing New York City's congestion pricing scheme. The potential for many of the surrounding neighborhoods of Manhattan's Central Business District to be even more congested with vehicles looking to avoid paying additional fares is a concern for the people living in those areas.

The selected site needed to satisfy a number of particular urban criteria. Both park-and-ride and urban agriculture desire to be located adjacent to a major transportation nexus with a large amount of vacant land. This adjacency facilitates the consumption of fresh produce at that multi-modal transfer point. An ideal such site is located north of the Sunnyside Rail Yards between Long Island City and Sunnyside, Queens; two growing neighborhoods east of Manhattan.

This site connects the urban farm to a larger existing network of food distribution within NYC, specifically City Harvest, which is an organization that collects and redistributes surplus food from NYC's restaurants to the hungry. The connection with City Harvest is the exchange of wasted food scraps, to be used as compost in the farm, for the delivery of fresh food to local markets and the enhancement of food security. The

^ Regional green spaces and walking distances.

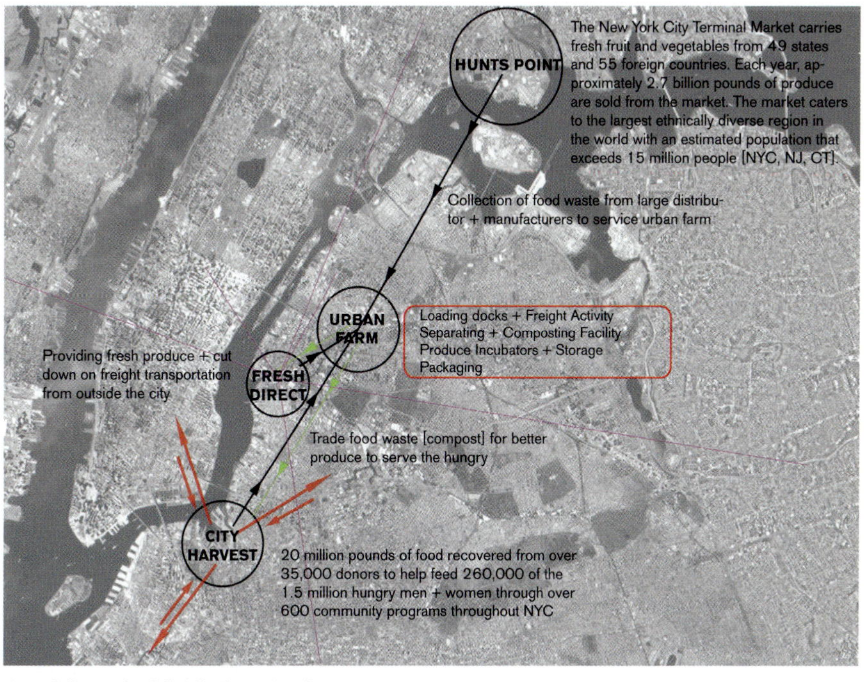

^ Adjacent food distribution network.

community-based infrastructure centralizes food production and creates a continuous productive landscape out of this enormous vacant land. The site also presents an opportunity to reconnect Long Island City and Sunnyside, which are divided by the Rail Yard's huge swath of open land, and provide a much-needed public green space. The existing green spaces surrounding these two communities are large cemeteries that do not function as successful public spaces.

Currently, Long Island City houses a number of cultural institutions that establish a unique corridor and place for the farm to be successful as a public gathering space. The site becomes the latest (agri)cultural institution to be included within the larger context.

The parking facility is hovering over the Rail Yards and is adjacent to Queens Boulevard, providing access for vehicles in both east- and west-bound directions. The constructed landscape is supported by a series of trusses. Depending on the span, the depth of the truss is deep enough to service the parking facilities, and shallow enough to support the soil of the growing beds. The site will be farmed using a biointensive method of double-dug raised beds, intensive planting, and companion planting with a range of vegetables to take advantage of a continuous seasonal harvest.

The landscape penetrates the elevated Number 7 train facility to bring a public green market space into the building to become a point of food consumption for those traveling between car and train. The proposed vertical green towers on site are localized gathering facilities acting as large indoor greenhouses above with public amenities below such as cafes and a farmers' market. These towers act as "farm frames" to suggest the stitching of the two adjacent neighborhoods by bringing visitors to the site along its corridors. The project suggests a multiplicity of landscape layers that provide a completely new experience of the simultaneous occupation of a parking lot, a park, a farm, and a transit station.

>>> **Austin Tragni, a Tokyo-based designer, is passionate about the natural and constructed landscapes within u rban environments.**

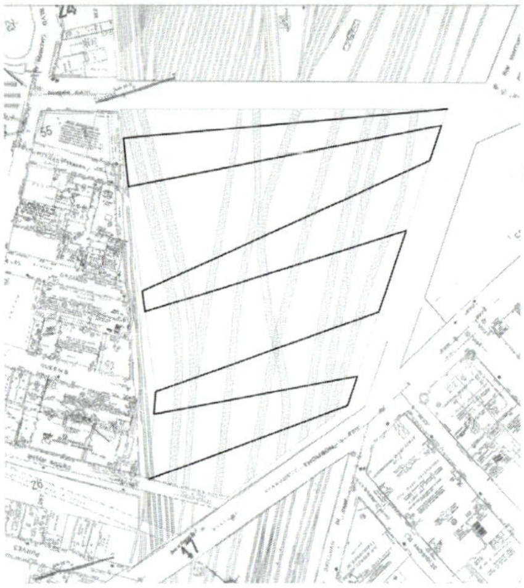

^ **Urban corridor.**

^ **Urban pattern.**

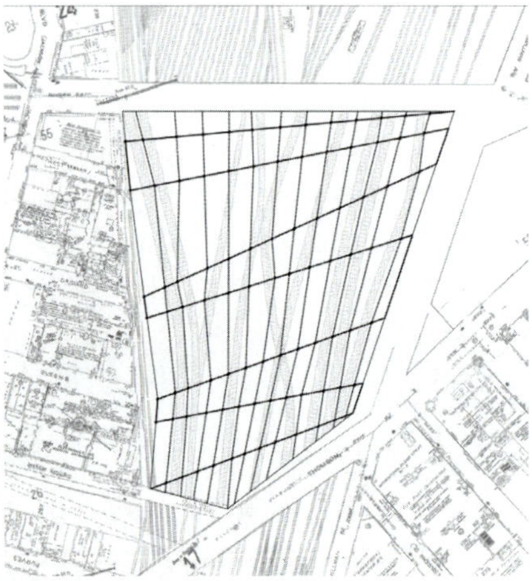

^ **Structural organization.**

^ **Parking layout.**

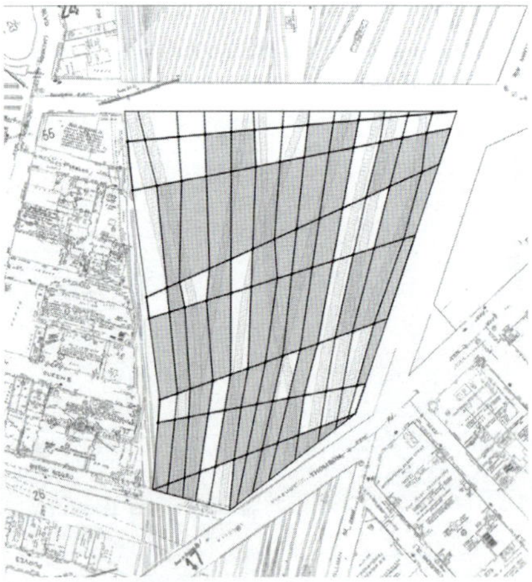

^ **Fragmentation.**

^ **Crop layout.**

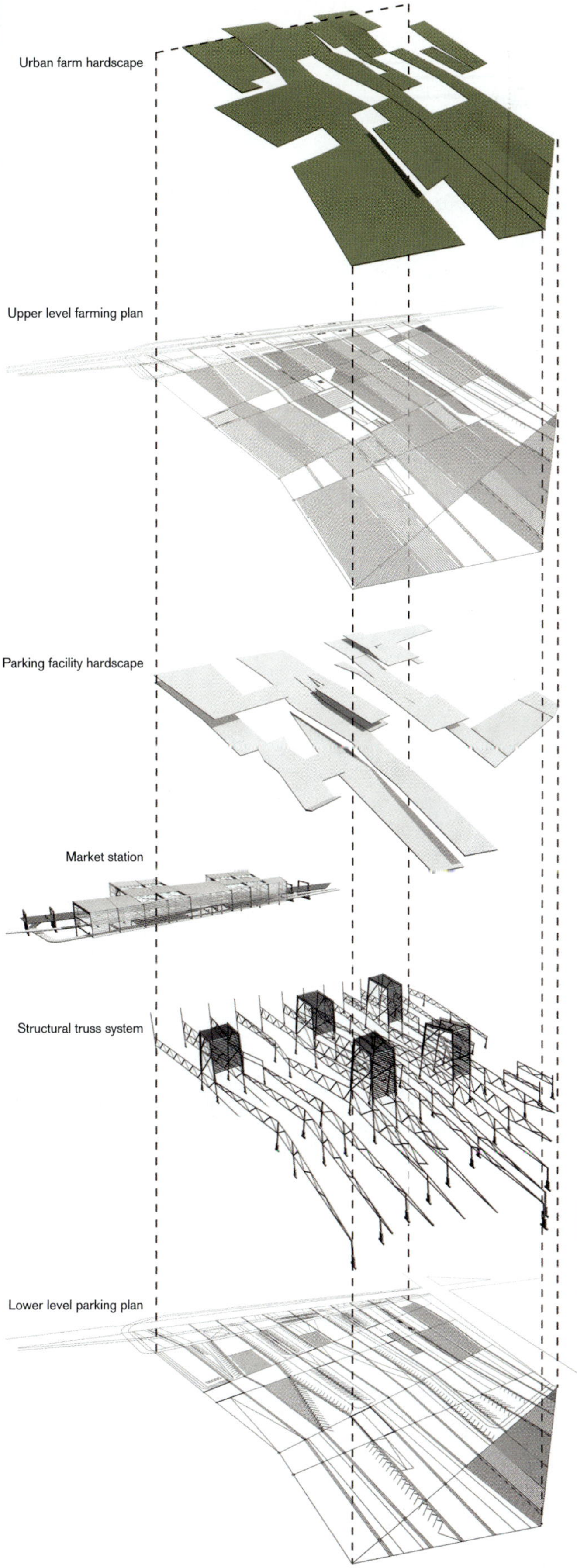

^ **Exploded perspective showing project layers.**

^ **Horizontal farm.**

^ **Transportation hub.**

^ **Farmers' market and parking.**

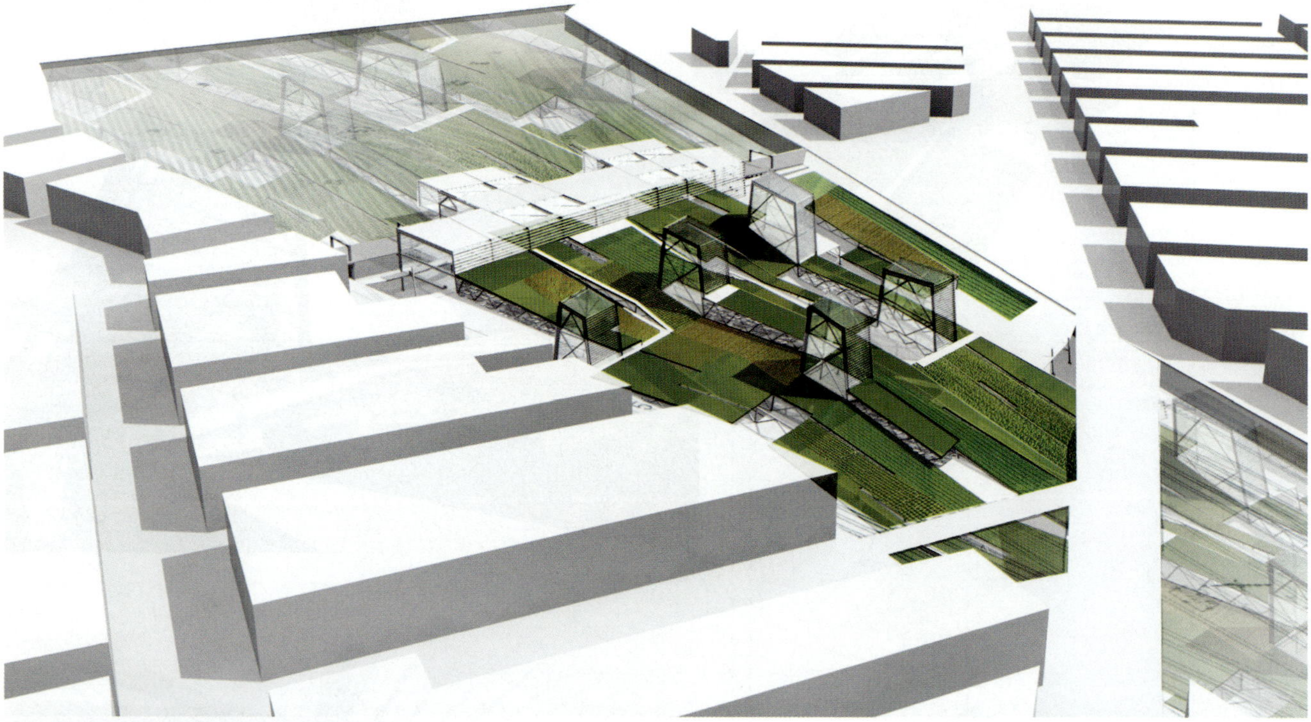

^ Daytime aerial view.

^ View of farm from public lawn.

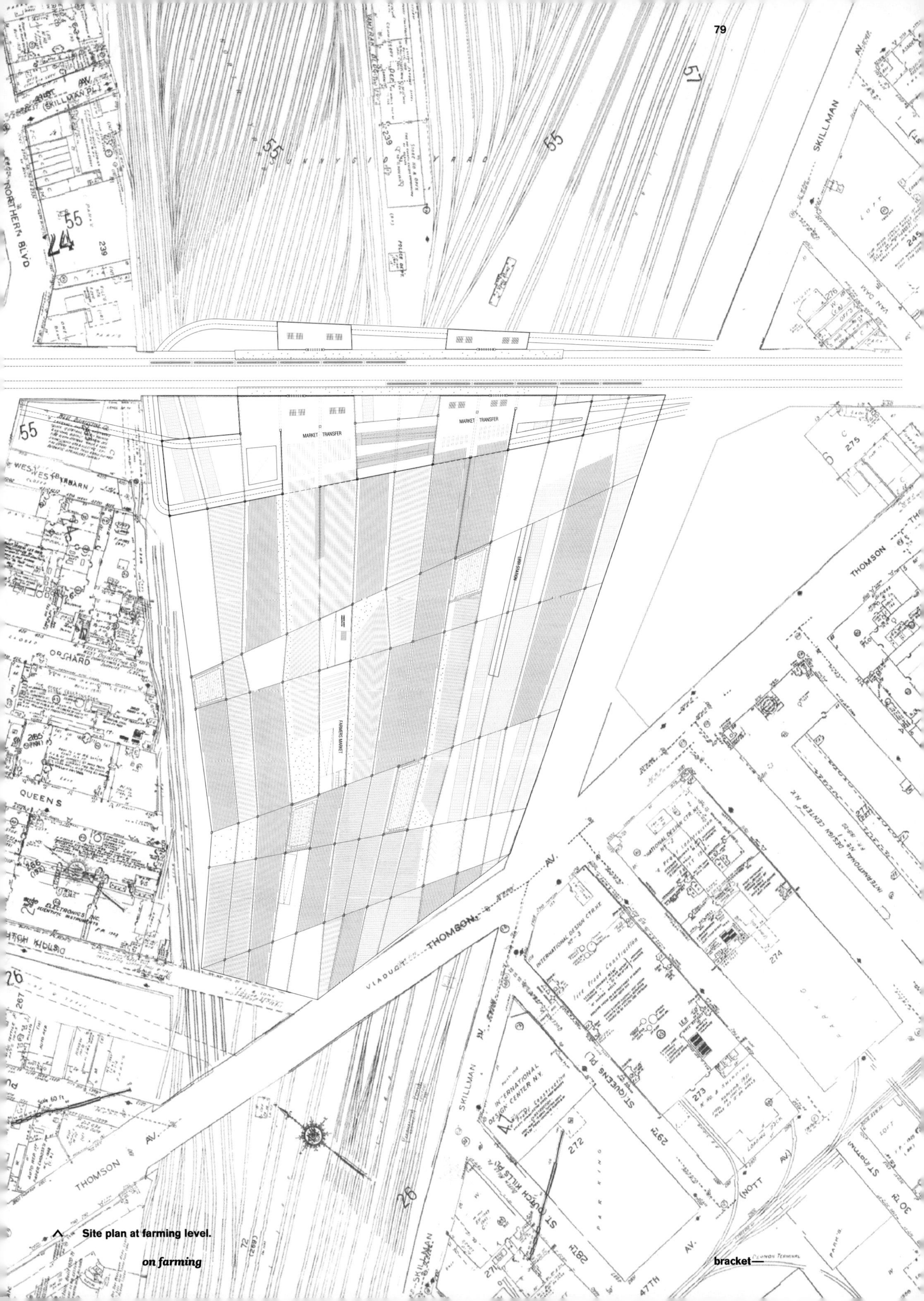

^ **Site plan at farming level.**

^ Sectional perspective showing intersection of food and transport infrastructure.

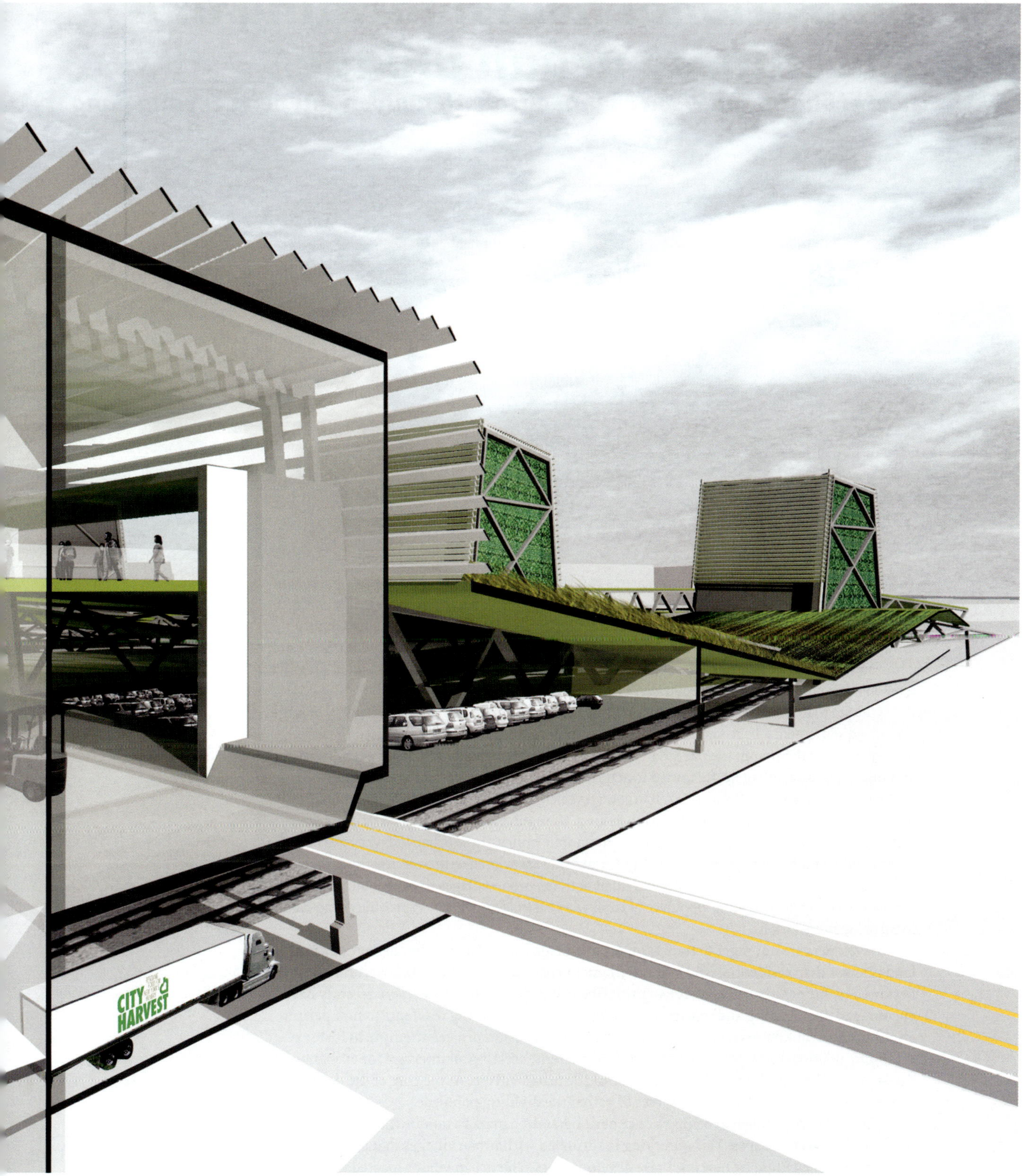
CITY HARVEST

migrational fields: farming and the chinese urban village

neeraj bhatia, marissa cheng, elizabeth nguyen, liu peng & yang jiang

^ Growth of Beijing from (left to right) 1959, 1983, 1991, 2000 and 2006. As the urban density sprawls, it engulfs the rural villages and forms urban villages that are disconnected from their source of income—the agrarian land that once surrounded them.

To be an urban or rural citizen in China carries vast distinctions in rights and duties. The *hukou* system was implemented by the Chinese Government as a means to register and classify each citizen as either rural (*nonghu*) or urban (*chenghu*), effectively polarizing the populace into lower-class farmers and upper-class urban citizens. The establishment of stringent policies to deter rural citizens from acquiring urban status and the associated benefits has resulted in grave tensions that culminate in the formation of an 'urban village.'

Unique to China, urban villages are rural villages that once occupied the periphery of large urbanized settlements. Post-1978 sprawling urban metropolises eventually encircled these villages, detaching them from their agrarian source of income. Surrounded by a wall of skyscrapers and infrastructure, villagers are forced to acquire 'urban' jobs but cannot do so legally due to their rural citizenship. A census taken in 2000 revealed that 3.8 million rural-urban migrants were living in over 300 urban villages within Beijing. Unregulated and untouched by centralized urban planning, infrastructure construction, and public policy, these urban villages have become *de facto* independent enclaves of informality. Commonly associated with overcrowding, inadequate infrastructure, social problems and illicit activity, urban villages have become proletariat sponges, soaking up a 'floating' populace of rural workers (*liudong renkou*) to provide cheap labor in urban agglomerations.

Developers are not eager to transform urban villages, as compensation for 'agricultural' land is lower than residential land and resettlement costs are too decadent. Governments have bestowed blame on urban villages for stigmatizing urban development and more recently have encouraged demolishing and replacing the villages, often relocating residents to new peripheries hundreds of kilometers away. The proposed urban design project attempts to foster a symbiotic relationship between urban and rural citizens, creating an interface for exchange as a design template to address the critical issue of the urban village. At the core of the proposal resides a productive landscape that serves to gather the dividing populace.

The planned greenbelt at Beijing's once peripheral zone marks a unique transition between urban and rural landscapes. This outer ring is sprinkled with urban villages that are quickly being encroached upon by new high-rise developments. The urban village located at the intersection of the planned Sun Palace transport station, the greenbelt, and the Bathe River is emblematic of several urban villages, imbued with a tenuous future as the reality of modern infrastructure construction threatens to destroy the settlement. The proposal begins with modifying Beijing's regional greenbelt plan, differentiating the vast landscape into productive zones—areas to farm clean water, air, food, and energy. These inject productive characteristics into the land that can be monitored, maintained and harvested by rural citizens. The re-establishment of productive land is further explored in the urban village. The morphology of the village as it urbanized reveals traces of its past—dimensions of the rice paddy fields established a critical planning block. Moreover, irrigation canals that once ran transversely through

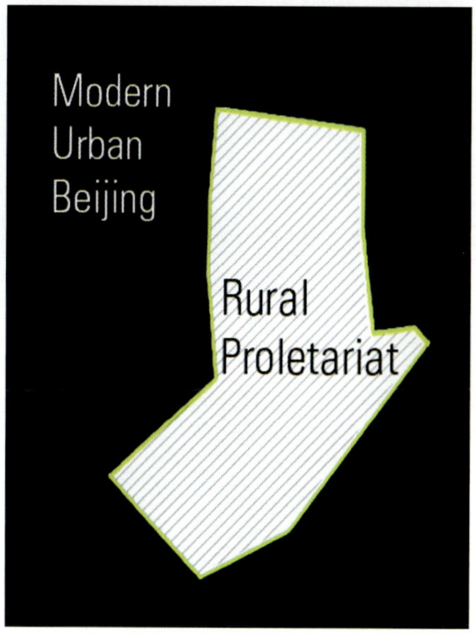

^ **Typical Urban Village (left) and Sun Palace Urban Village (center): Urban islands of the rural proletariat surrounded by modern China.**

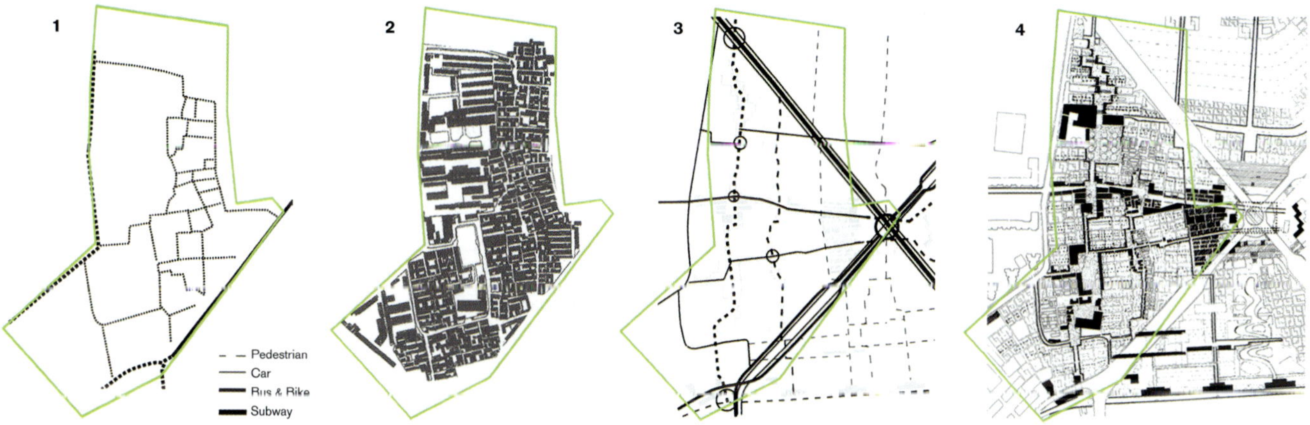

^ **1. Existing circulation; 2. Existing textural form—reveals rice paddy configuration and irrigation canals; 3. Proposed circulation—hierarchy of primary, secondary and tertiary routes based on the farming structure; 4. Proposed textural form and civic structure—textural form is integrated into the rice paddy module while the public civic structure links the existing north-south axis with the new transport station.**

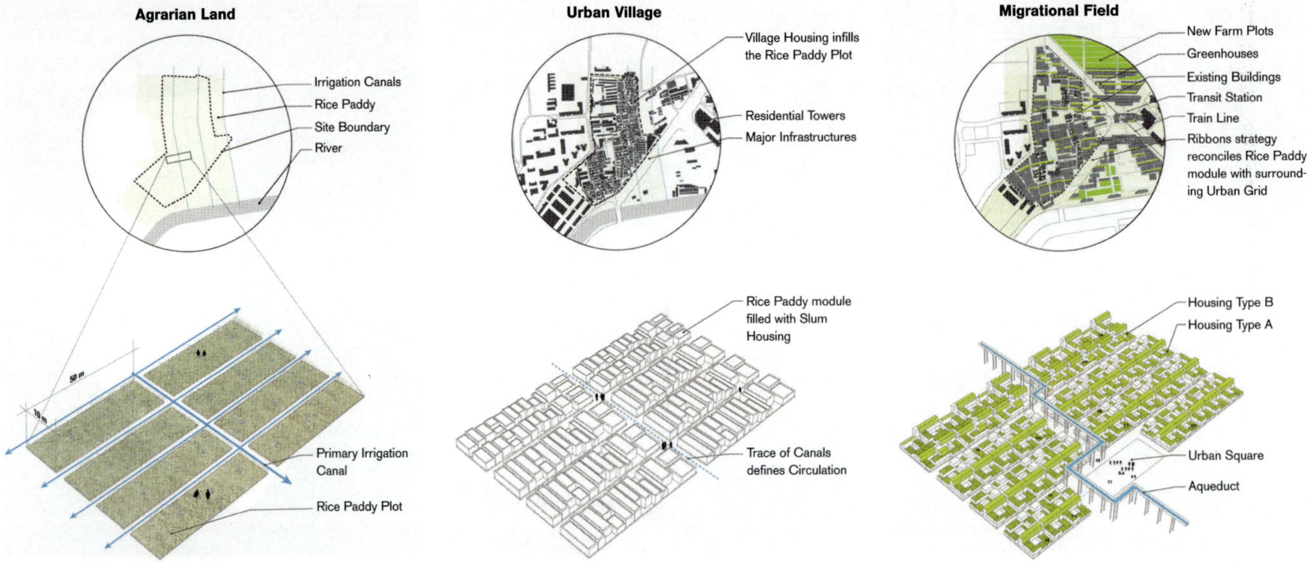

^ Evolution of the site from agrarian fields to urban village (left to right). The proposed scheme hybridizes the productive landscape with the urban structure to create a migrational field of exchange.

the fields have been replaced by circulatory streets. The proposal utilizes a series of 'ribbons,' which mimic the location and scale of the rice paddy configuration, to run throughout the site and mutate to link into the adjacent contemporary block configurations as well as the productive greenbelt. This framework is cut transversely by elevated irrigation canals that organize a series of public streets and plazas into a civic structure of major and minor axes. Laudable existing buildings are given new public functions and incorporated into this hierarchical framework. Three new housing typologies are developed that integrate rooftop farming, high-tech greenhouse farming, photovoltaic energy farming, and living quarters around traditional courts. These types are dimensioned to the block pattern and can be combined in numerous ways to create a series of topographic conditions—reminiscent of the terraced farms in rural China. The morphology of the settlement hybridizes the natural and artificial, the productive and consumptive, the living and working. One iconic moment, the transport station, serves as a civic indicator of the project, housing a large greenhouse, education, and research center.

New forms of farming and education provide a venue for migrant workers to increase their skills, inviting their potential urbanization. Urban citizens are able to consume the products of the farmers while scientists and horticulturalists learn from those who have vast intuitive knowledge of the land. The urban village is reinvented as a Migrational Field, a place of symbiotic exchange between traditional and contemporary methodologies, production and consumption, natural and artificial, and past and future. China's growing population requires massive amounts of water, energy, and food—and there is no reason why this cannot be done within the rural villages that have survived the sprawl of modern cities.

Research supported by Paul Sun Travel Grant, the Massachusetts Institute of Technology and Tsinghua University.

>>> Neeraj Bhatia is an architect and urban designer, partner of The Open Workshop and co-director of InfraNet Lab. He teaches at the University of Toronto and University of Waterloo.

>>> Marissa Cheng is a Boston-based architect and urban designer, interested in the social and spatial phenomena of food and consumption.

>>> Elizabeth Nguyen is an architect working towards licensure at Epstein Joslin Architects in Cambridge, Massachusetts.

>>> Liu Peng is an architect who graduated from Tsinghua University, Beijing.

>>> Yang Jiang is an urban planner with degrees from MIT, Cambridge and Tsinghua University, Beijing.

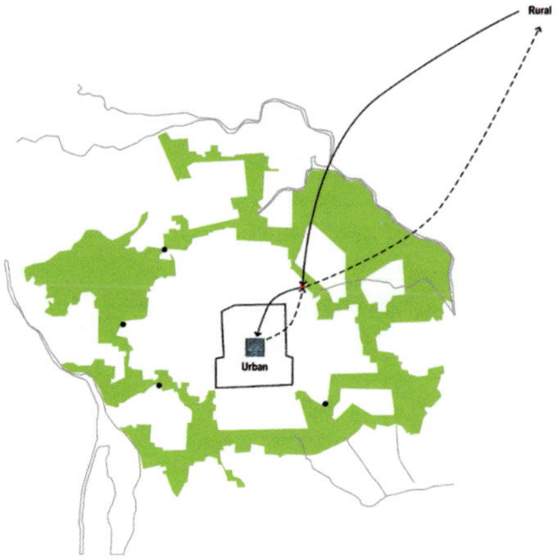

^ 1. Transform urban and rural lifestyles.
The site acts as an incubator for the integration of urban and rural lifestyles. It provides a forum for city residents (i.e., artists, sociologists, ecologists, etc.) and migrant workers to share aspects of their lifestyles (economic and social), creating a hybrid between rural and urban forms of living and working.

^ 2. Sustainable infrastructure.
The site is part of a larger system of sustainable infrastructure tied to the greenbelt, which has the potential to be productive—providing space to farm food, water, air, and energy. The site draws on the greenbelt, the canal system, the wetland, and housing, to maximize that potential. These systems are integrated into the housing typology at the unit, cluster and neighborhood scales.

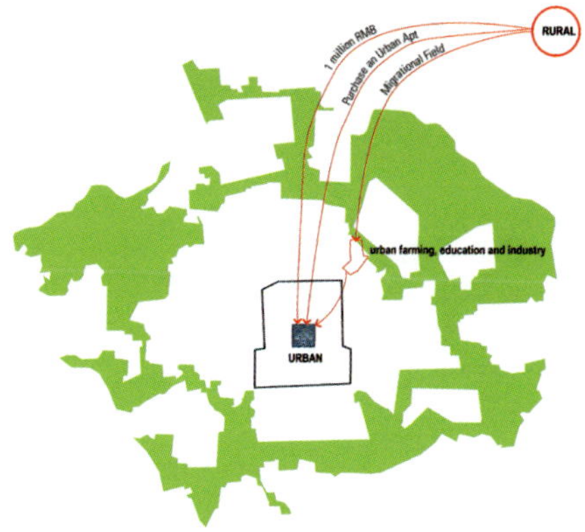

^ 3. Phasing and renewal.
a. Phasing of people entering and leaving the site: The site offers an alternative route for the urbanization of rural families. Through a process of "urbanization"—business education, light manufacturing and selling of goods—the economic activities of the city become available to the rural migrant.

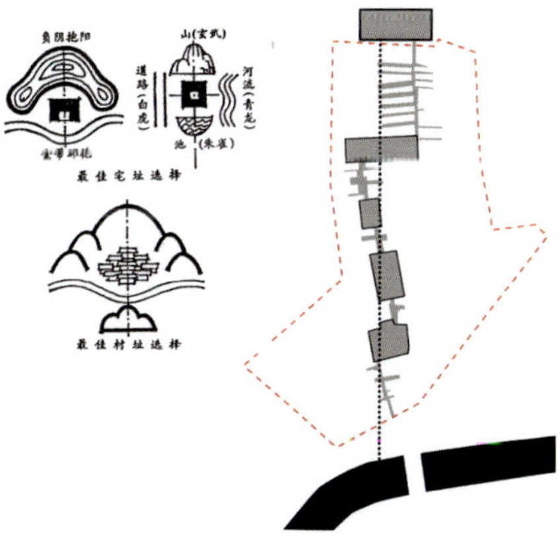

^ 4. Redefine the axis.
Formerly, a north-south axis linked temples within the site and aligned with the Temple of the Sun. The excavation of this axis and the reinvigoration of several existing buildings along it re-establish a local public way that becomes the central armature for the local community, containing a library, school, health care facilities, shops and restaurants. A series of fountains emphasize the aesthetic and productive value of water along the main axis, distributed from a high-rise water tower at the top of the axis to the rooftop farms via an aqueduct.

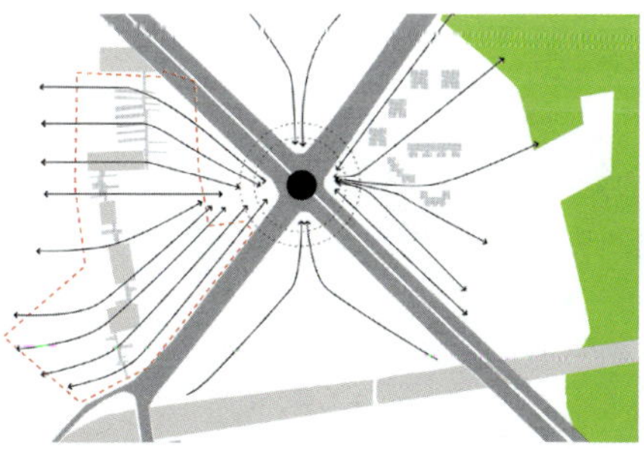

^ 5. Station as collector and distributor.
The station is an essential exchange point, drawing and distributing traffic to and from the road and subway systems, parks, water, historic axis, and local community. A market area linked to light industry mediates the space between the station and housing. The educational center and greenhouse located above the station create a regional marker and densify activity at the transport hub. The roads are lifted to allow the fabric and key layers of the project to flow through the station. This allows natural light and activity to occur in close proximity to the train platforms, embedding the station into the farming community.

^ 6. Reinterpretation of the low-rise, high density housing typology.
The site's existing housing will be replaced with low-rise and high-rise housing, integrated into the grain of existing housing and organized into linear "ribbons." The housing typology allows for flexibility and growth, while providing variability. Sustainable infrastructures—farmland, aqueducts, and greenhouses—are integrated into each unit.

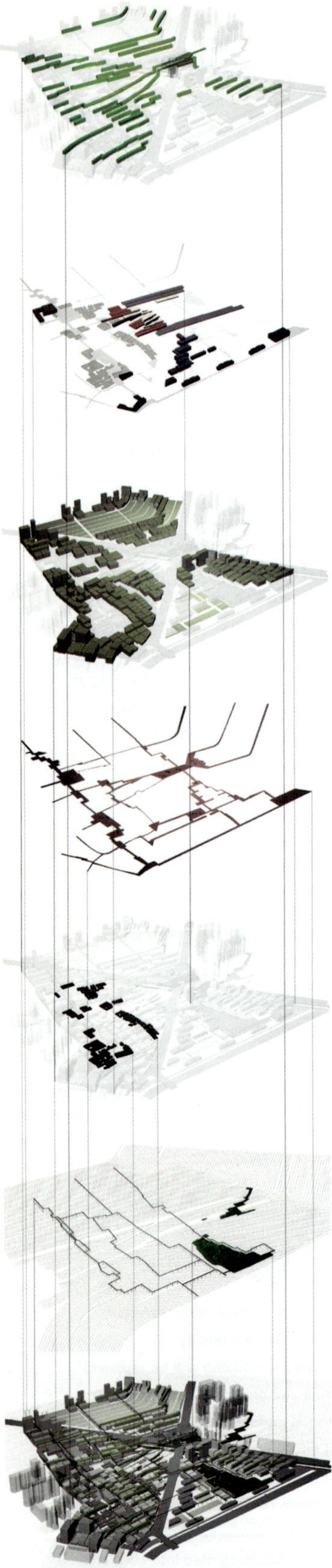

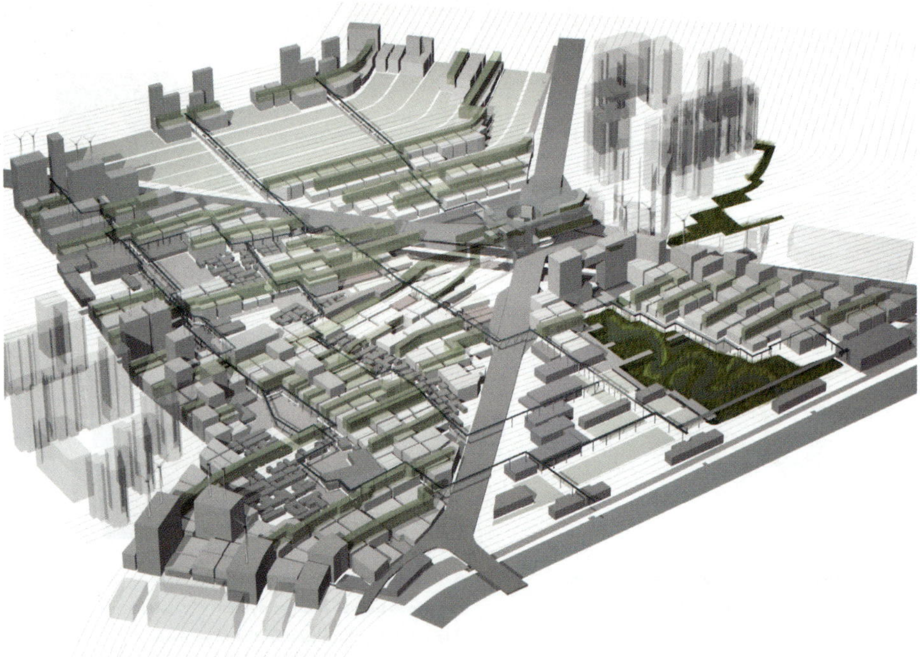

^ Full build out of the site, showing the relationship between the farming and public infrastructure.

^ Exploded layers of the proposition reveal traces of the rice paddy structure. Bottom to top: Full build out; Aqueducts and primary irrigation system; Existing buildings; Civic structure; Housing types; Public buildings; Greenhouses.

^ The weaving of new housing, retained buildings, greenhouses and irrigation canals into a new landscape.

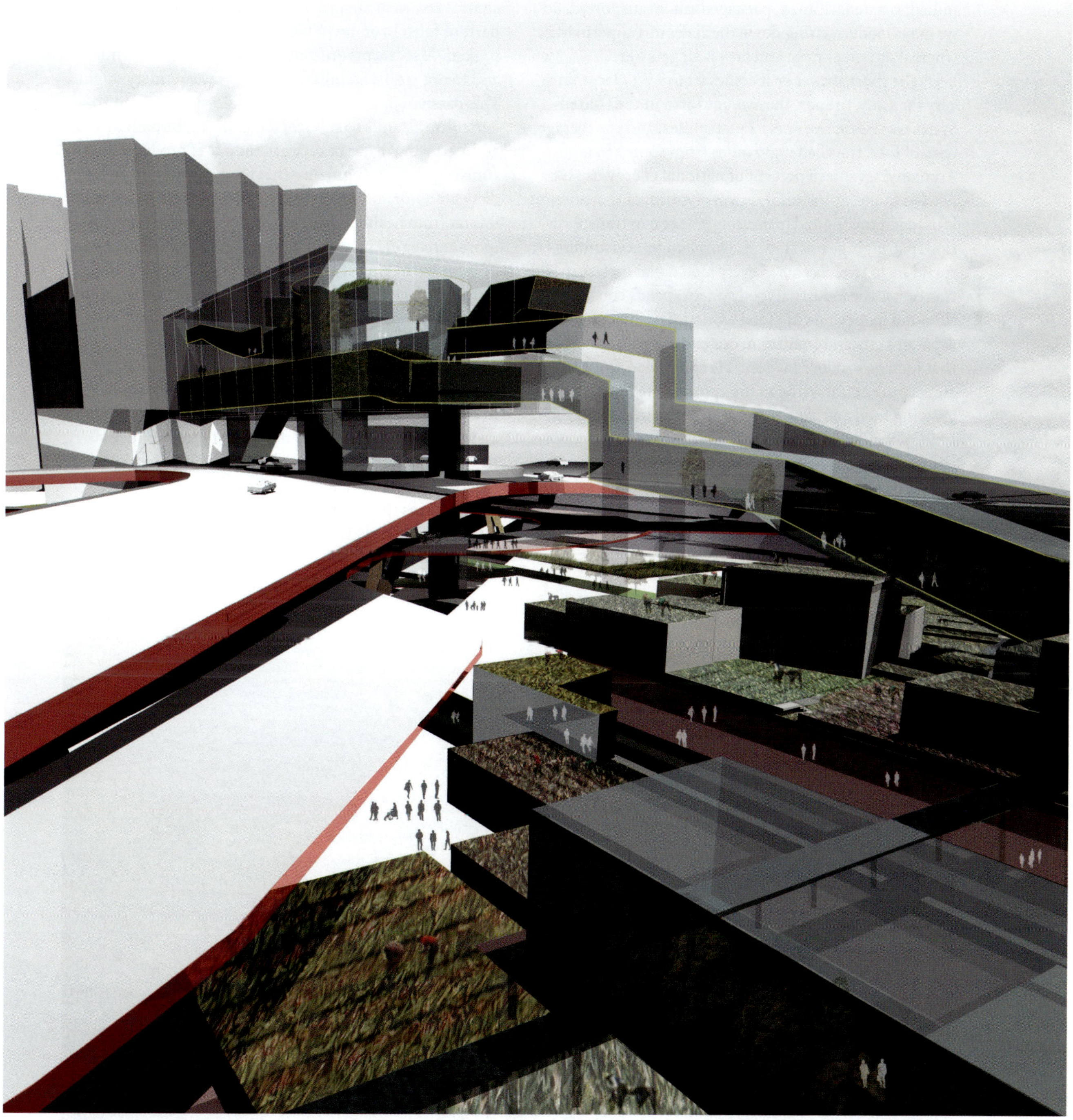

^ View of transport station and research centre with farming housing terraces in foreground.

farm plus: hybrid agricultural landscapes

ellen burke

The landscapes that produce our food are invisible in the supermarket. We cannot see the cornfield in the cereal aisle or the vegetable rows in the produce aisle, nor can we see the means by which the raw materials are processed, milled, roasted, cleaned, packaged, and transported. So we may wonder, staring down the aisles of a supermarket, where it all comes from and how it all gets here.

The short answer is: it comes from a very large farm, very far away. In 2007, the average farm in the United States was 449 acres, or 0.7 square miles, and the average item of food travelled 1,500 miles from source to table. Annually, seventeen percent of national energy use is expended in the agricultural sector; one-fifth of that amount is used in growing and harvesting, the rest in transportation, packaging, and storage. On average, consumers spend six to twelve cents of every food dollar on transportation costs. The industrial food system spends considerable time and money moving food around. The dualism that separates city and country in our culture resists the idea that food be produced where it is consumed. We envision a pure countryside as the appropriate location for growing food and cannot imagine farms in metropolitan centers of culture, commerce, and consumption. Should cheap supplies of fuel fail us, both growers and consumers would be put in a vulnerable situation by this system. If sustainability is a measure of the balance of a system's inputs and outputs, the current food system that expends eighty-seven calories of fuel to transport one calorie of food coast to coast is wildly unsustainable.

The design professions have not focused on access to food, as much as they have on clean air or water, as central issues of sustainable design. This is a result, in part, of faith in an invisible system that does not appear broken. Also, farms are generally perceived as harmless and to not require regulation in the way a refinery might. This passive approach has led to wholesale conversion of agricultural lands to suburbia in many metropolitan areas and a growing distance between the site of production and consumption. Recognizing that food is as important to daily life as air and water, and that the system that supplies it is far from healthy, how can the design professions take a more active role in shaping the sphere in which farming occurs? Farm Plus is an investigation of mechanisms by which designers can create and preserve farmland in urbanized areas.

Creating Farmland

Suggesting that food be produced where it is consumed, in cities and suburbs, seems somehow romantic, foolish, and utopian. While we may never be able to produce all the food we need in our backyard, there is potential for insertions, acts of radical agriculture, or cooperative partnerships that supplement the industrial system while questioning its wisdom. Evidence that the public supports

^ **Alemany Farm, four-acre farm on public land, San Francisco, CA.**

this approach exists in the fully-subscribed community supported agriculture (CSA) programs across the country, in the growing number of farmer's markets, and in the renewed popularity of victory gardens in residential yards. These movements indicate the public will to transform the way we eat.

Several major obstacles to locating farms in urbanized areas exist: high real estate values make it unrealistic for farmers to buy or rent land at market prices; there is limited availability of adequate parcels of land, as a working farm will require at least one acre; perceived nuisances of farm operations may create resistance from neighbors; and zoning laws restrict agricultural uses in urbanized areas. Mechanisms for countering these obstacles include conservation easements, right-to-farm agreements, and lease agreements with multiple private owners.

Another strategy, examined closer here, is to foster partnerships between landholders and farmers, creating hybrid agricultural landscapes. There are precedents for cooperative ventures, of farms inserted in housing and parks, bordering playgrounds and wetlands, producing food for their immediate communities. Successful examples include: a CSA in Portland, Oregon that operates from two locations—16 acres on city-managed open lands and 12 acres in a public park; and a four-acre farm on city park lands in San Francisco, California that sells at the local farmer's market.

In the cooperative venture model, the landholders may be private developers, non-profit organizations, or land trusts; or they may be public entities, such as cities, counties, or utility companies. The farmers may be employees of the land manager, tenants, part of a collective lease, or leasing several small parcels in an urban area. As a result of these partnerships, the agricultural operations act as part of hybrid landscapes, in conjunction with other landscape programs, such as public parks, planned housing communities, city-owned open space and university or museum campuses.

The projects examined here are working farms, small businesses with a goal of production and profit, rather than demonstration or educational projects. They are small in size, anywhere from seven to fifty acres. Markets for their produce include CSAs, farmer's markets, restaurants and produce brokers. Typically, as part of the partnerships, farmers are provided with basic infrastructure, including irrigation and fencing. Often a business plan has been prepared by the landowner or a non-profit to guide more inexperienced farmers to financial self-sufficiency. In some cases, farmers are provided with start-up capital. Permits and variances are acquired for agricultural use by the land owner or manager. Following below are brief case studies of three successful partnerships, each representing a different configuration of the owner-manager-farmer relationship.

Sandhill Organics

Sandhill Organics is a 16-acre farm located in the planned subdivision of Prairie Crossing, about an hours drive from Chicago. Prairie Crossing is a conservation community of 400 homes, set in a restored native grassland. The farm was conceived as part of the master plan for the community, and a sales fee from new homes funded the farm's infrastructure. The current farmer has a long-term lease for the land with the developer. Established in 2004, Sandhill Organics sells at farmer's markets and runs a 280-member CSA. One-fifth of the CSA members live in

^ **Sandhill Organics, Greyslake, IL.**

the surrounding development, and more residents shop at the on-site farmer's market. The farm has three year-round employees, and at the height of the season employs ten people.

An important part of the success of this project is because the farmer owns the business. Long-term management of an agricultural operation by a committee of non-farmers, such as the homeowners association, could result in poor business decisions. Initially, in this case study, a farmer was employed by the developer, but arbitrary rules made it difficult for the business to succeed. Leasing to an experienced farmer allowed the business to be run as needed more successfully.

Careful boundary design also contributes to the success of the overall landscape. Bordered by homes on three sides, the farm has never received complaints from the residents about noise or other disturbances. Topography and plant material are employed to separate homes from the farm. Sitting on a rise above the farms, the adjoining homes get views to distant fields, and a walking trail that connects throughout the development offers closer views of the farm. The unexpected location also contributes to the success of the project. The farm is near stores, schools, and neighbors and is within an hours drive of customers in the city. Yet the farm is rural enough to benefit from fuel and fertilizer deliveries from the agricultural co-op.

Appleton Farms CSA

Appleton Farms CSA is located on 25 acres in a public park an hour east of Boston. The Trustees for Reservations (TFR), a land trust, manages the park, which is approximately 700 acres and features walking trails, horseback riding and bird watching. Prior to being deeded to TFR, the land was an active dairy farm with operations dating to 1638. Given the history of the land, TFR committed to keeping active agricultural uses on the property, along with recreation and conservation programs.

This is an example of the employee model in which TFR employs the farmers and provides start-up capital with a business plan. Established in 2002, the Appleton Farms took four years to become financially self-sufficient. It is now fully subscribed—growing produce for 530 families in the area—and operates with revenues.

Given the rural setting, TFR understood that not competing with the neighboring farmers was crucial to the success of the project. Research indicated that no organic CSA was operating in the area, and this directed the choice of farm type. This strategy has proven to be successful, as there has been no complaint over privatized enterprise on public lands. The unique quality of public interaction is also a strength of the project. Rather than receiving produce deliveries, subscribers come to the farm to pick up their share. Produce is laid out in the barn for people to choose or customers may go into the fields to harvest some items themselves. This strategy can be implemented in urban or suburban farms to create strong links between consumers and the agricultural landscape through direct experience.

Sunol Ag Park

Based on the business park model where separate entities rent parcels of a larger development, Sunol Ag Park hosts 6 farmers on 18 acres. Developed by SAGE, a local non-profit, the park is located on a public utility easement, about an hour east of San Francisco. SAGE holds a long-term lease with the utility company and provides

^ Pick-your-own fields at Appleton Farms CSA, Ipswich, MA.

ongoing management of the Ag Park. Sunol Ag Park is also a recreational park with access to a small architectural icon at one end and a bike trail around the perimeter of the farm.

The tenants include a co-operative of immigrant farmers, a CSA for low-income subscribers and a specialty tomato farm. Each farmer leases an area from two to seven acres that are separated from each other by rows of sunflowers. The infrastructure, including irrigation, deer fencing, and a small office, was developed by SAGE prior to leasing to the farmers. Markets for the produce grown in the Ag Park include farmers markets, restaurants, produce brokers and CSA shares. Farmers pay rent and water costs to SAGE.

Identifying underutilized open land in metropolitan areas for conversion to farming is an important strategy for increasing access to land. Utility easements are a good prospect, given the stable nature of the land ownership and allowed uses. This type of project requires the presence of a third party, such as a non-profit, to handle the permitting, negotiations and management of the tenants and to act as the developer.

Parcelization of the larger acreage increased the number of sites available to farmers, which is of particular benefit in urban and suburban areas. Experienced farmers tend to own land in more rural areas, while farmers interested in city and suburban sites tend to be less experienced. With a greater number of smaller parcels available, the farmers can begin with a small operation, requiring less start-up investment. As they become more experienced at growing and marketing, they can expand their business, perhaps eventually finding a full-scale farm off-site. Working under this model also means that farmers can share equipment and avoid the burden of purchasing expensive items on their own.

Conclusion

These agricultural landscapes are important insertions in the urban and suburban context at many levels. As a process by which land can be used for agricultural operations near the source of consumption, these landscapes enhance the food security system of a community. By placing farms on land accessible to the public, either physically or visually, they provide an important spiritual link between consumers, the food they eat, and the processes by which it is produced. Finally, by reducing the distance between consumers and producers, these landscapes contribute to a more environmentally sustainable food system.

The farms examined here represent one segment of a wide field of agricultural experiments—small farms practicing outside the norm of very big and very far away. This is a form of agriculture that is not behind the scenes, but is very much part of the visible community. They serve as examples of the mechanisms by which landscape architects and planners can create and preserve farmland in urbanized areas.

>>> **Ellen Burke is a San Francisco-based landscape designer with a particular interest in productive landscapes and urban ecologies.**

^ **Zenger Farm, Portland, OR.**

45°50'8"n 119°41'57"w: hybrid-poplar farm

ryan lingard

Potlatch Hybrid-Poplar Farm

At 65 mph, a solid 5 minutes of Oregon's I-84 freeway travel is spent adjacent to this farm. A wall of green lines the Interstate's southern edge between the sleepy town of Boardman and its nearest neighbor, Hermiston.

With most crops, this event would go largely unnoticed. The rigorous cadence and uniform order of a farm typically leaves a traveler to find more exciting or less predictable available stimulus. But at a vertical height of almost 100 feet, standing proudly amongst the rolling sage brushed plains that tumble into the Columbia River Gorge, and a span of nearly 6 miles in each direction, the Potlatch Hybrid-Poplar Farm cannot be missed. Invoking mystery and astonishment, this stand of crop not only affects its vehicular passersby, those who take the time to walk its fields, experience how landscape can impact human experience.

In-Between Landscape

In 1993, the Potlatch Hybrid-Poplar Farm was planted in an otherwise common, arid landscape. This kind of terrain is found along Interstates and highways all across the western United States and is often experienced only as travel background. This is largely due to the difficulty people have in finding comfort in roadside landscapes such as these. Their open exposure, and plant and animal life is foreign in comparison to the developed, domestic environment.

In the American West, these landscapes are seen as both disposable and generative. It is this often forgotten, in-between landscape that bomb-testing grounds, nuclear waste sites, and landfills call home. Ironically, by introducing minute quantities of water and infrastructure, this property can become a highly productive and desirable commodity. In pursuit of growth, its original character is frequently disregarded and its image remanufactured. This discarded land has long been claimed by suburban developers as sites for master planning their

speculative communities or by farmers to house their expansive fields of crops. Its low land value and lack of human sentiment leaves it vulnerable to either terrible scars of human intervention or, in the case of the Poplar Farm, gloriously exhibit our ingenuity in cultivating life.

Growth and Order

These in-between regions of our country frequently experience heavy modification on behalf of growing crops as well as dwellings. This terrain is seen as a sort of empty, multi-disciplinary canvas for civil engineers and farmers alike to inject order and life. A once wild, untamed landscape, through planning and planting is brought under the control and order necessary for efficiently propagating life.

The trees' regulated 10-foot row offset allows machinery to easily access for harvest and provides equal distribution of sunlight and water. Computer controlled systems throughout the farm provide automated diagnosis and management processes. This conforming, scientific control of nature, creates a spectacular collection of regularized trees, but, in turn, emphasizes the subtle differences in each plant. Without the regular spacing, linear distribution, and controlled branch-trimming, it would difficult to distinguish the difference in bark patterns, variations in coloring, or the damaged branches of the trees. The machine designed to control nature in turn emphasizes the very natural phenomena of chaos.

Scale

A farm's scale is typically understood as the area its fields claim, a largely two-dimensional experience. When crops develop through growth cycle, their height adds a perceivable, yet nominal dimension relative to their horizontal area.

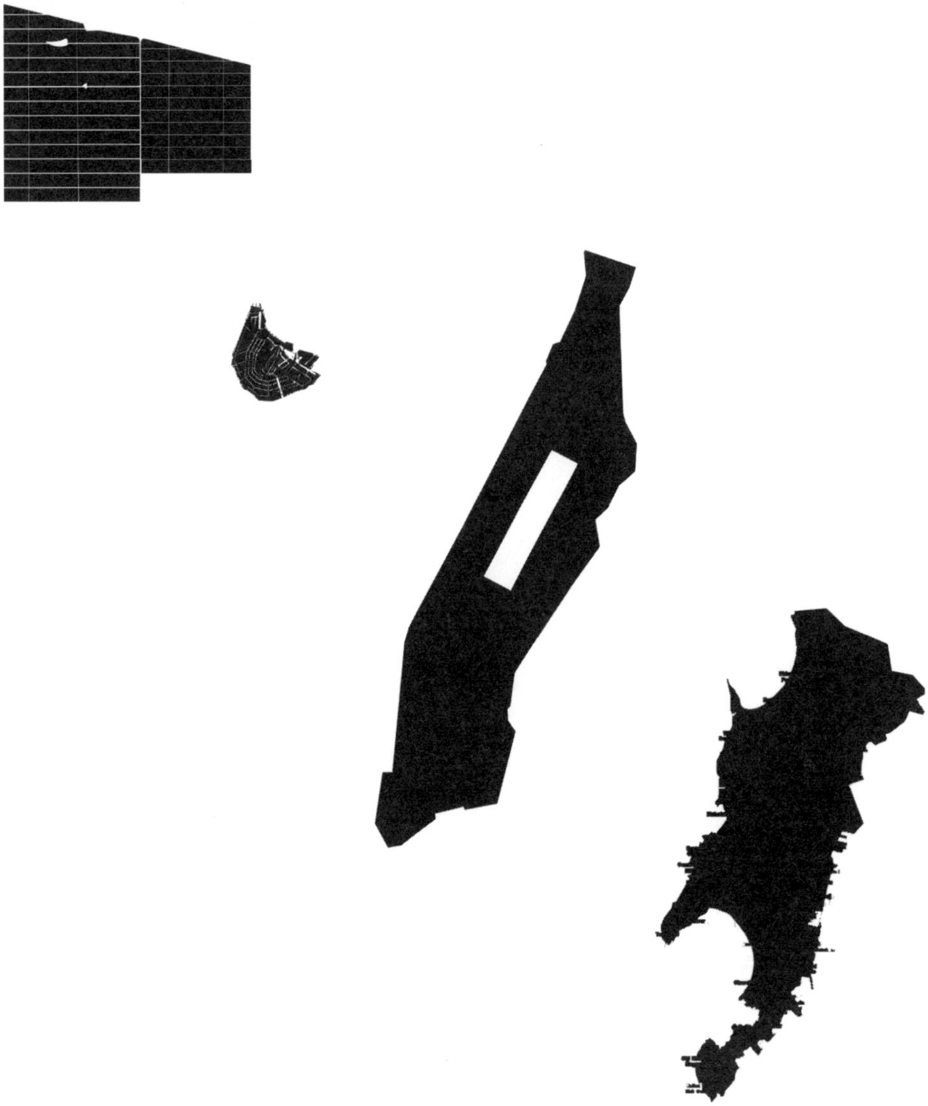

Potlatch Hybrid-Poplar Farm 26.6 square miles	274,780 trees per square mile	26.6 square miles = 17,000 acres = 68 square kilometers	7,171,760 trees 400 deer 2-3 cougars
Urban Center and Size	**Density**	**Area of Poplar Farm**	**Supportable Population**
Amsterdam City Center 4 square miles	11,549 people per square mile		301,430 people
Manhattan Island, New York City 34 square miles	27,147 people per square mile	x 26.6 square miles	708,540 people
Mumbai City 26 square miles	56,669 people per square mile		1,480,060 people

Ten-by-ten foot spacing between each tree yields 7,171,760 trees on this 17,000 acre farm. While this is far denser than the most compact of human settlements, a significant population could be supported on this property. When applying some of the world's urban population densities, calculations yield ample area to provide for entire urban communities.

Standing amongst the poplar trees feels like standing in an enlarged scale model of a rye grass field, or a 100:1 scale diorama of a wheat pasture. One can be removed from the human perspective, to imagine the distinct perception of scale of other places and other creatures. But at the 1:1 scale, with heights of almost 100 feet, the poplar tree's size puts humans in a humbling position to plant life.

Volume

For 16 years, this mass has occupied the landscape. But while adjacent fields have experienced a rotation of agricultural uses, this soil has supported a monoculture of plant life; a bound plot dedicated to the hybrid poplar. This plantation abuts the Interstate on one side and a landscape otherwise free of visible property lines on the other three sides. In the context of the open range, the tree farm holds its Interstate frontage as if it's a pioneering urban development. Its back edges, equally as abrupt, appear to either anticipate the arrival of similar developments, or push up against an invisible or imagined boundary. The farm sits in the landscape like a maverick, stubborn speculator fixed on its claim.

Yet what appears stationary and solid is in fact porous and constantly changing. While perceived as opaque from afar, the mass becomes porous upon approach. When aligned with the crop rows, view corridors stretch to undetermined terminuses bathed in light. Within the diaphanous solid, a series of roads and organically shaped voids provide land for the farm's work areas and outbuildings. Further, the fabric of the seemingly stoic landscape mass is constantly

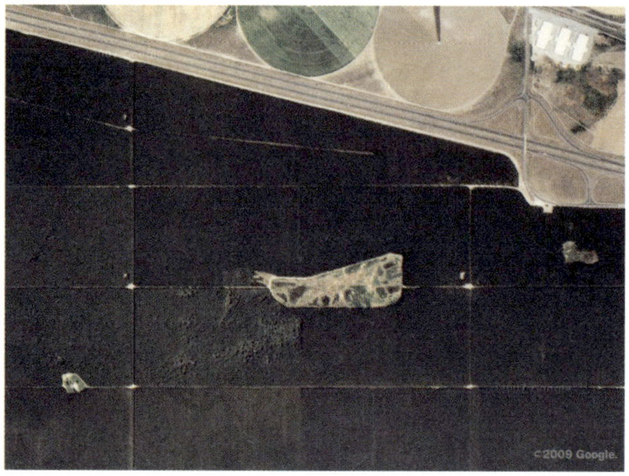

changing. Rectangular sections are harvested at a rate of 1 tree every 16 seconds. Tree heights are constantly increasing, and seasons shed and bring volume to its foliage. This mass, though maintaining its edges and contents, is constantly morphing.

In looking for parallel phenomenon in similar crops, corn comes to mind. The density and vertical dimension of corn supports the phenomenon of seasonal crop mazes. What are otherwise read as solid blocks in the landscape, act as hidden, internal playgrounds. Carving circulation and space out of these fields creates imprints of space. A field supporting a playful environment one day can be a plowed swath of soil the next. The following year, the field becomes a new imprinting canvas.

Harvest

Just over a decade ago, all across America, tree farms were planted to supply the paper industry with pulp. Hybridized poplar was common because of the ability to engineer accelerated growth cycles, producing quick, renewable yields. Patches of these crops can be found throughout the country. Some are located amongst rural dwellings and owned by individual families, and others as expansive farms run by corporations such as the Potlatch Hybrid-Poplar Farm.

Potlatch, a wood products company specializing in paper, started this farm in 1993. Planted adjacent a long arid stretch of Interstate 84 in Morrow Country, Oregon, the farm neighbors expansive open terrain, Umatilla weapons depot (a burying ground for retired ammunition), a mix of agricultural crops and the sleepy town of Boardman. Expansive computer infrastructure was built to facilitate the farm, automating all the farm's operations except its harvesting process. Its magnitude and success can be attributed to its proximity to the Columbia River. Extensive irrigation infrastructure supplies the Potlatch Farm with upwards of 175,000 gallons per minute of water.

In the 1980s and 90s, such significant pulp crops were justified because the paper industry primarily used virgin wood pulp for production. Nevertheless, the farm could not compete with the increasing popularity of recycled paper. While 90% of paper worldwide is still made from virgin wood pulp, the use of recycled, deinked paper pulp rendered the Potlatch Hybrid-Poplar Farm financially impractical. While Potlatch continued to maintain the farm and harvested trees for pulp for over a decade, it began looking for alternative uses for its highly developed crop. The Farm remained largely underutilized for years, waiting to find a new consumer and caretaker.

In 2008, the Potlatch Hybrid-Poplar Farm was acquired by GreenWood Resources, a supplier of sustainable wood products to the building industry, who has established it as a Forest Stewardship Council certified tree farm. The wood is commercially available and being used for interior trim, casework, sheet material and furniture through wood distributor Collins Wood. Remarketing the crop, farmers now trim the trees lowest branches to yield knotless lumber and have established a new trademarked name for the crop, Pacific Albus™. The farm's staying power rests on its engineering, infrastructure, and quantity. The crop's consistency, quality, and ease of harvesting, which once benefited the thriving paper pulp industry, is now serving its contemporary counterpart, the construction industry.

Cycle

Irony and poetry lies in the rise and fall of the Potlatch Hybrid-Poplar Farm's various caretakers. First, the ambition to grow a renewable resource in an adversarial landscape to produce something as common as paper. Then, the viability of its mission is undermined by its more sustainable counterpart, recycling. Next, the ambition to improve the highest consumer and waster of wood products in the country, the building industry, reclaims the farm's purpose to

make interior, residential wood finish materials, seen almost as frequently as paper. And, now, the success of this product hinges on the fate of a failing housing market.

Yet, against all odds, what has remained constant is the care and cultivation of a crop, no matter its purpose. Despite the change in consumer demands and the improbability of the undertaking, the infrastructure and quality of the crop has remained relevant. Driven by both the human desire to grow and the compassion to sustain life, this farm has continued to weather the storm.

>>> **Ryan Lingard is a designer living in Portland, Oregon. His work includes environments, objects, and images.**

the productive surface

mason white

Out of a legacy of high modernism, a history of form, technique, function, and technology has dominated the representation of architecture's and landscape's surfaces. This history has perpetuated a binary interpretation of the role of the architectural surface—either form and technique driven or function and technology determined. Form privileges the expressive, a response to the question of what the surface could be; while function privileges the operational, a response to the question of what the surface should be. Certainly these hard-lined distinctions have not always been so clear, though they have organized camps of thought in twentieth-century design histories and practices. Yet, external to this pairing is an overlooked history of what could be called the "productive surface." The productive surface is a constructed terrain that has the ability to, simply put, yield something. In other words, it has a tangible, positive byproduct—energy, biotic, or abiotic components, for example. The productive surface is premised on an intimate understanding of context, climate, and natural processes. It may operate at the scale of a building, a region, or scales between—because of its networked and scalable logic. This text seeks to reveal a twentieth-century history of the productive surface, establish its key progenitors, and offer an interpretation and assessment of its recent resurgence.

What is the productive?

Prior considerations of the productive within architecture and landscape have remained firmly embedded within the functionalist camp. This is for good reason, given that the primary instigators have been agriculturists, technologists, botanists, material scientists, and engineers. However, contemporary design ambitions—as evidenced in the almost forty projects and texts found within this volume of *Bracket*—and an increased interest in critical sustainability, ecology, and biological analogies have posited new possibilities. Taken literally, the productive surface refers to the capacity for a designed surface to generate a usable component—agriculture, renewable energy systems, and water harvesting systems, among others, do this. Much as Reyner Banham is synonymous with the exposure of a previously suppressed lineage of environmental engineering in modern architecture, the field requires, once again, exposure of a sociological and methodological account of environmental technology's impact on design.[1] This account, though challenging within this brief space, is timely and urgent in a contemporary design context increasingly littered with ecological metaphors, sustainability as design justification, and divergent qualifications of the role of infrastructure.[2]

To begin, an account of a productive surface must be made. What does architecture and landscape already produce—intentionally or otherwise? And how is that component managed by design? Possible inadvertent byproducts of any new construction include heat, light, and various matters of waste; but also shifts in air patterns, hydrology, species patterns, and land value. The envelope, or outermost dermis of a volume, is the principal plane of contact between a regulated interior and a fluctuating exterior. The primary objective of this surface, in the case of architecture, has typically been preventative—keep cool or warm air in or out. And in landscape, the primary objective has typically been to perform at the level of nature. The productive surface embraces the possibility of a wider range of possibilities. Increasingly, designs seek to convert its very liability (an exposed surface) not only into a low-impact element but also into an economic and climatological asset. Returning to a central question: How has the role of the horizontal surface—with its land use, properties, and resources—evolved toward the productive?

Zoning Production

Significant to an understanding about the rise of the productive is its arrival at a blurred, yet empowering, distinction between ecological and experiential ambitions. The argument here is that the productive is an extension and evolution of sustainability without the dubious empirical and technological determinism. In order to isolate a pre-existing history of the productive surface, the role that zoning and land-use regulation have had upon larger urban assemblies is essential. The rapid corporatization of land during the 1970s and 1980s and the subsequent deindustrialization of cities are well-documented and has provoked inventive propositions for new forms of urbanism.[3] As one of the urban conditions that provoked the response of sustainability and now, new forms of productivity, what are the origins of the industrial city?

One of the initial urgencies of early twentieth-century urban planning was to manage the complexity and unhygienic conditions of the expanding city and its relationship to both rising industrial production and strains on agriculture. Leading up to the turn of the century, several models of urbanism

1. In Reyner Banham's 1984 re-issue of *The Architecture of the Well-tempered Environment* (The University of Chicago Press: Chicago, 1969), he chastises readers (and librarians) who "have put the first edition of this book among general introductions to technology—in which role it would leave much to be desired." (p.9) Instead, Banham sought an integrated history of architecture and technology.

2. For a recent critique of sustainability see *Log*, vol. 8, Summer 2006, especially Mark Jarzombek's "Sustainability: Fuzzy Systems and Wicked Problems," in which he argues that "sustainability as a field is dangerously afloat in ambiguity and indeterminacy." (p.11)

3. See especially Alan Berger, *Drosscape: Wasting Land in Urban America* (New York: Princeton Architectural Press, 2007) and Berger, *Reclaiming the American West* (New York: Princeton Architectural Press, 2002). Also Lars Lerup, *After the City* (Cambridge: The MIT Press, 2001) for the particular case of Houston, Texas—a city without zoning ordinances—as well as clarification of the terms "stim" and "dross."

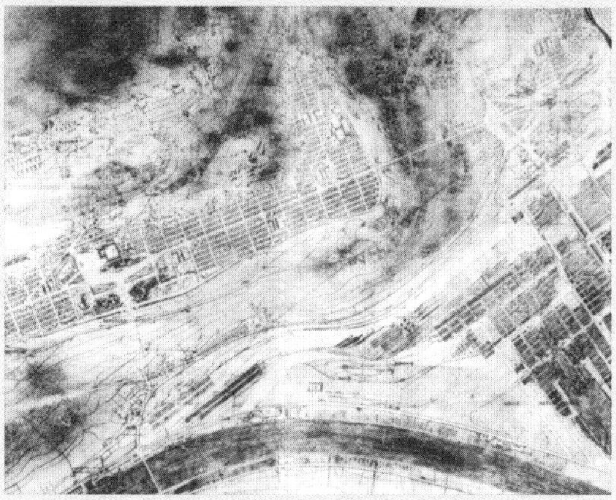

^ **Aerial Plan of the Cité Industrielle project as presented at the exhibit of Envois from Rome in Rome and in Paris in 1904. From Tony Garnier, *Une Cité Industrielle* (New York: Princeton Architectural Press, 1989), 35.**

were expounded that sought to coordinate these functions. Ebenezer Howard, Patrick Geddes, and Tony Garnier are part of a lineage of thinkers, planners, and designers that confronted this difficult transition. Geddes offered a biologists view of urbanism, citing cities as forms of life.[4] Howard, trained as a stenographer, offered an ultimatum found in his "three magnets" diagram in which he positioned the garden city as an ideal alternative to the current town form. Extending these provocations into a situated and detailed design proposition, Garnier developed the *Cité Industrielle* over a period of about 15 years, finally publishing his treatise *Une cité industrielle* in 1918. These three visionaries were seeking the form and format that the nascent industrial city might take. It is perhaps Garnier's proposition that most accurately reflects an early modern ambition of a model that incorporates the productive amidst the residential.

Known for its Fourierian overtones, the *Cité Industrielle* emerged within the context of early forays into regionalism.[5] *Cité Industrielle*, planned for a plateau in southeast France, incorporated residential, cultural, governmental, manufacturing, and agricultural facilities into a grided and zoned city. Additionally Garnier embraced the pre-condition of site as both opportunity and obstruction for development. Unlike Howard's Garden City concept, Garnier's vision anticipated growth and evolution through an expandable linear framework. Programs were located into dense homogenous clusters though strategically proximate to each other. Divided into either residential or public zones, the plan used circulation, green belts, and park areas to modulate separations as required. To address the industrial component, the plan specifically called for a metallurgic factory, a silk manufacturing factory, mining, a cattle farm, abattoirs, a vineyard, and a hydroelectric station and dam. Garnier's city was conceived and strategically programmed to absorb the productive capacities of industry for its own benefit—for the production of energy, food, and a strong economy.

Garnier's preoccupation with the hygienic nature of the *Cité Industrielle* determined many aspects of the city's configuration: its east-west alignment, the height of building masses, its intricate drainage systems, and the separation of functions. The overall zoning strategy was also intended to better facilitate future expansion, though it also espoused a distinct separation of functions, which became increasingly more commonplace throughout the twentieth-century. It can be argued that this structure prompted the current condition of segregated, enclaved urban environments, and the very catalyst for the dominance of the suburban, peri-urban, and exurban typologies. Later, Frank Lloyd Wright would extend and critique Garnier and others with the spatial order found in his (suburban -utopian) *Broadacre City* (1932-59).[6]

Howard, Geddes, and Garnier struggled to realize even partial developments of their visions within their time—not unusual for most planners—and the course that modern planning would eventually take did little to adopt essential components of their plans. Today, a renewed interest in urban agriculture, renewable energy technologies, and positive-energy developments invites a re-reading of early modern planning with regard to the productive urban surface.

Resource Terrains

More than four decades after Garnier and the early regionalists, another significant provocateur propelled and expanded an understanding of engineering, design, and the role of the productive surface. Operating at a significantly larger scale, and armed with the propensity of a comprehensivist, R. Buckminster Fuller's vast body of work is evidence of one of the most complex figures in architecture of the twentieth-century. His transdisciplinary practice, which had many labeling him a dilettante, led to a pursuit of a more holistic worldview of energy, resources, and urbanization. As John McHale, a stalwart supporter, wrote: Fuller was a "phenomenon which lies outside the customary canons of architectural judgment."[7] Two projects within his substantial oeuvre stand out as particularly relevant here, yet remain suppressed in a more common historical account of Fuller. The first is the 1946 "Profile of the Industrial Revolution" as a visualization of a global condition and the second is his research initiative *World Design Science Decade 1965-75*, which led to a study and proposition called "World Game."

Fuller's "Profile of the Industrial Revolution" (1946, later updated in 1964) charted the acquisition of the 92 chemical elements found naturally in the environment.

4. Patrick Geddes' *Cities in Evolution* (1915) was a seminal publication documenting his claim to a shift toward a "neotechnic" city. For more on Patrick Geddes, see the excellent Volker M. Welter, *Biopolis: Patrick Geddes and the City of Life*, (Cambridge: The MIT Press, 2002).

5. For a more thorough account of regionalism and its relation to Garnier's *Cité Industrielle*, see Dora Wiebenson's *Tony Garnier: The Cité Industrielle*, (New York: George Braziller, 1969), 16-20.

6. For more on Frank Lloyd Wright, Ludwig Hilberseimer, and Andrea Branzi and the positions of an 'agrarian urbanism,' see Charles Waldheim, "Notes Towards a History of Agrarian Urbanism" within this volume.

7. John McHale, "Buckminster Fuller," *Architectural Review* 120, no. 714 (July 1956): 12-20.

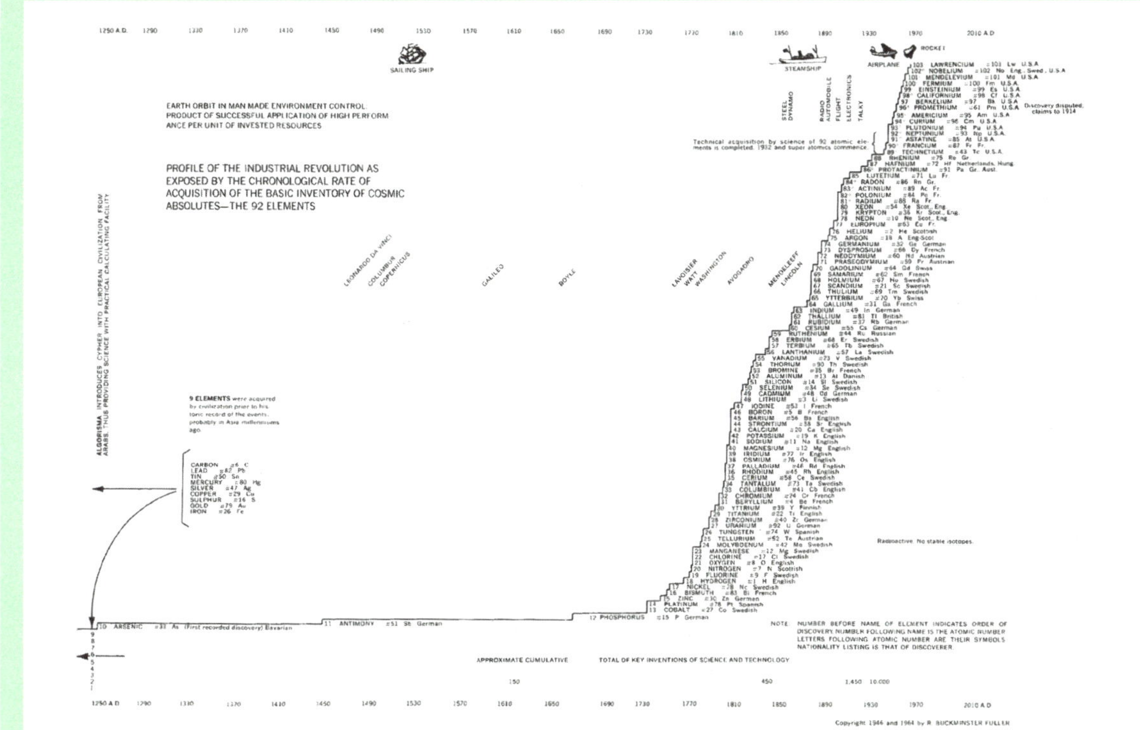

^ **Buckminster Fuller. Profile of the Industrial Revolution, 1946 (updated 1964). Courtesy, The Estate of R. Buckminster Fuller.**

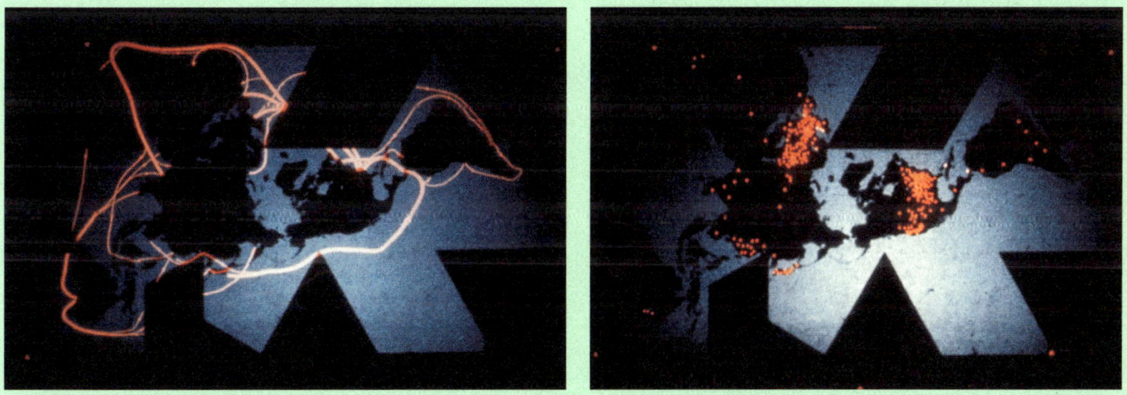

^ **Buckminster Fuller. World Game, seminar 1972. "Food Movement—Bread Grains," (left); "Energy Slaves," (right). Inventorying of world resources, development trends and world population needs. Courtesy, The Estate of R. Buckminster Fuller.**

Beginning in 1250 with the discovery and isolation of arsenic and culminating with lawrencium in 1961, Fuller leverages the chart as evidence of humankind in "command of the complete inventory of building components with which the universe is structured."[8] Generally, the elements occur organically and inorganically and are recovered through mining and other earth-altering techniques. In short, they are sourced and extracted from soil, water, air, and biologically; they are farmed from the earth's surface for conversion and transformation. Alongside the pragmatic extraction of biological resources (resulting in food, energy), elemental resources are cultivated more out of human curiosity (resulting in science, materials). The observations and intention of this chart can be found throughout later research by Fuller and his collaborators.

Fuller's preoccupation with a total inventory re-emerged in his 1961 prompting to the International Union of Architects at their VIIth Congress in London, England that architecture schools "invest the next ten years in a continuing problem of how to make the total world's resources which now only

8. Buckminster Fuller, "Earth, Inc.," from *The Buckminster Fuller Reader*, edited by James Meller (London: Jonathan Cape, 1970), 223.

^ 360° panorama of the San Gorgonio Pass Wind Farm in San Gorgonio Pass, California. Photo by Gregg M. Erickson.

serve 40% serve 100% through competent design."[9] This venture spawned a series of documents under the banner of *World Design Science Decade, 1965-1975*. Fuller worked closely with John McHale, a sociologist-artist, in developing ambitious documents such as an "Inventory of World Resources, Human Trends, and Needs" co-authored by Fuller and McHale in 1963. McHale's increasing involvement in the WDSD project led to one document that he authored individually, "The Ecological Context: Energy and Materials" (1967). This research was later expanded into a publication of the same title. McHale argued that industrial and agricultural undertakings needed to be redesigned as "ecologically operating systems rather than piecemeal aggregates of unrelated processes."[10] McHale wrote that "[o]ur concern here is to more fully appraise the role of man-made systems which are also natural systems in the overall integral functioning of the system."

The WDSD culminated in a project called "World Game," which was also developed in anticipation of the 1967 World's Fair in Montreal. Already embedded in its title is the ambition of its scale of thinking. And with World Wars, the World Bank, and World Health Organization, a 1960s citizen was increasingly familiar with the scale of thinking needed to play Fuller's Game. If Reynar Banham's *Architecture of the Well-Tempered Environment* (1969) charted an overlooked environmental interior, Fuller and McHale's WDSD documents and "World Game" charted an overlooked environmental exterior. Both undertakings posited a more efficient use of resources, land, and materials—both preceding the science of climate change. Specifically, the "World Game" located current and potential energy sources, current and potential resources, food trade, mobility patterns, and education, among others, at the scale of the globe. Fuller wrote that "a great world logistics game [could] be played by introducing into the computers all the known inventory and whereabouts of the various metaphysical and physical resources of the earth."[11] Therefore—though certainly no easy task—the inventory of known human needs would need to be calculated and mapped. Fuller intended "World Game" to operate as a piece of software that could compute "answers to the mounting social and ecological crisis, which he had been predicting since around 1950."[12]

It is significant that Fuller's proposition of comprehensive thinking of the Earth's resources coincided with contemporaneous space travel that facilitated a view of the Earth not witnessed before. Both the computationally-inventoried and the visually-documented Earth afforded a new comprehension of large-scale thinking. These observations can also be said to have launched the back-to-the-land counterculture movement of the 1970s, as chronicled by Stewart Brand's *Whole Earth Catalog*.[13] A century earlier, George P. Marsh published *The Earth as Modified by Human Action* (1874), an impressively

9. Buckminster Fuller and John McHale, "Phase I (1963) Document 1: Inventory of World Resources Human Trends and Needs," (Carbondale: Southern Illinois University, 1963).

10. John McHale, "Phase II (1967) Document 6: The Ecological Context: Energy and Materials," (Carbondale: Southern Illinois University, 1967), 7.

11. Joachim Krausse, Claude Lichtenstein, editors, *Your Private Sky: R. Buckminster Fuller, The Art of Design Science*, (Baden: Lars Müller Publishers, 1999), 473.

12. Ibid., *Your Private Sky*, 473.

13. For more on the *Whole Earth Catalog* and cataloging see Maya Przybylski's essay "The Catalog: From Ploughs to Clouds" within this volume.

complete 644-page survey of current physical modifications to the environment and their quantifiable consequences. In it, Marsh traces "the history of man's industry as exerted upon Animal and Vegetable Life, upon the Woods, upon the Waters, and upon Sands."[14]

The incorporation of resources, and the industries dependent on them, into the design disciplines is significant in the shift from mitigative sustainability to the productive surface. The inventories underlying the "World Game," for example, were meant to offer not a single solution, but a resources framework for possible solutions. Fuller wrote in 1969 that to "accomplish the game's objective, the resources, pathways and dwelling points around the surface of our eight thousand mile diameter, spherical Spaceship Earth must be employed by the players in such a way that the world's individual humans would each be able to exercise complete actional discretion."[15] Much as aerial photography modified perceptions of the twentieth-century landscape and its use, satellite imagery and Geographic Information Systems (GIS) are changing twenty-first century perceptions of land use and interpretation. These processes materialize new techniques of managing the constructed surface, something that John May argues are not infrastructures but "statistical-electrical control spaces."[16] In fact, several contemporary practices and thinkers have emerged as a proponent of these issues and methods.[17]

Generative Plots

In more recent developments of the last decade we find a host of referents for the productive surface. Massive undertakings in energy and water harvesting, agribusiness, and other productive surfaces are transforming architecture, land use, and the larger landscapes—turning Fuller's inventory into sites of intervention, development, and generation. These new surfaces harvest, yield, and farm the environment, generating byproducts that are then integrated or distributed.

There is the groundbreaking 2003 renovation of Ford River Rouge Plant in Dearborn, Michigan by William McDonough + Partners, which converted the liability of the massive roof into a water harvesting system. The living roof, the largest of its kind in North America, now cleans about 10 billion gallons of water annually in what they call an "industrial strength" landscape. A similar intervention was completed in 2009 on the roof of the Atlantic City Convention Center in New Jersey. 13,486 photovoltaic panels produce an average of 26 percent of the convention center's energy. Savings from the project are estimated at $4.4 million US over the next 20 years. In addition to water and solar, there is wind farming.[18] For example, there is the San Gorgonio Pass Wind Farm, installed during the 1990s and 2000s in southern California, is an array of 3,218 turbine units delivering 615 MW. Combined with the Altamont Pass and Tehachapi Pass

14. George P. Marsh, *The Earth as Modified by Human Action*, (New York: Scribner, Armstrong, & Co., 1874), vii.

15. Ibid., *Your Private Sky*, 473-479.

16. John May, "The Logic of the Managerial Surface," (University of Toronto) lecture, April 7, 2010.

17. In particular, the Center for Land Use Interpretation (CLUI) is most notable here, but also the work and research of Keller Easterling, Albert Pope, Charles Waldheim, Alan Berger and P-REX, Pierre Belanger, Christopher Hight, Urban Lab, InfraNet Lab, and many others, especially within this volume.

18. For more on wind farms and agriculture, see Jason Sowell's "Cash Crops, Energy Landscapes," within this volume.

^ Thanet Earth in Kent, United Kingdom. The hydroponically-farmed tomato glasshouse, January 2009. Courtesy, Thanet Earth.

wind farms, they account for 11% of wind energy generated worldwide. Approvals and development of these farms were initiated as early as 1981 in response to the 1970s energy crisis, meaning that several of the turbines are now significantly out of date. A renewal proposal that broke ground in 2008 at Tehachapi seeks to expand the field further with ambitions to supply power to 3 million homes by 2013.

Then there are propositions and implementations of agriculture and the productive surface. Eurofresh Farms in Wilcox, Arizona—a place with no shortage of sunshine—houses some 318 acres of hydroponically grown tomatoes. In the UK, Thanet Earth, majority-owned by Fresca Group Ltd., has recently completed phase one of a massive greenhouse buildout in Kent. There tomatoes, peppers, and cucumbers are grown in glasshouses in which everything, according to their website, is "computer controlled—from the blinds in the ceilings to opening the windows, the liquid feed make-up, the heating, lighting and carbon dioxide levels." The site was selected for its light levels, proximity to the National Grid line to upload electricity, high level of local unemployment, and good transport links. Most notably, though more modest in scale, the US White House installed a vegetable garden on the South Lawn in 2009, which is the first on that site since Eleanor Roosevelt's victory garden during World War II. In many ways, this is more telling than any industry or science development. While operating more didactically—though it does provide produce for White House chef Sam Kass's cooking—the garden plots yield vegetables, fruits, herbs, and honey.[19] In controlling harmful bugs in the garden, even ladybugs and praying mantises are employed by the President.

Producing Design

Productive surfaces articulate a new public realm, and with that a new public; A public not characterized by its degree of urban, suburban, or rural, but by its ability to participate in the cultivation of its consumables. With a dramatic increase in clean technology, and with design practices seeking innovative responses to a project's operating cost, the shift from Modernism's "function" toward the contemporary's "production" can be charted through architecture's relationship to its larger environment. The productive surface yields, making it not only responsive to its environment, but indeed operational because of it. This is not a sustainability mantra as much as a biological one; An architecture of synthetic surfaces servicing variously scaled constructed environments—its roof, its site plot, and its wider climatological and ecological territory.

The productive surface acknowledges and capitalizes on its innate potential for a seasonal or cyclical yield. It is dynamic, responsive, and yet occupiable and tangible. As a history of the productive surface emerges, its key provocateurs are often corporations, venture capitalists, farmers, and technologists. How might the productive surface generate new economies, programs, typologies, and public realms? The role of the architect and landscape architect here is ripe for opportunity.

19. There was some controversy around the Obama's vegetable garden being constructed on the "people's lawn." Many felt that the symbolic value of the lawn had been compromised. The irony of this suggestion and controversy is paralleled in the research and projects of both Fritz Haeg and Heather Ring, found within this volume.

plots / allotments

Daniel Kariko and John Raulerson

Anthony Acciavatti

Bryan Boyer

Nick Glase

Dimitris Argyros

Joseph Grima, Jeffrey Johnson, and José Esparza

beyond disney: the vanishing florida family farm culture and the decline of florida's last cash crop, the housing market

daniel kariko & john raulerson

^ Plastic fruit stand, HWY 27, Florida.
Image, Daniel Kariko.

^^ **Bringing in the cattle, Albritton Farm, Belle Glade, Florida. Image, Daniel Kariko.**
^ **Citrus hedging machine, Central Florida. Image, John Raulerson (left). Farmall tractor, Central Florida. Image, John Raulerson (right).**

^ **Cowboy sunrise, Albritton Farm, Belle Glade, Florida. Image, Daniel Kariko.**

The aim of "Beyond Disney" is to capture the qualities of a vanishing subculture, the Florida Family Farm, in the face of a changing Florida landscape. Over the years, Florida's farmlands have been bought out by developers. In light of the current financial crisis, developers have not been able to secure financing to complete their sprawling subdivisions and commercial properties, rendering much of what were fertile farmlands useless.

As early as 1925, Florida's farm and ranch lands have been steadily decreasing. According to the Florida Department of State in 2004, approximately 1,000 people were moving to Florida every day. In addition, Larry Arrington, University of Florida, Dean for Extension, indicated in 2007 that over the next 50 years Florida's population is expected to double to 35 million. During this time, Pierce Jones, director of the Extension's Services Program indicated more than 11 million new homes, millions of square feet of commercial space, and thousands of miles of additional roadways will be needed to accommodate the influx of new residents. According to a University of Florida study, 150,000 acres of farmland are rezoned for commercial and residential use each year. Florida farmland is disappearing at an alarming rate and with that the culture of the native farmer.

Evidence from statistical data offers one level of understanding, but to see a visual document is entirely different. Although information on this topic exists in other formats, these sources lack poignant visual imagery that captures the feelings and thoughts of families experiencing the disappearance of "Old Florida" as it happens. "Beyond Disney" is designed to explore the cultural significance of the family farm to the state of Florida through documentation using multiple media. The project goal includes the collection of material through firsthand observation and documentation using photography and film/video. The film and video content of "Beyond Disney" represents personal life experiences of the farmers, their opinions of current farming issues, and the future of Florida farming. Thematically related segments of the documentary are shown along with still imagery as a contextual framework in which the photographs are displayed.

^ **Simms farm house, homesteaded land, Brooksville, Florida. Image, Daniel Kariko.**

The background and significance of this project can be traced to the earliest uses of the photographic medium, specifically the role of the photographer during the American depression era. During the 1930s, the Farm Security Administration mounted efforts to combat rural poverty. The FSA is famous for its small but highly influential photography program that portrayed the experiences of the rural poor and provided educational materials to the press and public.

During the same time period the Resettlement Administration funded two documentary films by Pare Lorentz, "The Plow That Broke the Plains" (1936), which focused on the creation of the Dust Bowl and "The River" (1938), which was about the importance of the Mississippi River. These films were deemed culturally significant by the United States Library of Congress and selected for preservation in the National Film Registry.

Parallel to the farm investigation, Kariko and Raulerson started documenting Florida's suburban developments and their influence on the environment. This work developed as Florida's housing crisis became more and more evident. It was necessary to document the change in the Florida landscape from agrarian to residential and commercial.

In this tangent work, the navigation through our contemporary built environment is investigated, drawing on ideas and earlier work of such highly regarded photographers as Robert Adams, Lewis Baltz, and those represented in the "New Topographics" exhibition. The aerial photographs of central Florida's arrested suburban developments create patterns of human habitats in crisis. From the air, the negative imprint of suburbia and its effect on nearby farmland and forests is evident. By contrast, the photographs from within these abandoned developments create post-apocalyptic sentiments of nature taking back. The juxtaposition of the two bodies of work generates a dialogue between the suburban and agrarian landscape, drawing a contrast between the population growth in Florida and sustainability depicted in agricultural resources.

Preliminary research for this project began with the privilege Kariko had of studying with William Jenkins, curator of "New Topographics: Photographs of a Man-Altered Landscape" exhibition in the International Museum of Photography at the George Eastman House, Rochester, NY, in 1975. The exhibition featured ten photographers exploring and recording the expansion of residential and commercial development and the rapidly increasing human imprint on the landscape. This work was a reaction to the utopian, pictorialist landscape work of such 1930s and 1940s photographers as Edward Weston and Ansel Adams, because it objectively represented the

effects of the industrial culture. "New Topographics" is still considered one of the most influential photography projects, reshaping the image-maker's approach to contemporary landscape. The impact of this exhibition can be seen in the healthy renewal and growth of interest in all things environmental more than thirty years later and particularly in the seminal contemporary photographic work of Edward Burtynsky's "Manufactured Landscapes."

"Beyond Disney" provides a visual expression of the ideas and cultural concerns in understanding the disappearance of "Old Florida." The exhibit consists of images of farming processes, equipment, and aerial photographs of the agricultural landscape encroached upon by urban sprawl. It examines the possibilities of preserving the family farm, while illustrating the symbiotic relationship between farmer, land, and lifestyle. This social, anthropological record provides an in-depth examination of the context in which the agrarian environment influences the residents and farming industry, and vice versa.

Currently, in Florida, family-owned agricultural endeavors such as cattle ranching, dairy farming, and citrus production are disappearing. The families' attachment to their lands, which in some cases have been owned for generations, and the impact of selling their properties to large corporations or developers are a part of the new cultural environment in Florida.

>>> **Daniel Kariko is a native Yugoslavian photographer and digital media artist.**

>>> **John Raulerson is a native Floridian, sculptor, photographer, and digital media artist.**

^^ One tree, King's Ridge, Clermont, Florida. Image, Daniel Kariko.
^ Land development billboard, Central Florida. Image, John Raulerson.

^^ Frontlines, HWY 27, Florida. Image, Daniel Kariko.
^ Land for sale, HWY 27, Florida. Image, Daniel Kariko.

^ Cul-de-sacs, unfinished subdivision, HWY 27, Florida. Image, Daniel Kariko.

precipitating a productive countryside: a renewed company town model

anthony acciavatti

Within the last half of the twentieth century, technology and print media made it possible for designers to vividly imagine the instantaneous propagation of urban environments that previously took centuries to produce. The campus community remains one of the few defensible vestiges of the desire to immediately construct context and community as a whole. The need for people, governments, and corporations to ascribe meaning to the landscape is paramount to understanding the countryside through the lens of the corporate campus model.

Whereas it has proven exceedingly difficult to bring new life to existing urban conditions, the company town and new corporate campuses, because of their initial size and a priori goals, allow for the possibility of materializing social interaction into a physical and marketable reality. This provides a total landscape. Paired, dwelling and working synthesizes both the physical and social benefits of urban infrastructure and access to the seemingly rural and autonomous countryside. It is the Trojan horse of development.

Situated in the valleys and foothills of the Mournes and Cooley Mountain ranges of Ireland, County Lough and adjacent counties are undergoing unprecedented changes in terms of industry, domestic space, and infrastructure. Due to European Union subsidies and a competitive global food market, this tightly fractured agricultural landscape is no longer financially profitable for agrarian labor. This proposal takes advantage of the unique corporate tax system within Ireland, the abundance of rain, and the island's strategic placement between continental Europe and North America to construct a new company town model that makes use of the metabolic rates of change in the landscape and human appetite.

Ireland's picturesque landscapes and corporate tax incentives make it a desirable place to establish a corporate community with immediate access to recreation and diverse outdoor activities. Similarly, the abundance of rainwater throughout the year and convenient access to Carlingford Lough (one of the busiest ports on the island) makes it possible to establish small- and medium-scaled fish farms that service Europe and North America. Unlike traditional agricultural practices, fish farming consumes less energy to produce more food and employs the same number of people as conventional field agriculture. These company town models allow for economic change to take place without ruining the highly articulated and constructed ecologies of the region in order to precipitate a productive countryside.

>>> **Anthony Acciavatti is a partner in Somatic-Collaborative, an architecture and urbanism office based in Cambridge, MA. He is currently a Frederick Sheldon Fellow at Harvard University.**

^ Company Town Case Study: Zlin, Czech Republic. Perhaps one of the most attractive aspects of the company town model in terms of dwelling and working is the ability to bring all of the physical and social benefits of urban infrastructure and access to the seemingly rural and autonomous countryside.

^ Googleplex. Today the corporate campuses of companies like Google attempt to construct environments that integrate work with recreation and efficiently consolidate many of the services found within cities. Unlike many of the company towns of the early twentieth century, these corporations do not include housing within their architectural and planning schemes. ©2010 Google, ©2010 Europa Technologies.

Alluvial Corridors and Channels

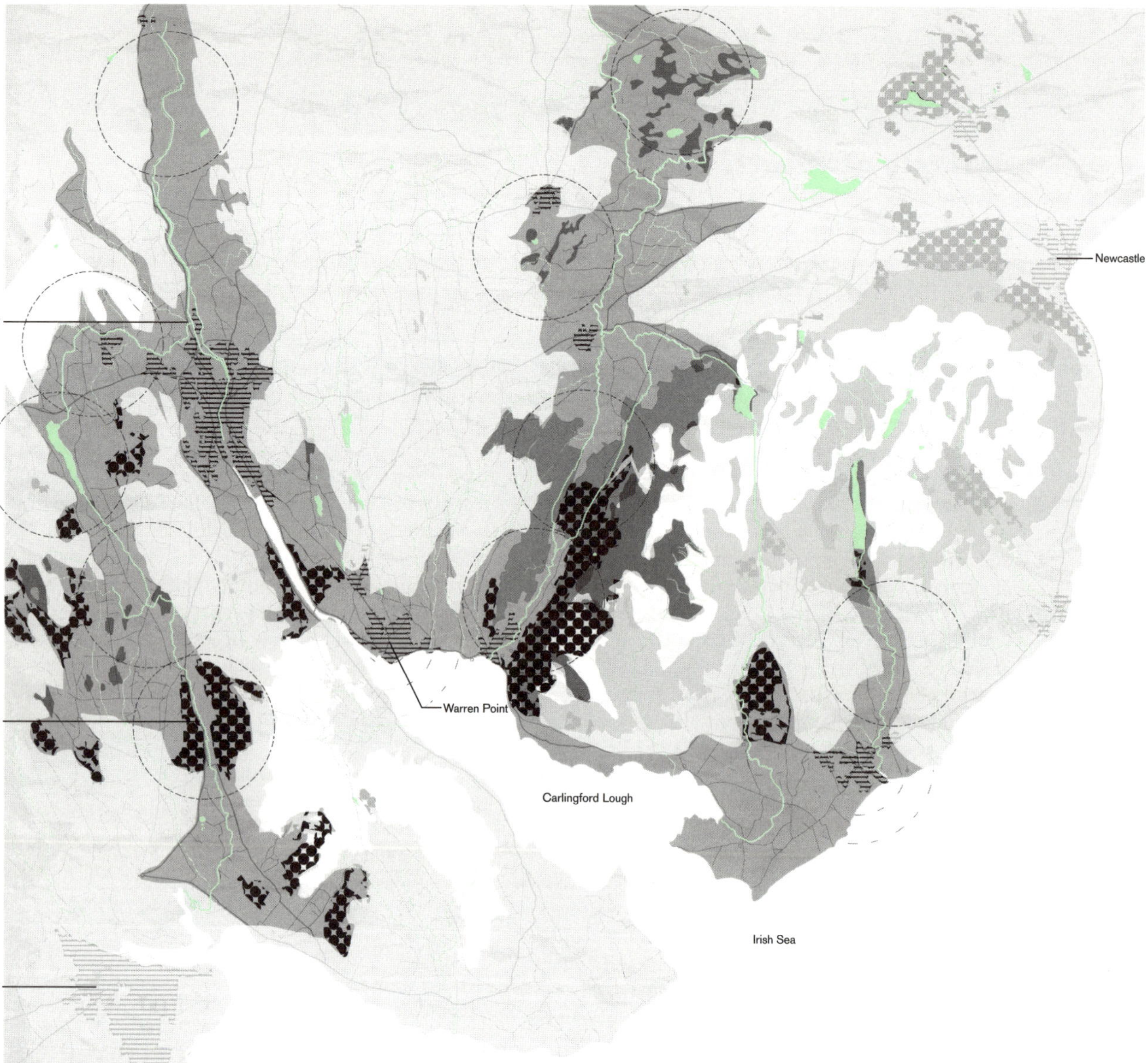

^ **Corridors of potential development and advantageous sites.**

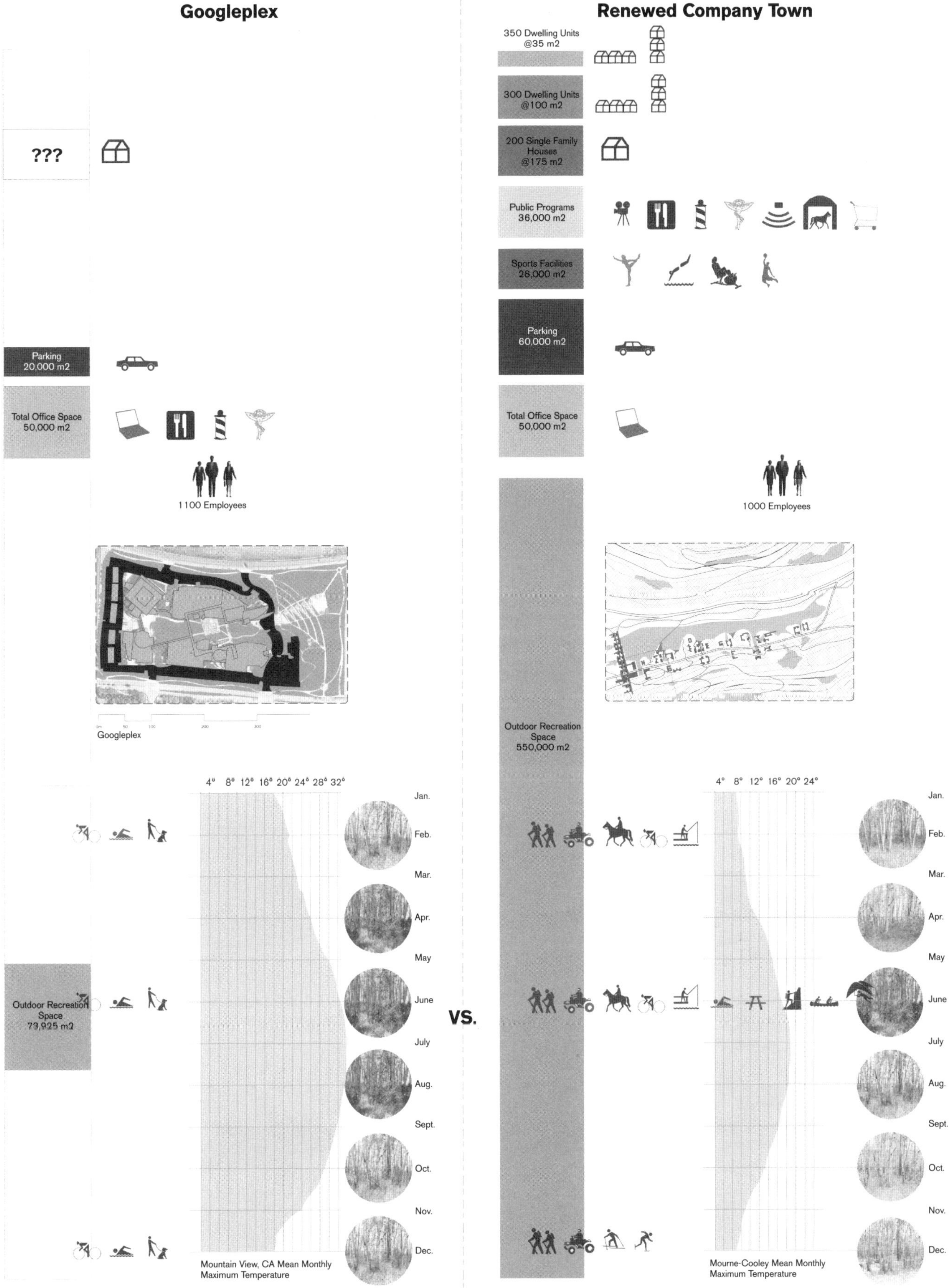

^ **Googleplex Amenities vs. New Irish Company Town Model.**

Emergent Morphologies

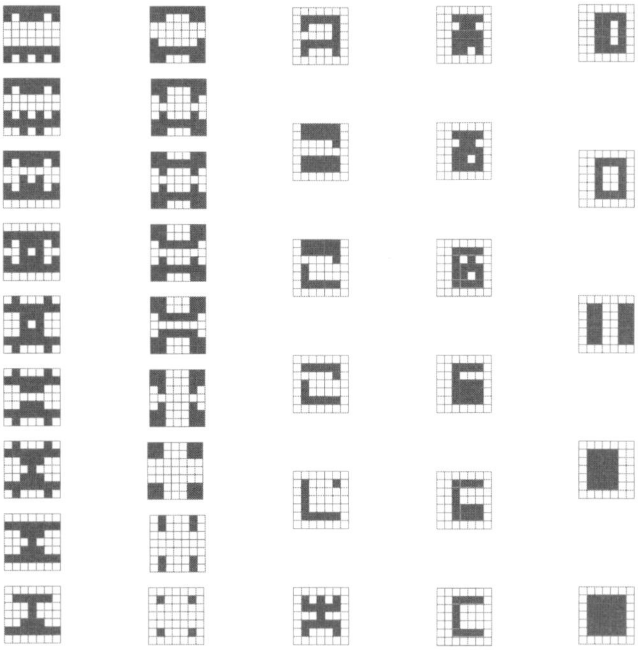

Ground Permutations in 50m x 50m Parcels

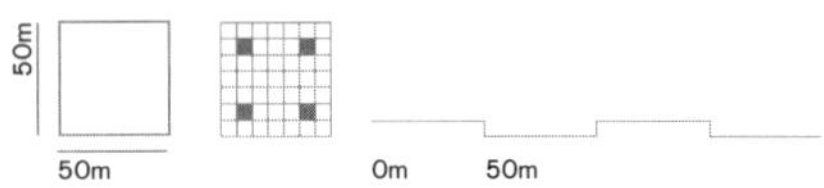

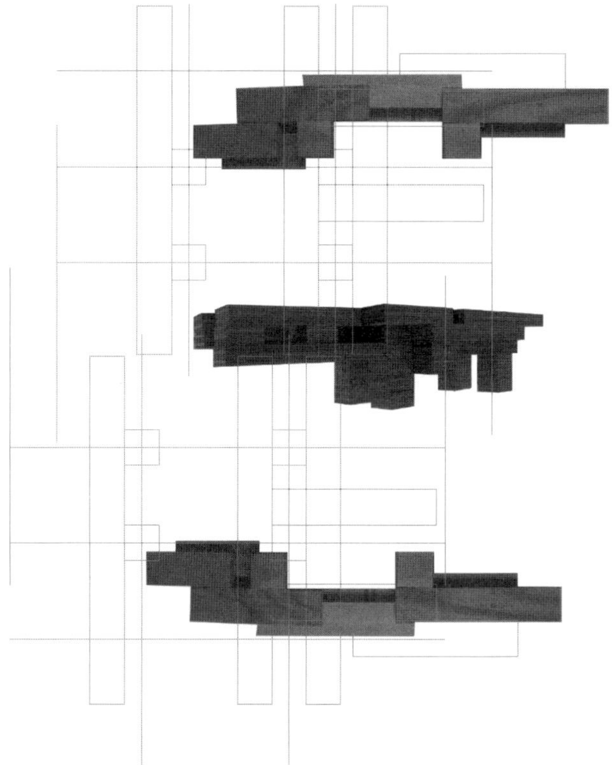

Potential Bar Building Combinations

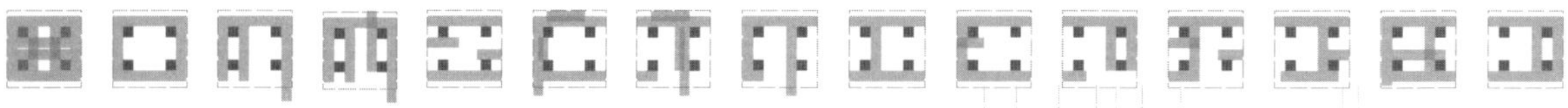

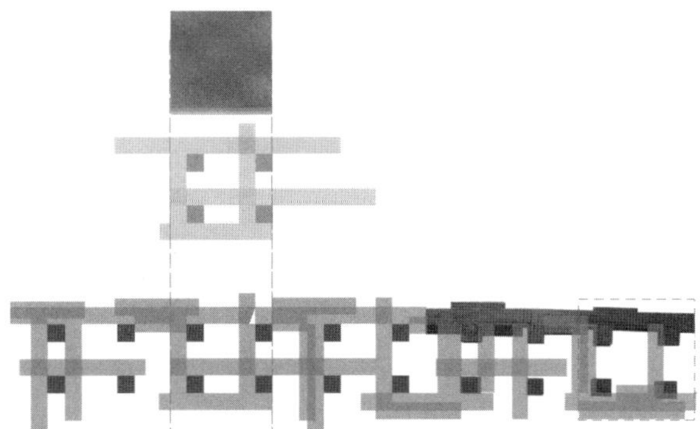

Bar Building Plan 1.0

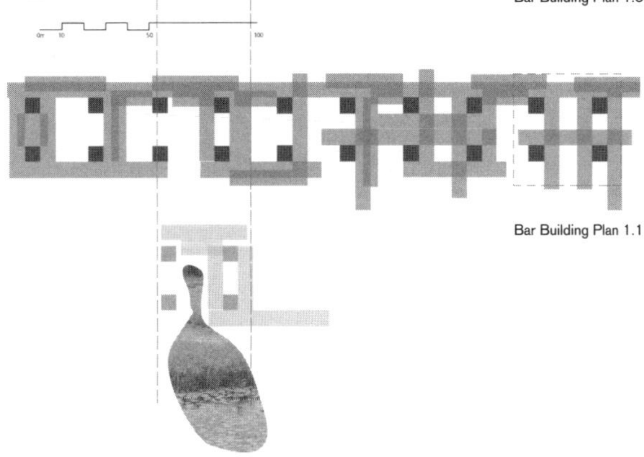

Bar Building Plan 1.1

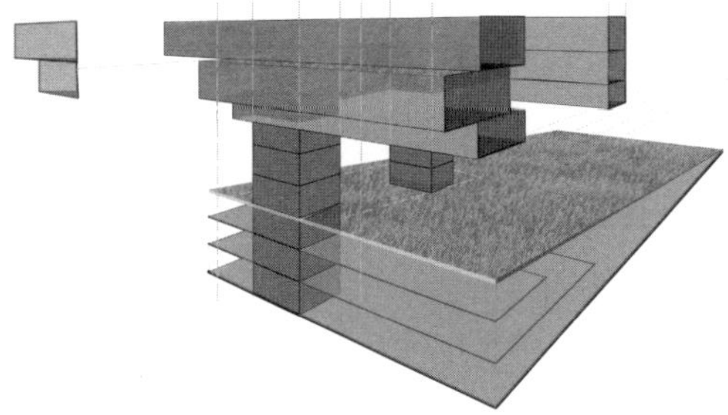

Sectional Perspectives

^ **Emergent morphologies of development.**

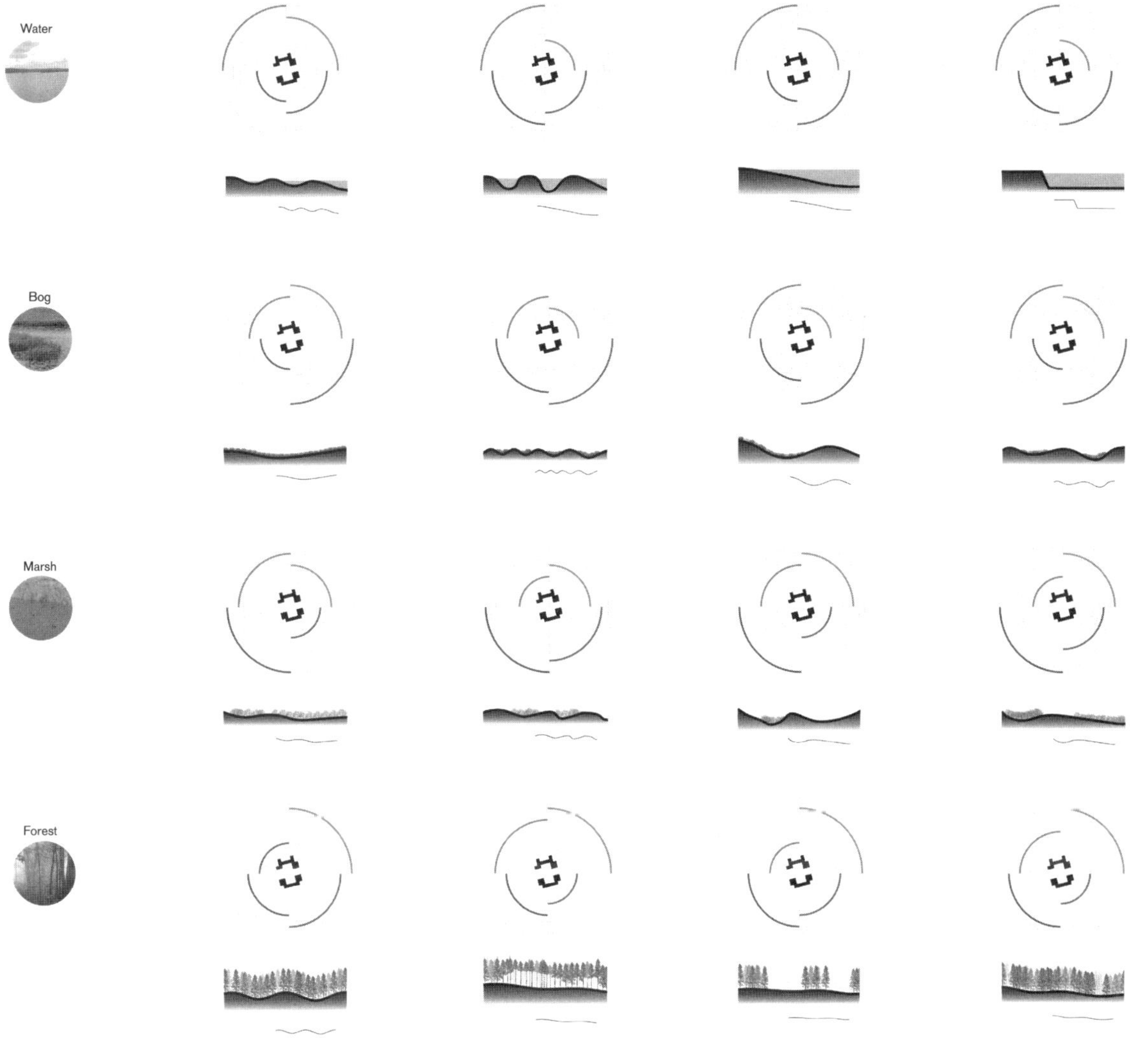

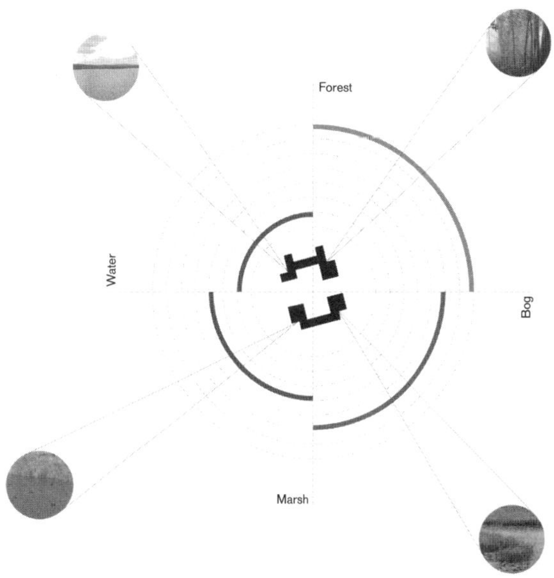

^ **The Botanical Matrix constructs measured relationships between the featu res of the ground and building envelope to produce spatial and environmental affects.**

The Bar Building houses many of the large programs, infrastructures, and primary work spaces of the company town campus. Sectional differences allow for numerous ground and floor plate adaptations.

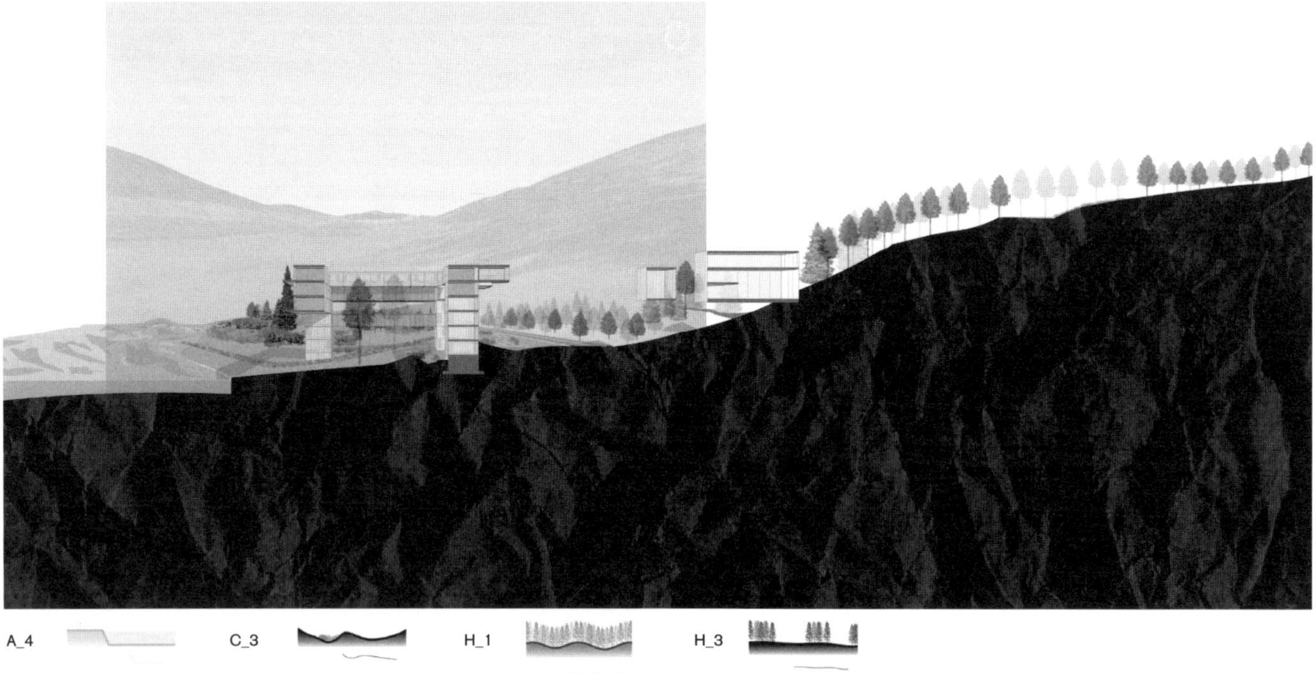

^ Each section contains a series of permutations of the Botanical Matrix. Within each drawing there is an index of the typologies of construction.

learning from salinas (hopefully)

bryan boyer

^ An 80mph billboard glance.

Driving north on Highway 101 though the Salinas Valley, one of America's most fertile regions, you will find Dennis Caprara kneeling in his field with a wide grin as he holds a head of lettuce freshly plucked from the soil of his farm. You're free to drive by on any day and at any time, Dennis will be there proud as can be. This is because Caprara happens to be 12 feet tall and 3/4" thick: he's a scroll-cut billboard painted by John Cerney erected with minimal structure on the edge of a field. What business does a farmer have putting a billboard in his field?

After all, this is not the sort of farm where you are enticed to pull off and purchase fresh victuals. So the sign is not performing a traditional marketing roll, nor is there much of an imperative to cultivate a brand identity. In this part of the country, growing contracts run for long terms and a reputation is something that develops over decades and beers rather than branding. It would be hard to deny at least a touch of hubris in an act such as this, but that's not enough of an answer. According to the farmer behind the billboard, this was his way of putting a face to the behemoth that is industrialized farming and in doing so honoring those who work the land. The necessity of giving a face to something that seems so essential is a funny problem to have. Shouldn't everyone know what lettuce looks like and thus be able to appreciate it on their own?

The revolution of industrial farming vastly increased our ability to produce foodstuffs but it required the concentration of farmlands on a scale that is hard to imagine. While our culture perpetuates a bucolic vision of food being grown on 40-acre family farms, these have been relegated to an ancillary role in contemporary food production. Organic, local, and fair trade may be the buzzwords but they're far from the predominant condition. As farmlands have been concentrated into vast single-grower tracts stretching for miles, so too has the mental distance from field to consumer increased. Answer truthfully: Do you know what a field of lettuce looks like? Do you know what kind of plant walnuts are harvested from? Does asparagus flower?

This gap is what Dennis Caprara's billboard quietly addresses by trading the horizontal, unknowable flatness of the agricultural field for the vertical, illustrated flatness of the billboard. After all, Caprara finds himself in an interesting position: he's growing something that 99% percent of the people who pass by his farm consume and yet most of them would not be able to identify exactly what is going on in his field. Since recognition and a hint of knowledge are essential preconditions to appreciation, it's in Caprara's best interest to help passers-by understand what he is doing: to illustrate for them the very literal process of farming, which has become abstract and largely isolated from our lives by the gasket that is the grocery store. In the parlance of economics, this is a shrewd competitive technique described as elevating a needs-based decision driven by price ("I just need some fresh vegetables for nutrition") to an ideological decision based on value-add ("I will pay more for locally grown lettuce because it's a responsible choice"). Using a billboard such as this is an act of depiction rather than deception: Caprara's lettuce may be equivalent to his competitors', but by depicting his product in a new light he helps the public to understand both farming as a general concern and the specific nature of his product—locally grown-produce—so that they may better appreciate (and buy!) both.

What would the architectural version of this billboard look like? Would it be the stale Randian image of an architect standing proud on the job site, blueprints slung over their shoulder? Maybe a team of designers sitting around a table hashing out ways to bring the budget back into reasonable territory? A CNC (Computer Numerical Control) robot happily milling parts for a curtain wall? A student pouring over airflow and daylight simulations? How about someone shepherding a group of community members towards a mutually beneficial decision about land use? It could be any of these, of course, because the contributions of the architect have expanded dramatically since the profession evolved away from cottage craft to play an increasingly strategic role in the development of cities, campuses, business plans, and more.

Architecture suffers from its deceptive literality. Like the farmer who uses a billboard to add transparency to his otherwise opaque soil, architects must seek ways to make visible the specific benefits of their cultural contributions. In the United States (and perhaps to a lesser degree in other countries), the profession of architecture suffers from an inability to communicate its cultural value. If the ambition is to develop new and better ways of living, why do job site billboards just show renderings of buildings?

The profession of architecture suffers because its public image is equated with its built output rather than being merely symbolized by those creations. If the true value of architecture

1. Unfortunately, Betsky's introduction to the Biennial was the pinnacle of the event. Apparently in the context of the Biennial, "Beyond Building" = furniture. This is not a step forward.

is "beyond buildings," as Aaron Betsky would have it, the profession must find ways to educate the public as to the nature of their contribution.[1] At its most basic level this means that architects must help the public to understand space the way designers are trained to: something palpable, manipulable, and active. While any architect will agree that the development and disposition of buildings and the spaces between them really matters because it changes the patterns and quality of life, this is not necessarily something that a layperson detects. Non-architects seldom have any training to help them perceive the potentials of spatial relationships despite existing in a spatial, visual world and thus feeling some obvious kinship with it. Most children will have at least one measly course in art appreciation and music in elementary school, but what about spatial reasoning?

This is similar to the situation with farming: food is something we all experience on a day-to-day basis, but few people actually have the knowledge or care to think critically about its production and potentials. Naturally, the products of farming enjoy a special relationship with humanity and food is celebrated in countless ways by every culture on earth, but it's important to note that graduated knowledge about food (especially cooking it) is available to the public in a variety of different formats. From billboard to magazines to television shows, food knowledge is dispensed in a public-friendly manner such that one may enter into the world of cooking as a novice and reasonably expect expertise to be available in many different graduated forms whenever one is ready to elevate their knowledge level. This is a simplistic comparison but imagine a situation where a layperson decides they want to learn something more about architecture so that it may enrich their lives or better their business. Where do they turn? Does one have to go to architecture school to properly appreciate the work of an architect?

>>> **Bryan Boyer is a designer working on strategy issues at Sitra—The Finnish Innovaion Fund—an independent public fund under the supervision of Parliament.**

seasoned pasture: a demonstration range and public park

nick glase

This proposal for a new public park in the grassland hills in San Francisco's East Bay, transforms a degraded rangeland into a unique type of public space. Seasoned Pasture is a place for teaching ranchers new, ecologically restorative ways to manage their herds and a place where visitors are exposed to an emerging, highly-symbiotic form of environmental stewardship. Almost 30% of the world's total surface area is devoted to livestock. Livestock production systems have long been cited as the world's single largest environmental threat and contributor to greenhouse gas emissions. And, when taking a closer look at the problem, it becomes clear that there is no simple solution.

For thousands of years, California grasslands have depended on free ranging herbivores to stabilize their ecosystem against colonization by shrubs and trees. When the Spanish missionaries settled the region in the 19th century, fenced paddocks and cattle grazing came to dominate the landscape, while resident elk populations were slowly eliminated. When looking at the region from the air, the ranching tradition still defines much of the landscape. A culture of farming has preserved vast areas of developable land that otherwise might be consumed by urban sprawl. However, cattle farming has simultaneously caused widespread environmental degradation.

In contrast to wild grazing systems, where large tightly-packed herds rampage through the landscape devouring everything in their path, cattle ranching has traditionally been a laissez-faire endeavor: put up a perimeter fence and let cows pick and choose whatever and wherever they want to eat. In many contexts, this type of management has been responsible for significant ecological destruction. In California, where winter rains are followed by a long summer drought, traditional management is not only inefficient in maximizing the forage potential of the grasses, but it also has devastating

^ Site photo looking west. Once the home to perennial grasslands and Tule Elk most grasslands in California are now covered with exotic annual grass and cattle.

^ Land use map highlighting range land. In the Oak grasslands surrounding the San Francisco Bay, cattle ranching has helped ward off development, but has destabilized ecological health.

^ Ranch parcel plan with proposed park shown dark gray.

1. Steinfeld, H., P. Gerber, T. Wassenaar, V. Castel, M. Rosales, and C. de Haan. 2006. U.N. Food and Agriculture Organization. Rome, Italy "Livestock's Long Shadow—Environmental Issues and Options." Retrieved December 5, 2008.

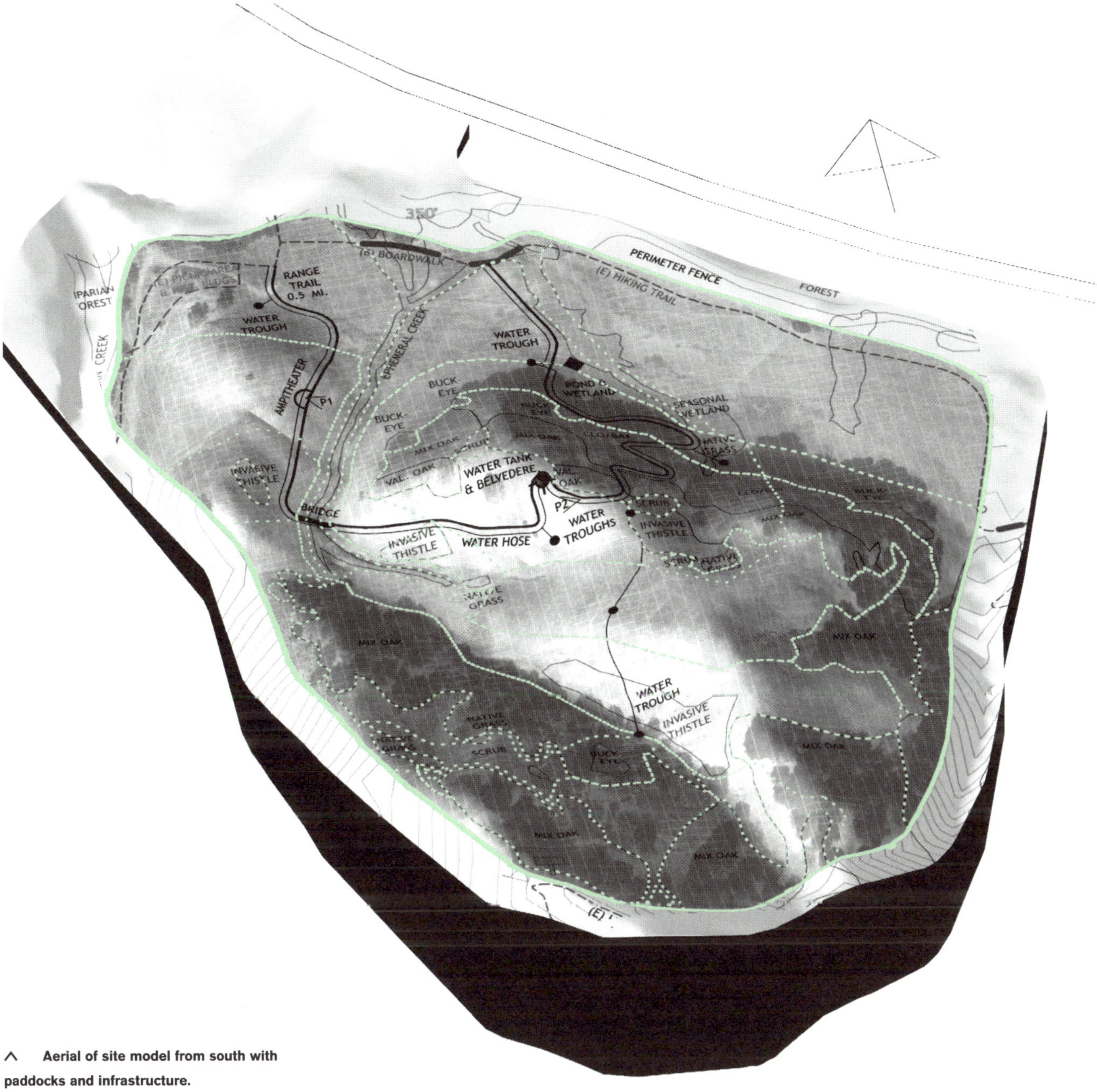

^ Aerial of site model from south with paddocks and infrastructure.

downstream effects: heavy hoofs on wet soil leads to a proliferation of invasive species, erosion, stream contamination, and habitat loss.

There is no simple solution to this big-hoofed problem. Taking cows off the landscape would lead to a complete transformation of the region's ecology. In California, that would mean the emergence of a shrubby landscape susceptible to frequent flash fires—a huge liability near an urban population.

By replacing existing wire fences with 'fenceless' GPS-controlled shock collars (a technology currently being pioneered in Australia), cows can be sequentially corralled from paddock to paddock according to grass growth cycles, soil saturation, and special ecological events. With this system, cows are always directed to fresh palatable forage, while avoiding sensitive land.

In order to construct an interpretive framework for visitors to view this otherwise invisible landscape, "Seasoned Pasture" introduces a series of transformable fence-post markers. Through color and compositional patterns, the posts become a blatant way to read the nuances of the landscape: saturated slopes, growth phases, and seasonality become more visible. Otherwise elusive areas—such as vernal pools, valley oak saplings, and native grasses—are highlighted. Carefully choreographing the movement of cows while allowing visitors and wildlife to move freely across the landscape, creates a liberating type of public space. In contrast to private farms, where animal management is seldom visible, or conventional parks, where trails determine the experience, this proposal counters traditional notions by re-weaving productive, human, and ecological landscapes into a more visible, experiential, and symbiotic relationship.

>>> Nick Glase is a Los Angeles-based architect and landscape architect whose work seeks to synthesize rural and urban technologies, values, and ordering systems.

Fall Grazing Plan
- target one
- target two
- graze invasive
- oak woodlands

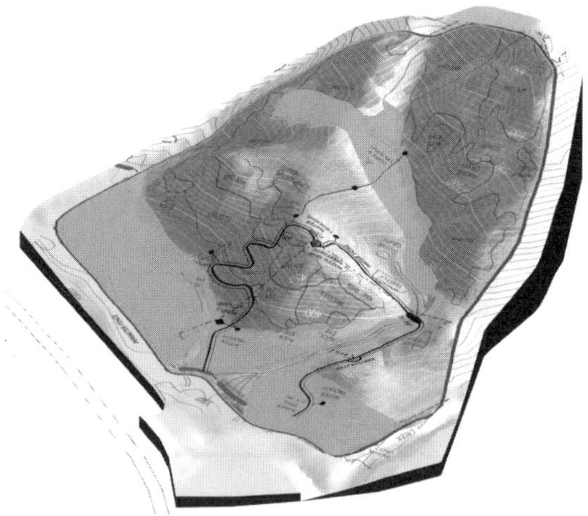

Winter Grazing Plan
- target one
- target two
- graze invasive
- perrenial grasses
- oak woodlands

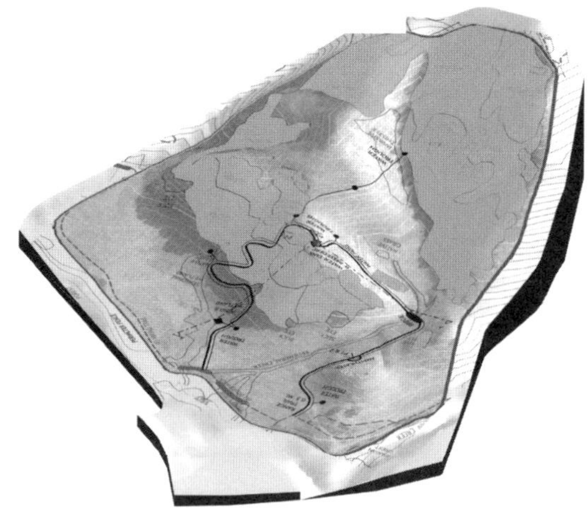

Spring Grazing Plan
- target one
- target two
- graze invasive
- perrenial grasses
- oak woodlands

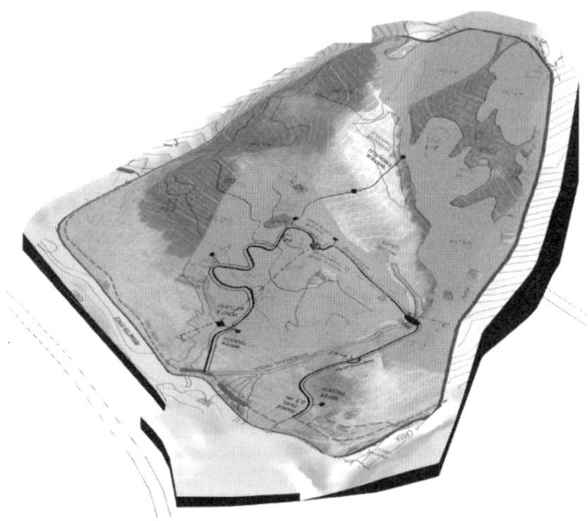

Summer Grazing Plan
- target one
- target two
- graze invasive
- perrenial grasses
- oak woodlands

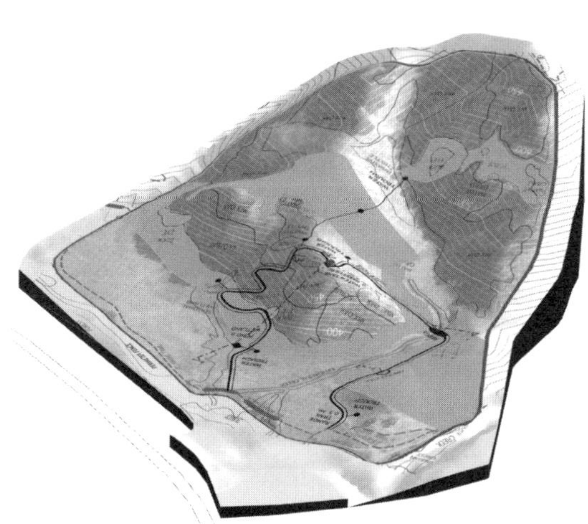

^ Proposed seasonal grazing plans. Throughout the year cows are moved to preferred paddocks based on sun exposure, wildlife vulnerability, wetland cycles, and soil moisture.

^ **Renewed ecological order. In contrast to typical grazing lands, where cows scatter and selectively feed, at Fernandez Ranch, lines of colored posts reestablish an ecological order reminiscent of the region's native Tule Elk.**

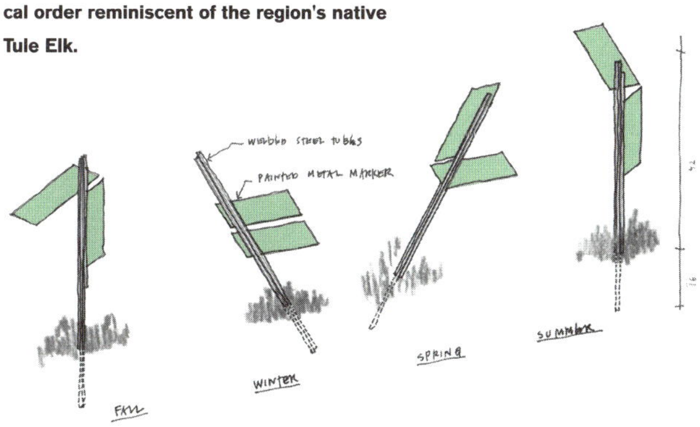

^ **Detail of plastic and steel sign posts. Colored markers indicate seasonally appropriate areas to graze, while the signs shapes can be reconfigured to communicate additional information, such as seeding, flowering, drought, or over saturation.**

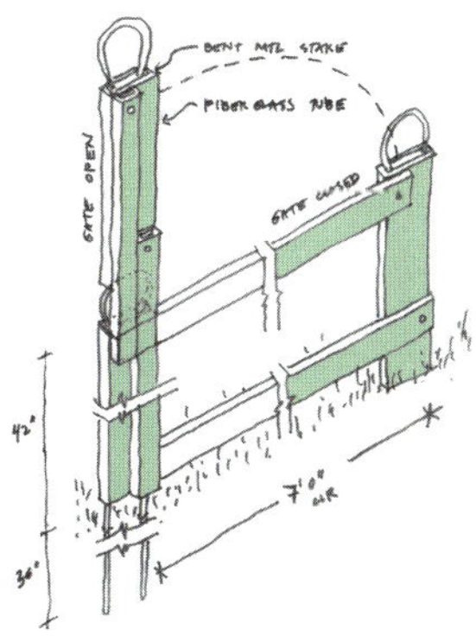

> **Cattle gates. In a 'fenceless' grazing system, gates can provide a useful visual cue for cattle. Gates are designed to be relocated easily.**

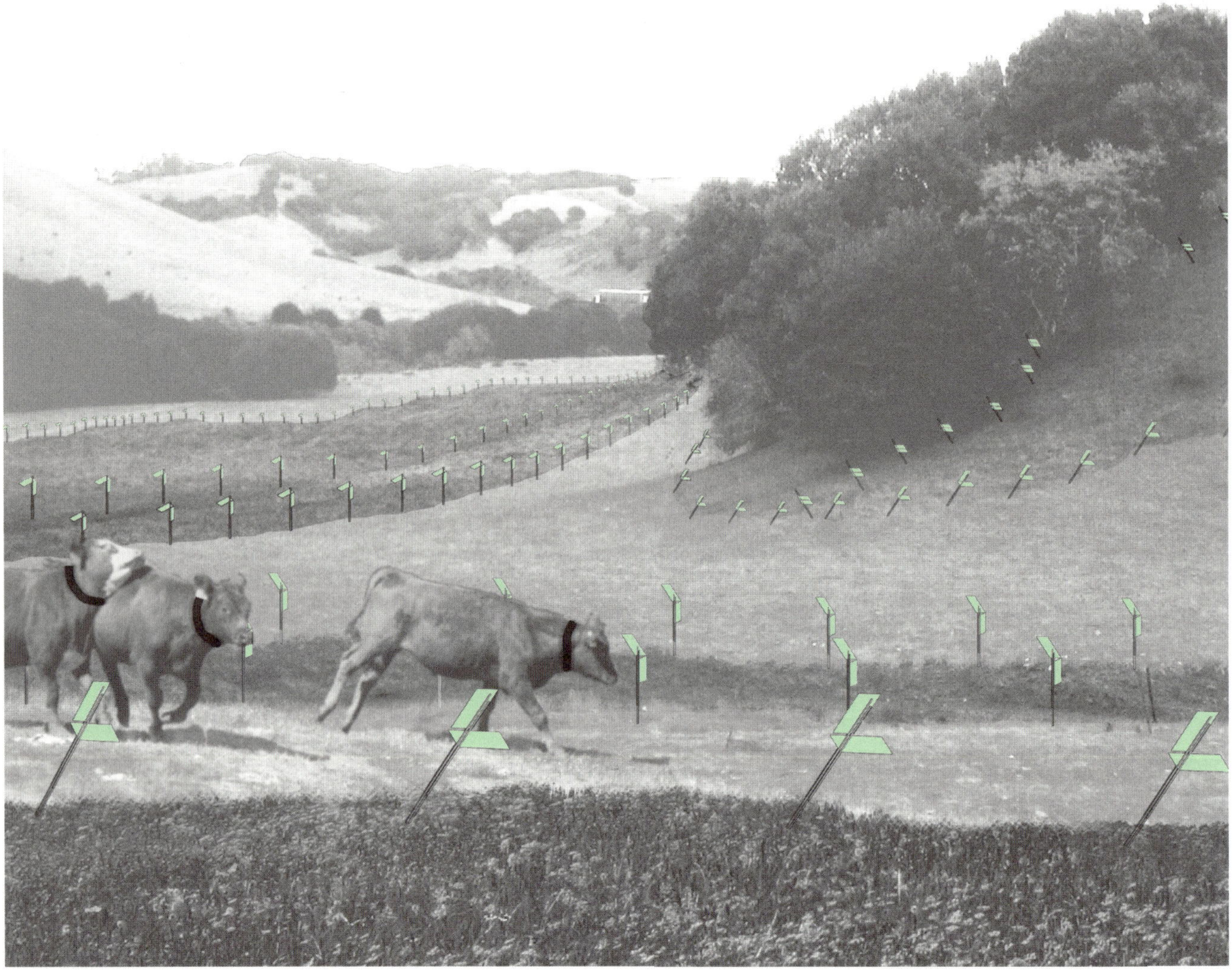

^ Cattle in paddock with interpretive signage. With the help of GPS-controlled 'invisible fence' collars, cows are rotated to new fresh grazing areas daily. Colored sign posts provide visual cues to park visitors, highlighting the grazing circuit and sensitive habitat areas.

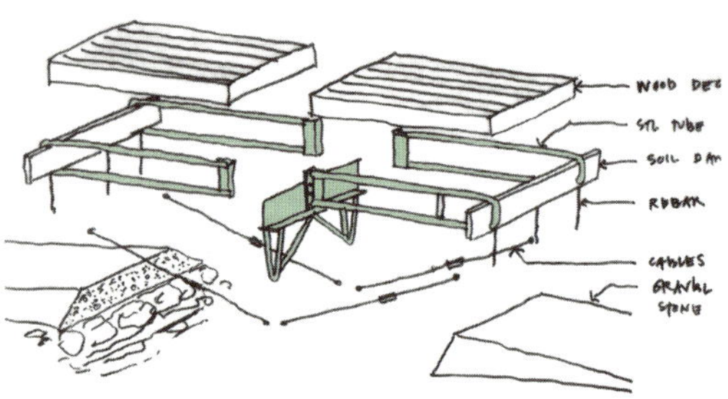

^ Sketch of cattle bridge. Cattle movement in wetland areas causes widespread habitat destruction. New relocatable bridges enable cows to bypass sensitive wetland areas.

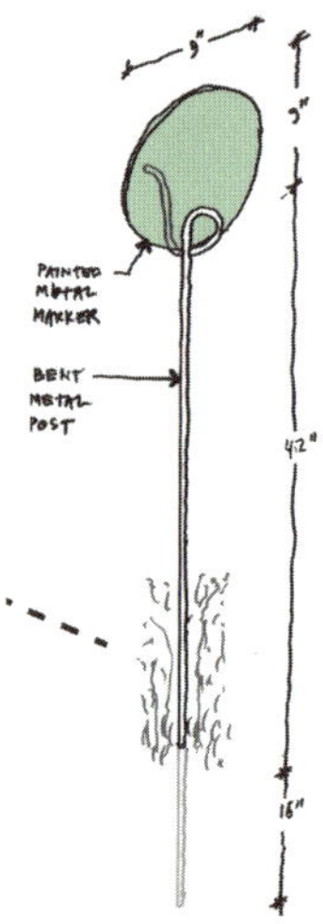

^ Detail of sign posts. Sign posts made from simple painted metal materials.

living tower: a vertical horse stable for luxor

dimitris argyro

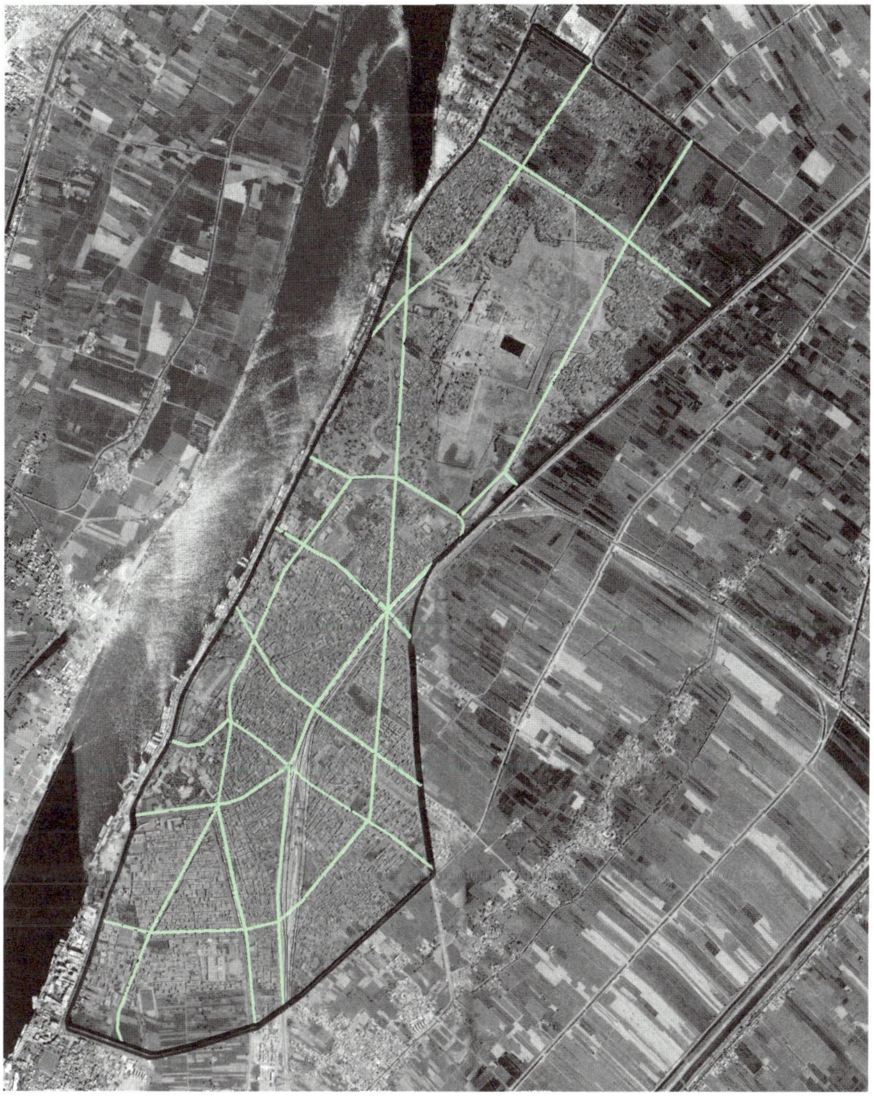

^ Aerial map of Luxor with proposed pedestrian zoning and bus routes.

^ Early collage studies for proposed horse-and-carriage public and goods buses.

A sustainable development pilot scheme seeks to alleviate Egypt's Luxor Central of its urban congestion as well as problems of land deprivation and poor horse care, all while sustaining Luxor's economic growth and future prospects. This is achieved through the re-invention of the transport system owned by the government using only horses and carriages. The proposed scheme consequently operates at three scales: urban, community, and building scales.

Urban Scale

"Living Tower" prompts a master plan in which the horse reshapes the city just as the car did, to adverse effects, for Western cities. The scheme proposes pedestrian and congestion zoning, bus stops, terminals, suburban car parks, and inner-city garden areas for equine and public use. The proposed system utilizes the existing tourist horse-and-carriage system in combination with technologically advanced methods of transport and waste recycling for a self-sufficient system of energy consumption. An annual festival in the form of a horse race is proposed to link together the peripheral isolated communities, bring in tourism, and encourage local participation in the city.

^ Collage model of brick market and food-storage wall.

^ Collage model of horse-accommodation towers.

Community Scale

A community-scale architectural intervention, called the Union, in the form of a transport terminal is proposed. The Union is strategically situated at the boundary of the city and the farming strip and is the key element for the success of the transport system. To address Luxor's shortage of available cultivatable land, a layered system of natural and artificial undulating landscapes, horse-accommodation towers, workshops, industrial silos, and horse and goods circulation routes is introduced to accommodate the 1,000 horses, 200 drivers and carriages, 60,000m^2 of required pasturelands, and huge amounts of food and waste needed for the entire operation.

Building Scale

The horse-accommodation towers seek to re-create a natural environment for the horses through a construction strategy based on a sustainable hybridisation of imported new technologies as the underlying structure and vernacular ones as the cladding. The vernacular is built over time by the locals. A proposed manure-adobe brick connects the process of food and waste with equine inhabitation and consumption of raw materials. The cladding acts not only as an insulating dry façade, but also as a sustainable responsive greening façade—requiring no added water, soil, fertilizer, or stabilizing structure—from which horses can snack on during their daily washing routine.

>>> **Dimitris Argyros is a London-based designer, urbanist and lecturer at the Bartlett School of Architecture, UCL.**

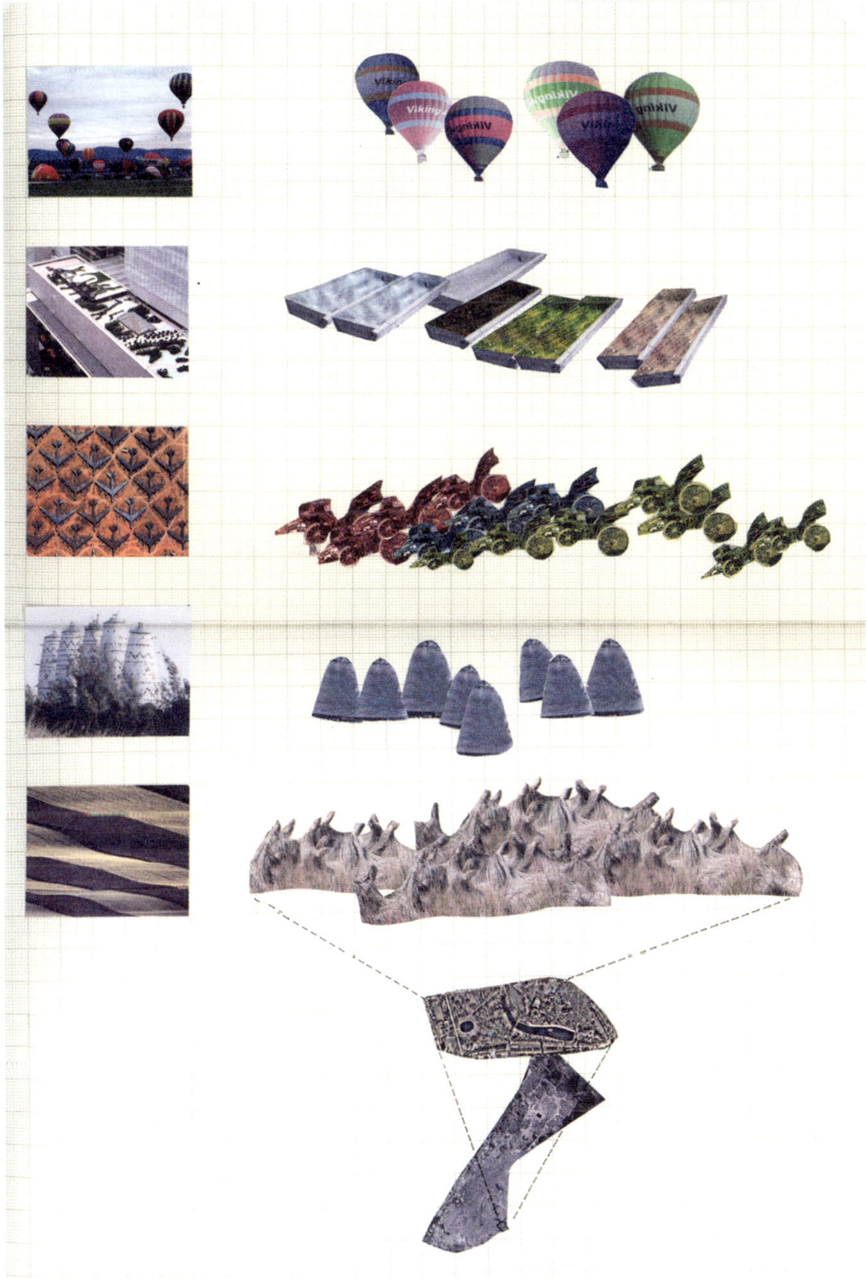

∧ **Building as a layered system of natural and artificial undulating landscapes.**

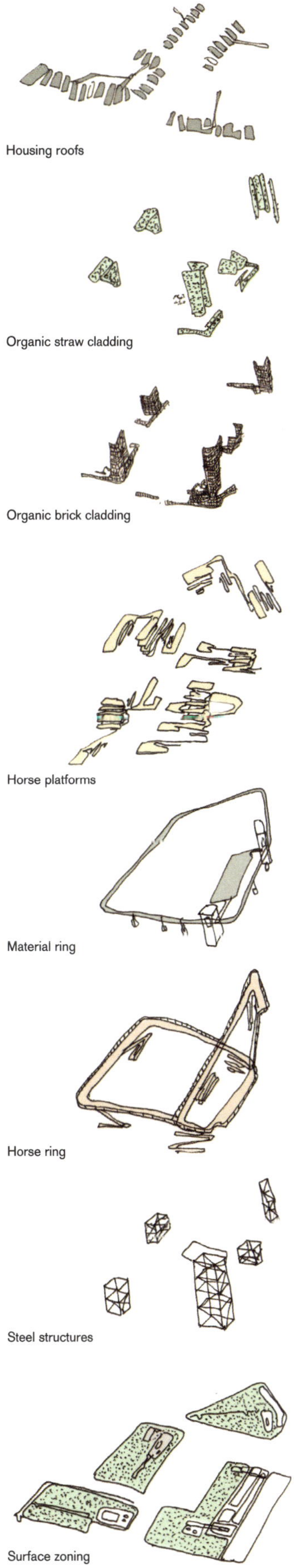

∧ **Exploded diagram of the building kit of parts.**

^ 1:1 prototyping of the tower's living wall, constructed of horse manure based bricks.

^ Paper and clay collage of tower northern balconies, where horses are cleaned, cooled and fed.

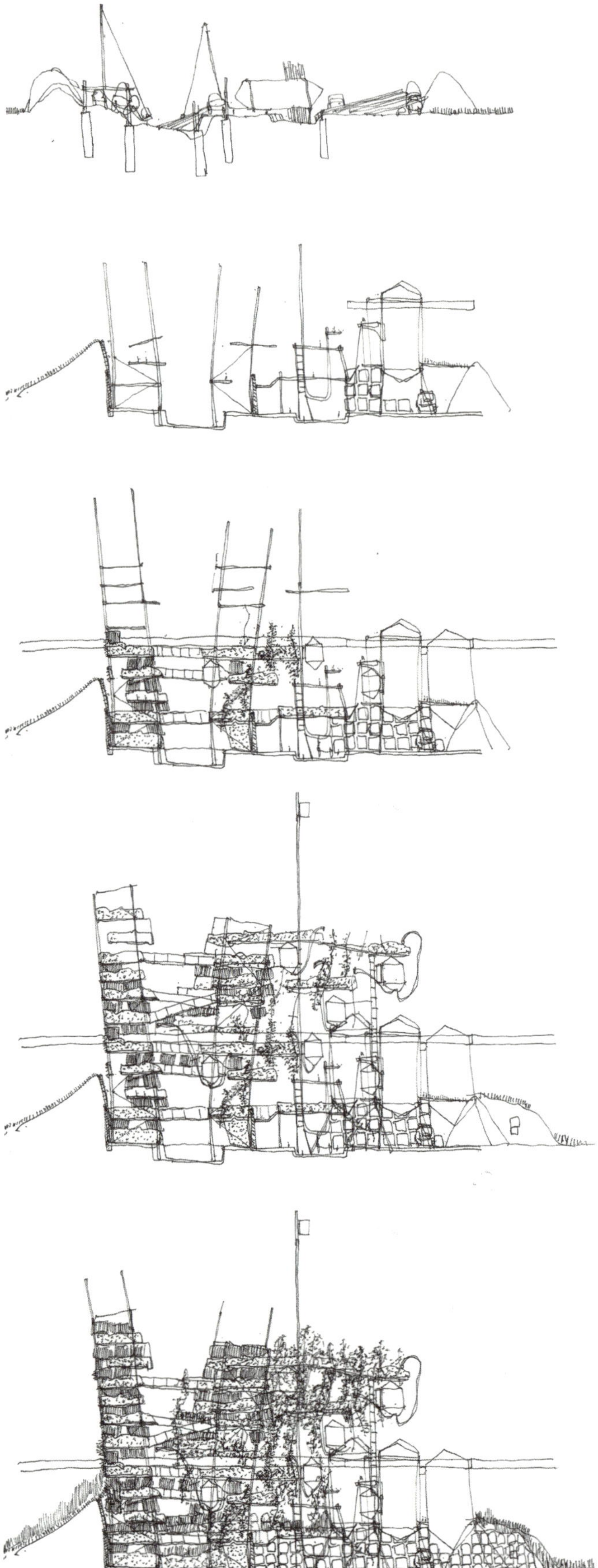

^ The locals build the building over time.

^ View of annual horse race through Luxor Central.

^ View of the "Living Tower" seen from the dessert. Building demarcates the farming industry and the new pedestrian city.

^ **Perspective model. The Tower mirrors in scale the Ancient Egyptian Karnak and Luxor Temples.**

> Following: Sectional collage model.

ORSE RACE
FESTIVAL

landgrab city

joseph grima, jeffrey johnson & josé esparza

"Landgrab City" is an installation commissioned by the Shenzhen/Hong Kong Biennale of Architecture/Urbanism 2010 and is located on Shenzhenwan Avenue (Nanshan), a busy shopping district in the city of Shenzhen. Conceived as an investigation of the full extent of Shenzhen's spatial footprint, the installation is comprised of two parts: a map of one of the city's dense downtown areas, home to approximately 4.5 million people, and a plot of cultivated land divided into small lots. This land is a representation, at the same scale as the map, of the amount of territory necessary to provide the food consumed by the inhabitants of the portion of city sampled in the map, projected to 2027—the year China is expected to overtake the US as the world's leading economy. Each lot represents the extent of a single food group's footprint: vegetables, cereals, fruits, and pasture (for livestock), among others.

In reality, of course, these agricultural territories are not clustered around Shenzhen, as in the installation, but scattered across China and contiguous regions. As China's political and economic apparatus grows in range and influence, significant proportions of these agricultural territories have, in fact, migrated to far-flung regions of the planet such as Africa, Latin America, South-East Asia, and Eastern Europe.

Spurred on by the volatility of food prices and a turbulent global economic climate, highly urbanized regions of the world are purchasing vast tracts of land in distant continents. Entire extra-urban regions of the developing world are being transformed into insurance policies against future food scarcity. Since 2008, China, South Korea, many Gulf states, and several other nations that have undergone explosive economic, demographic, and urban growth, have scrambled to purchase or lease vast quantities of land in foreign countries. Korean agribusiness, Qatari corporations, Chinese food traders, Saudi investment banks, and European private equity funds—all with government backing—have become the proprietors of entire regions of arable terrain in Argentina, Uganda, Brazil, Philippines, and Kenya, to name a few. Mega-farms of unprecedented productivity have sprung up in impoverished Sudan, while Mali has seen a spike in foreign acquisitions of agricultural land. The output of these territories bears no relation to the economies of the nations that surround them—they are in effect satellites of the distant urban entities to which they belong, and to which their output is shipped the very moment it is harvested.

By closely examining its spatial footprint, "Landgrab City" attempts to produce a complex representation of a 21st century metropolis. It proposes a new spatial definition of the city and thereby a more complex understanding of urbanism, one that no longer considers city limits as the boundary of its remit, but instead looks beyond—even across international borders—to the spatial, social, economic and political implications of rapid urbanization.

>>> Joseph Grima is a Milan-based architect, editor, curator and researcher, and the editorial director of *Domus* magazine. He is the former director of Storefront for Art and Architecture, and a PhD candidate at the Centre for Research Architecture, Goldsmiths College, London.

>>> Jeffrey Johnson is the founding director of China Lab, an experimental research unit at the Graduate School of Architecture, Planning and Preservation, at Columbia University, where he also teaches. He is also a co-founding principal of SLAB architecture based in New York City.

>>> José Esparza is an independent New York-based architect and researcher. He received his degree in Architecture from ITESO, Guadalajara, Mexico, and has since worked as a curatorial associate at Storefront for Art and Architecture, New York.

^ Landgrab City. Shenzhen/Hong Kong Biennale of Architecture/Urbanism 2010.

Landgrab Headlines

18 December 2003

3,000 Chinese Farmers To Farm Abroad (China Daily)

13 November 2007

Japan Mitsui Affiliate Buys Brazil Farmland (Reuters)

04 March 2008

Food Crisis Looming Over Korea (Chosun Ilbo)

04 May 2008

China Farms The World To Feed A Ravenous Economy (The Guardian)

15 May 2008

Jordan To Invest In Sudanese Agriculture (Sudan Tribune)

08 June 2008

UAE May Invest US $500 Million In Pakistan Farms (The National)

15 June 2008

Korea Will Grow Wheat In Sudan (Korea Times)

30 June 2008

Hopes And Strains In China's Oversea Farming Plan (Economic Observer)

04 July 2008

Africa At Large: China Eyes Idle Farmland In Continent (Business Daily)

21 July 2008

China And Mozambique Invest In The Zambezi Valley To Make Chinese “Grain Store” (Reuters)

22 July 2008

UAE Signs MOU With Philippines To Ensure Food Supply (Gulf News)

25 June 2008

Enter The New Farmers: Morgan Stanley Buys 40,000 Hectares Of Land In Ukraine (Reuters)

29 July 2008

S Korean Co To Invest US $2 Billion In Indonesian Maize Plantations (Antara)

07 August 2008

UAE Stepping Up Agricultural Investment In Sudan (Sudan Tribune)

07 August 2008

Saudi Arabia Joins In The Ethiopian Land Grab (Ethiopian Review)

13 August 2008

Indonesia, Bin Ladin Group Discussing $4 Billion Rice Project (Dow Jones Newswires)

22 August 2008

Goldman Sachs Buys Chinese Poultry Farms (Thepoultrysite News Desk)

24 August 2008

UAE's $4 Billion Food Bill Set To Climb (Bloomberg)

26 August 2008

Saudi Arabia Looks For Brazilian Land To Feed Saudi Population (Brazil Magazine)

30 August 2008

Egypt To Buy Kazakh Wheat, Grow Grain In Uganda (Reuters)

10 September 2008

Korean Land In Argentina To Be Developed (Chosun Ilbo)

11 September 2008

Banks, FIS May Get To Fund Overseas Fertile Land Buys (Economic Times)

14 September 2008

Blackrock Plans $200 Million Agriculture Hedge Fund, People Say (Bloomberg)

15 September 2008

Saudi Arabia Eyes Investment In Grain, Cattle-Breeding In Kazakhstan (Asia Agribusiness)

02 November 2008

3 Gulf Firms Sign Major Farm Deal With Turkey (Arab News)

20 November 2008

Madagascar Enthusiastic At Corn Deal With Daewoo (Associated Press)

02 December 2008

Qatar Looks To Grow Food In Kenya (The Guardian)

05 December 2008

Daewoo Unsure Of Madagascar Deal (Financial Times)

14 December 2008

Global Grain Rush Under Way As Rich Nations Snap Up Farmland Overseas (Chicago Tribune)

13 January 2009

Qatar Signs Food For Port Deal With Kenya (Arabian Business)

^ Landgrab City. Shenzhen/Hong Kong Biennale of Architecture/Urbanism 2010.

Grima, Johnson, Esparza

Landgrab City

the building that farms...

fritz haeg

^ "Edible Estate Regional Prototype Garden #1: Salina, Kansas," 2005. Photo by Dale Cole.

At my neighborhood public library in the early 1980s, I knew the architecture section by heart. They had a surprisingly good selection of books on environmentally conscious, low-impact, alternative building types, which I was particularly drawn to. Around that time I wrote my sixth grade research term paper on underground homes, and another the following year on passive solar strategies. The 1973 oil embargo, and a nascent modern environmental movement had produced a strain of outsider architecture that was becoming even more marginalized (though I wasn't aware of it at the time) during the Reagan era because of a dialog on style, what a building looked like, mostly on the outside, postmodern they called it. The buildings that intrigued me, however, seemed to care less for their appearance, with their exposed trombe walls of black painted oil drums, bulky thermal shades, and scavenged window frames. Performance or appearance, they seemed to be mutually exclusive, one had to choose.

I was born into a typical controlled homogenous Midwestern American suburban existence, shuttled on freeways between a sealed windowless mall surrounded by parking, an industrial school box, and a two-story "colonial" developer home that would be very familiar to many of you. In comparison these hand-made, heat-absorbing, water-collecting, south-facing, ferro-concrete, plant-covered, earth-sheltered, inflatable constructions thrilled me. They seemed so exotic and visceral in comparison. Here was the possibility of a different way of life, evidence of a more evolved existence, and it seemed like the people in those buildings were having so much more fun.

The idea of farming probably does not sound like fun to most people. The word farming may trigger images of vast fields of one crop. Some romantic notions of the farmer may persist, but we know that farming today primarily means an industrial landscape of monocultures on a life support system of chemicals and irrigation, in vast inaccessible spaces that are rarely even tended by humans. Like those sealed climate-controlled boxes that I grew up in, this is the modernist dream of global consistency without regard for local nuances of climate, place or people.

^ "Edible Estate Regional Prototype Garden #3: Maplewood, New Jersey," 2007. Photo by Curtis Hamilton and Canary Project.

I would argue that the most significant recent developments in architecture have nothing to do with the appearance of the physical shell of a building, but with the way that architects and landscape architects are beginning to intervene in the expansive local and global systems that the building is a part of. Architects are increasingly considering their task to involve some reconciliation between the impossibly complex physical, emotional, creative, intellectual needs and desires of the people they are accommodating with the equally complex and varied environmental resources and forces surrounding them. Architects are asking themselves questions about what we need, and then looking around to see how much of that might be provided right where we are. What an exhilarating feeling to think we might have everything we need already!

Let's expand the role of architect beyond builder and planner, to also include farmer. The farming architect, the building that farms. What would that look like?

The building that farms pays attention.
It takes careful stock of its situation. It understands what is uniquely possible and available in a particular location and figures out how to sustainably transform it into something useful. What do we have here? Algae, birds, clouds, compost, dead trees, data, fire, fresh air, fruit, geo-thermal activity, honey, hot springs, humidity, lightning, manure, meat, moss, mud, oceans, rain, rocks, seafood, salt, sand, seaweed, smog, sunlight, spices, tides, trash, traffic, vegetables, running water, wind, and worms? It is an inherently localized activity.

The building that farms takes responsibility for what it needs.
It takes steps towards removing itself from the corrupt short-sighted global supply chain that exploits people and places we will never see. Even though the building may not even come close to extricating itself from this remote support system, modest gestures in that direction start to provide evidence of what is needed for us to survive and be comfortable in that place. Staying warm by the heat radiated from walls exposed to the sun, computers powered by electricity generated from the wind, eating a tomato that grew on the roof, the people in the building become more aware of how their needs and desires are being satisfied.

The building that farms is sexy with a casual charm.
Buckminster Fuller was well known for scolding young architecture students, warning them that willfully trying to make a building beautiful was a ridiculous pursuit. If one were to follow the basic lessons, models, systems, and principles found in nature and in the environment, the building would be inherently beautiful in ways you couldn't anticipate. A building that really performs, that has a job to do, that is responsive, that is engaged in a complex feedback loop of localized farming activity develops a particularly charming appearance that reveals itself over time. It can be messy, rough, tactile, sensual, and sometimes hard to predict.

^ "Edible Estate Regional Prototype Garden #4: London, England," 2007. Photo by Heiko Prigge.

^ "Edible Estate Regional Prototype Garden #5: Austin, Texas," 2008. Photo courtesy Arthouse.

^ **"Edible Estate Regional Prototype Garden #6: Baltimore, Maryland," 2008. Photo by Leslie Furlong.**

The building that farms is alive.
It is constantly changing. Plants grow, the sun moves, the wind dies down, the rain comes, the leaves decompose, the food ripens, and we are there to be a part of it. Are the squash ready to harvest? Is the rain barrel full? Is the sun out? Should we open the windows? It is interactive. It requires attention and welcomes us to be in dialog with the building and each other. We are invited to become active participants in our environment and in our lives. The insidious zombie role of the passive consumer is not possible here.

The building that farms welcomes a new era of active participation.

harvest / yield

Studio Lindfors
atelier eem
Stephen Becker and Rob Holmes
Paul Schuette
Alexandros Avlonitis
Joyce Hwang

cloud skippers

studio lindfors (ostap rudakevych & gretchen stump)

^ Earth's surface has grown verdant as people migrated to the clouds.

> Fins keep the rigs aligned with prevailing winds.

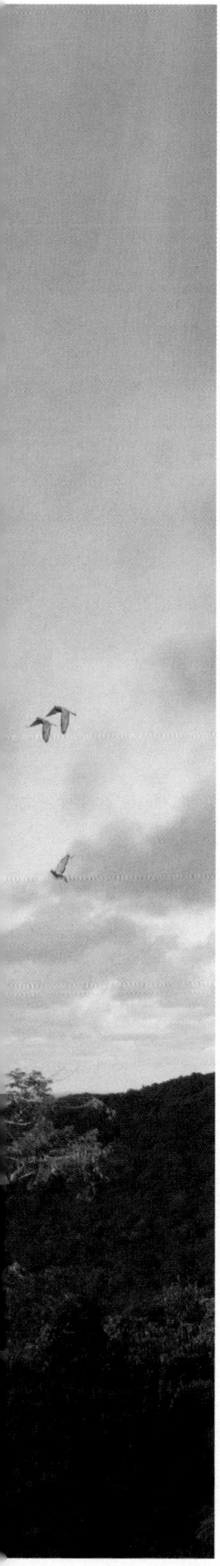

^ **Skippers cast lines into the ocean in hopes of catching fish.**

^ **Skipping along cloudscapes.**

^ Wings are studded with solar panels and sensors to maintain aloft autonomously.

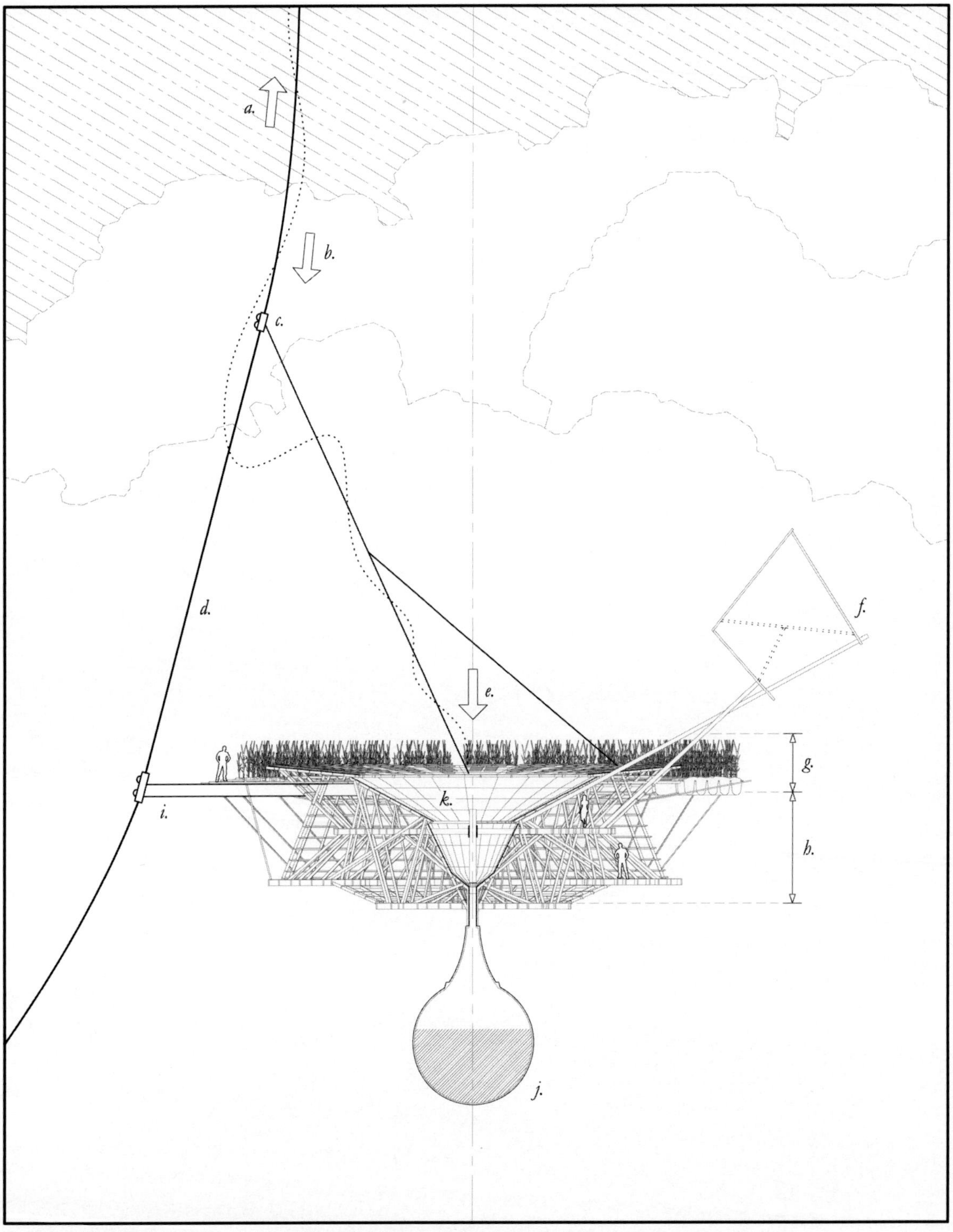

Cross Sectional Diagram of Typical Dwelling Componenets

a.) *Lift force pulling up equals weight suspended* b.) *Electrical power from wind turbines above* c.) *Cable truck allows for altitude adjustment along cable* d.) *Support cable* e.) *Rainwater collected, used for irrigation* f.) *Stabilization fin* g.) *Edible vegetation zone* h.) *Habitable zone* i.) *Compression strut* j.) *Expandable potable water storage bladder* k.) *Water collection funnel*

^ **A small cluster of rigs passes over undisturbed fields.**

The clouds drifted languidly across the grassy plain. Watching them made me feel as if the ground was sliding backwards. I studied their formation and structure for signs of erratic wind or pending bursts. The campfire at my feet was dying down, so I stoked it with some dry branches, waited a few minutes, then threw some fresh green grass onto the flames. The smoke thickened, rising in a tapered column up into the clouds. I scanned the horizon again, and then lay down with my hat over my eyes. I thought of returning to my home, to the city of Aether. I'd been gone almost two months—a month longer than planned. My packs were full of seeds and cuttings that I had gathered to expand the diet of my fellow Cloud Skippers back home. That extra month I was kept busy watering the shoots and tending to some rare toads, which the medics would use to extract a medicinal salve. Though I had never been to this region before, I was tired of camping in the same spot, waiting. I looked forward to returning to the city.

There are no walls surrounding the city of Aether—no gates or landmarks signaling its location to a returning traveler. The approach to the city is never the same. Its surrounding landscape is ever changing: one day, it may be nestled among the snow-capped peaks of a continental divide, and the following day sprawled out beneath the big sky of the grassy plains. By week's end, it may be a collection of silhouettes against the endless horizon of the ocean. A floating, unmoored community, Aether drifts over lands and regions freely.

Remembering how my legs shook when I first touched down on the surface, I dozed off. I was awakened by the vibration of my mobile. They must be near! I jumped up and started the signal fire back up. As I scanned the western horizon, my heart jumped when I spotted the first few dots growing larger as they approached. Soon, the whole of Aether was visible, a collection of rigs shifting through the sky like a swarm of insects. As the city approached, I ran over to the half-buried concrete structure from a bygone era to check the mooring rod once again. If the mooring rod was missed, I would have to spend another month or two waiting for the city to return. The rigs were upon me; I gathered my packs and brought them to the mooring station. As they passed overhead, I spotted my home and could barely see Nyxa's faint profile peering down over the edge, waving. The transfer rig approached, two thin cables with large grappling hooks at the ends swinging in its wake. Before I realized it, there was a loud clang of metal as the hooks grabbed on to the steel mooring bar. I watched as the trailing cables pulled taut, the springs at their ends cushioning the shock. I was relieved that the hooks held—I would be home at last. Before long, my fellow Cloud Skippers lowered a balloon along the length of the cable, and I climbed onboard after stowing my packs. We checked the weight and radioed up to the transfer rig to inform them of the additional load. Balloons are invaluable for surface outings as well as for traveling between rigs. We reached down and released the mooring hooks. Our balloon floated upward, still attached to the trailing ends of the cables, as the transfer rig gained the altitude and speed to rejoin the city. I looked down over the surface where I had just spent two months and watched it recede into the distance as our balloon was hoisted back towards the rig.

The Skippers onboard were simultaneously lofting a balloon with a payload of water equal to the weight of me and my packs, thus maintaining the always critical balance.

The following day I was summoned to the navigator's rig where I met with Lapetus, a direct descendant of the founder of Aether. Joined by Chief Akmon, we sat at an old wooden table from the bygone era. Flashing screens surrounded the table. The Chief informed me that I would be heading topside first thing in the morning to help with repairs. In my absence, several generating wind turbines had failed and Aether was running dangerously low on electrical power. While topside we were expected to conduct a thorough inspection of the wing foils and the sensor and lift apparatus, which could take several weeks. Lapetus apologized for the abrupt notice, empathizing with me about the short time I was allotted to spend with my family since my return.

As the sun broke over the horizon, we loaded the lift balloon with tools and provisions for two months. Our team checked the guide cable once again before climbing onboard. The mooring line was released and our balloon began its slow ascent up along one of the main cables, upward into the clouds. Soon we were shrouded in the cool white mist of the clouds. With visibility reduced, we struck up a conversation in the newfound intimacy. The sun blinded us as we broke through the top of the clouds, skipping over their frothy peaks for a brief moment. The rest of the ride was uneventful, as the earth shrank away from us on our way up to 30,000 feet. Halfway up we put on our pressure suits and engaged the supplemental oxygen tanks. Soon the airfoils came into view. Aether is founded on these giant wings; without them, the city could not exist.

The founding engineers had devised a way to glean the enormous energy of the jet stream[1] circling the planet in continuous bands of fast moving air. Huge kites with immense surface areas that provide the lift needed to suspend their various structures known as rigs were designed by Aether's engineers. Sensors embedded in sturdy airfoils constantly monitor air currents and send information to autonomously controlled flaps which adjust the pitch and yaw of the rig to keep it afloat. The surface of each wing is a patchwork of electricity producing panels; the cable structures below are studded with wind turbines to meet nighttime demand. Redundant high-strength cables are equipped with safety parachutes and antennae to facilitate communication between rigs. The absence of any mechanical noise is one of the striking aspects of the city; the low whoosh of rushing air is all one hears. Because the city runs on sun and wind power, and no combustion takes place, the city is free of engine noise and smoke. Running entirely on quiet electrical power, the city is like an ever-shifting swarm. The dwellings were designed and formed for maximum efficiency, functionality and self-sufficiency. Shaped like large funnels, the dwellings collect rainwater into elastic storage balloons. Rainwater is supplemented with cloud water extraction tubes.

As we finished our first day of work, I rummaged through my pack looking for my log book. It was nowhere to be found. I realized with a tinge of regret that I had left it laying by the campfire back on the surface. I would have to wait another few months with the hope that the shifting air currents would bring us back to the same spot, so I could retrieve my notes and sketches from the foraging journey.

>>> Studio Lindfors explores the contours of imagination.

1. Jet streams are fast flowing (constant speed of around 100 miles per hour), relatively narrow air currents found at the tropopause, the transition between the troposphere (where temperature decreases with height) and the stratosphere (where temperature increases with height), and are located at 10-15 kilometers above the surface of the Earth. They form near boundaries of adjacent air masses with significant differences in temperature, such as the polar region and the warmer air to the south. The path of the jet typically has a meandering shape, and these meanders are one manifestation of Rossby waves. The major jet streams are westerly winds (flowing west to east) in the Northern Hemisphere. (source: Wikipedia)

globalgaelisation

atelier eem (marc blume, estelle nicod & francesca liggieri)

Globalgaelisation is a consideration on the delicate balance of a social organization based on the exploitation of a single resource. Algae can be an extraordinary source of energy that satisfies multiple needs: alimentation, gas and electricity production, thermoplastic and textile manufacturing, among others. The multiple applications of this resource allow us to consider transferring the mechanisms of our contemporary society entirely based on oil towards an all-algae future.

Yet could the intelligence of these organisms, which produce photosynthesis and protect the ozone layer, sustain such a vast complexity of consumption on a massive global scale? And what if it was precisely today's main advantage offered by algae—its capacity of creating energy and oxygen from carbon dioxide—that would create a new imbalance in the ecosystem? Will there be a day when we need to artificially re-inject carbon dioxide in our atmosphere?

>>> atelier eem [iim] is a multidisciplinary office based in Paris, founded by Estelle Nicod (Switzerland, 1978), Francesca Liggieri (Italy, 1979) and Marc Blume (Germany, 1976).

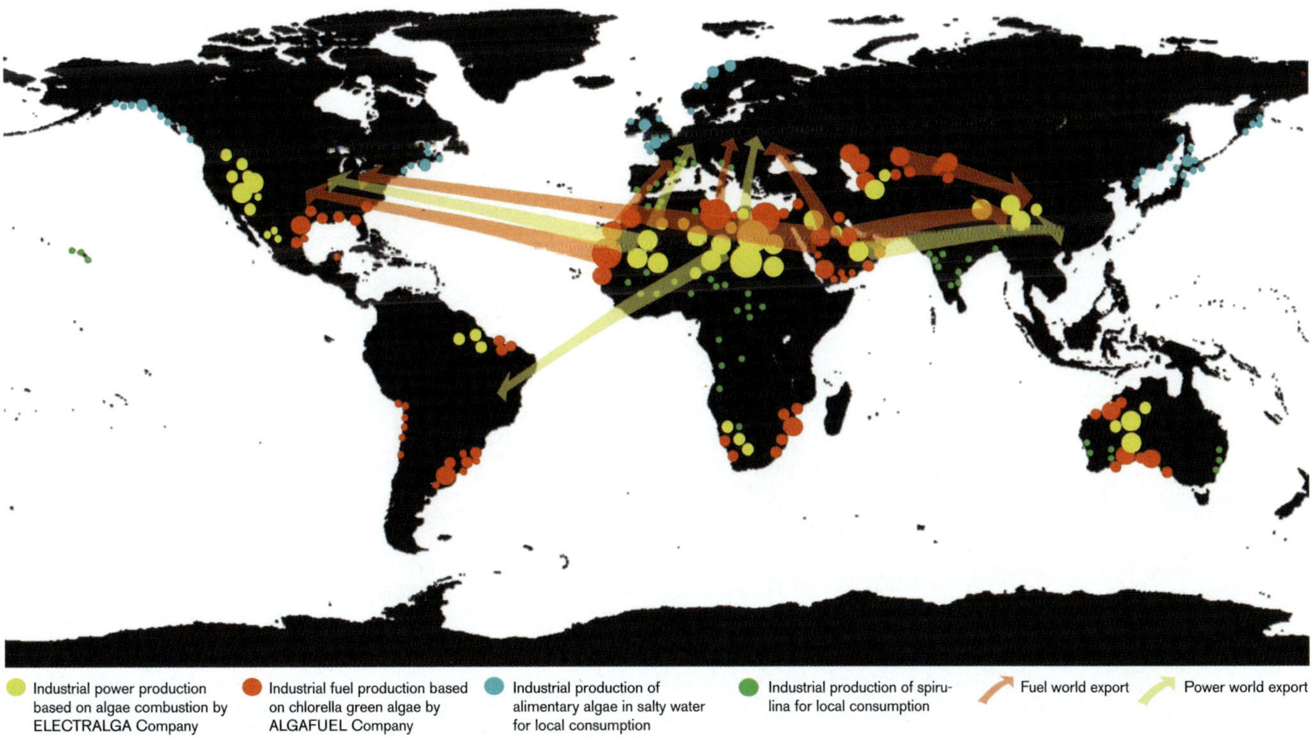

^ **Global algae-based production / 2030.**

^^ **Algae worker.**

^ **ALGAELECTRA algae bioreactor farm, Agadez, Nigeria.**

ΛΛ Advertisement for the New Algabella Colony, Plainview, Texas.
Λ Algal energy buildings in Belleville district, Paris.

hydrating luanda

stephen becker & rob holmes

Luanda, Angola, the fastest growing city in the world, is desperately short of clean water. Only one in six Luandan households has running water, forcing most of the inhabitants of the musseques (the vast slums that constitute the majority of Luanda's land area) to depend on contaminated water trucked in from rivers several hours north and south of the city. The price of water in the musseques can be as high as 12 cents a gallon, a huge burden on a populace that lives on an average of $2 per day. In 2006, the worst African cholera epidemic in a decade devastated the musseques, killing 1600, spread by contaminated drinking water and contact with sewage.

What if water, already inextricable from agricultural farming processes, was itself farmed? Beyond the direct benefits a renewable source of fresh, clean water would provide Luanda, farming water seeds the city with potential. By establishing an infrastructure to effect the farming of water, one may farm landscapes, societies, or production: in short, a city.

Farming water in parched terrain seems paradoxical, but an established process for doing so already exists: fog farming. As with most forms of farming, fog farming can be seen as a special instance of infrastructure with, as an infrastructure, both direct and indirect effects. It is directly responsible for condensing and collecting airborne moisture, but it also has many indirect effects on the growth and health of the urban system. The fundamental inquiry of our proposal is a unique treatment of infrastructure, considering not just how its programmed function can enable positive development of some urban condition, but how the thing in itself—the system which enables that function—can have other indirect effects, and how they can be understood and utilized.

Before exploring those indirect effects, though, a brief description of the techniques and requirements for fog farming may be helpful. This technology has been explored for several decades as a means to obtain potable water in arid environments, though until now it has been primarily confined to rural locales, owing to problems with airborne particulate pollution. A typical apparatus consists of a nylon or polypropylene mesh, at least a meter or two square, stretched across a metal or plastic frame, with condensed moisture dripping down the mesh into a collection pipe at the bottom of the net. The basic necessary conditions for the deployment of such an apparatus are an arid environment and strong fog. Such fogs are found primarily along particular ocean coasts (the Pacific coasts of Chile and Peru, the Atlantic Coasts of Namibia and Angola, or the Indian coast of the Arabian Peninsula) where certain ocean currents produce atmospheric moisture which is then confined and concentrated by obstructing mountains.

^ World Map.
Circles catalog the cities of the world projected to have five or more million inhabitants by 2025. Circles are proportionate to the projected population at that date. The blue fill within the circles indicates rate of growth; the darker the blue, the faster the rate of growth. Hatches indicate areas of water scarcity, with denser hatches indicating more severe scarcity. Red hatches mark those areas where topography, ocean currents, and prevailing winds combine to make fog farming practicable. Luanda lies on one of the three coasts where fog farming is most practicable, in an area of water scarcity, and has the highest rate of population growth of any city catalogued.

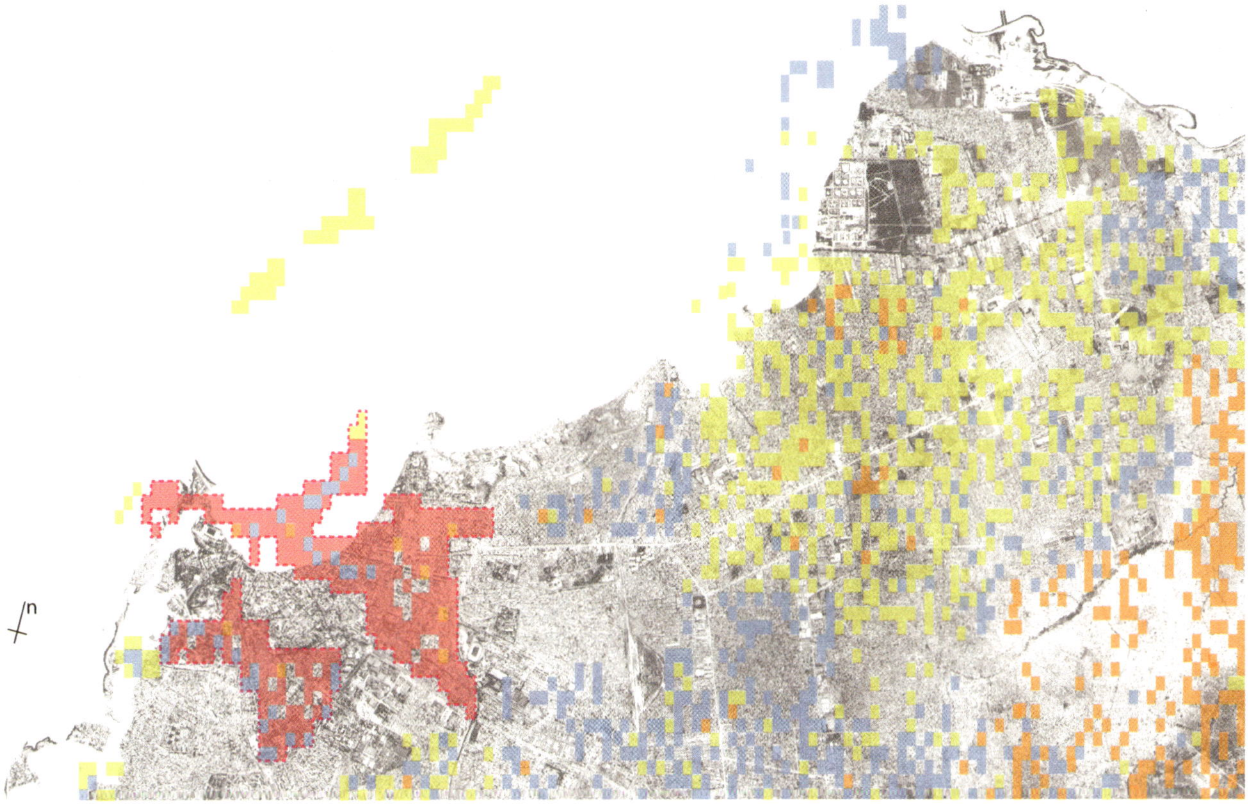

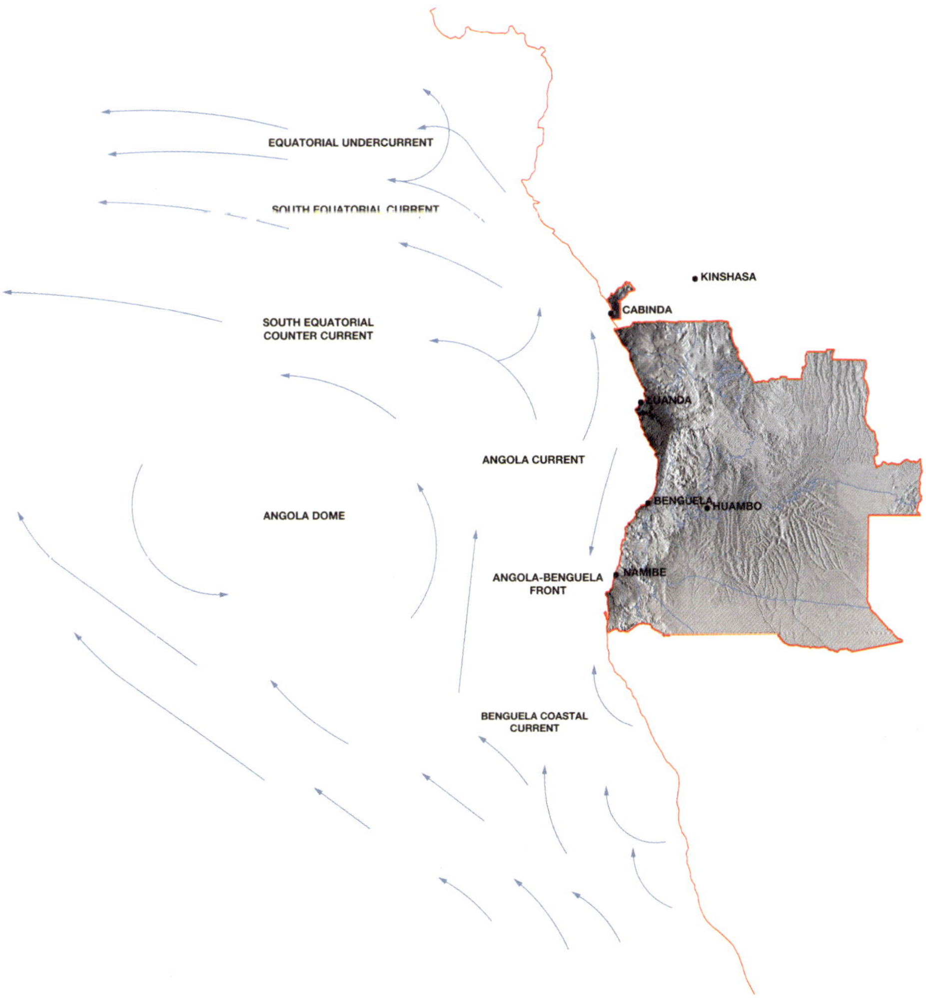

^ Net Distribution Snapshot— Ingombotas to Musseques.

Yellow: 85% of nets chlorinated: Used for harvesting water when disconnected from municipal supply.

Orange: 85% of nets with soil fertilizer: Used for aiding gardening and farming in the city limits, and rejuvenating the soil at the periphery.

Pink: 85% of nets with algae fertilizer: Used for increasing the capacity of the algae farming infrastructure.

Blue: 85% of nets un-treated: Used in Musseques when chlorinated nets are cost prohibitive, or when linked to municipal supply.

Pink dashed line: Denotes government owned, closed, centralized network.

< Regional Map.

Three geographical elements must overlap to provide an environment suitable for fog farming: winds and currents, topography, and aridity. Luanda possesses all three. The Angola and Benguela Currents meet off the coast of Angola, producing a humidity that hovers over the coast in dense fogs. The plain that Luanda lies in is one of the driest parts of Angola (in the diagram at left, the light blue hatches indicate precipitation, with denser hatches indicating heavier precipitation), and is pinned against the coast by a high range of cliffs and mountains that separate the interior plateau from the ocean.

^ **Ingombotas.**
At the outer edge of the district of Ingombotas, both direct and indirect effects of fog farming can be observed. Citizens erect private frameworks atop single-family homes to provide chlorinated drinking water and nutrient-rich water for plants, sharing or selling the surplus with/to neighbors. Government-built networks span the gaps between taller buildings, arranged perpendicular to the prevailing western winds to maximize productivity. These networks feed a secondary network, of tubes growing algae, which also hangs off buildings and across streets. The space beneath these networks can be appropriated by citizens as a public market, layering additional program onto the street.

Luanda is by far the largest city in the world in which fog farming is practicable, as it is projected to jump from a current population of four million to over eight million by 2025. Though Luanda experiences something of a rainy season between March and April, for most of the year precipitation is nearly absent—averaging 0 millimeters/month from June to August—as the Benguela and Angola currents of the Atlantic Ocean combine to prevent the humid air from condensing into rain. This combination of humidity, aridity, and explosive population growth creates a situation uniquely suited to an experiment in urban fog farming.

Fog farming has the direct potential to meet this need for potable water; but an investigation of this form of farming can also reveal important things about the role of infrastructures in urbanism, if we consider its indirect effects. We believe that this move—the equal valuation of indirect effects and direct effects—is essential to the development of an alternative modernist urbanism, for several reasons:

A. We argue (like many others have, particularly in the past decade or two) that the modernist planning movement has run its course; has been exposed as insufficiently flexible to react to the flux that characterizes a vital urban condition. This is because modern urbanism was concerned primarily with controlling the city, which inevitably produced conflict between the dynamic nature of the urban system and the static nature of the designer's reality.

B. We also take as a starting point the inadequacy of some alternate proposals, particularly those such as new urbanism and parametricism, that propose an alternate set of forms to modernism while failing to account for the fact that it was the overreliance on formalism itself—on the controlled expression of the rationality of the designer—that made modernist planning unable to cope with the reality of the urban condition.

C. We suggest that infrastructure is an appropriate object of design for the urbanist, the architect, and the landscape architect, as infrastructure can be embedded with some characteristics that provide definition (a means for the urbanist to have influence on the direction of change), as well as

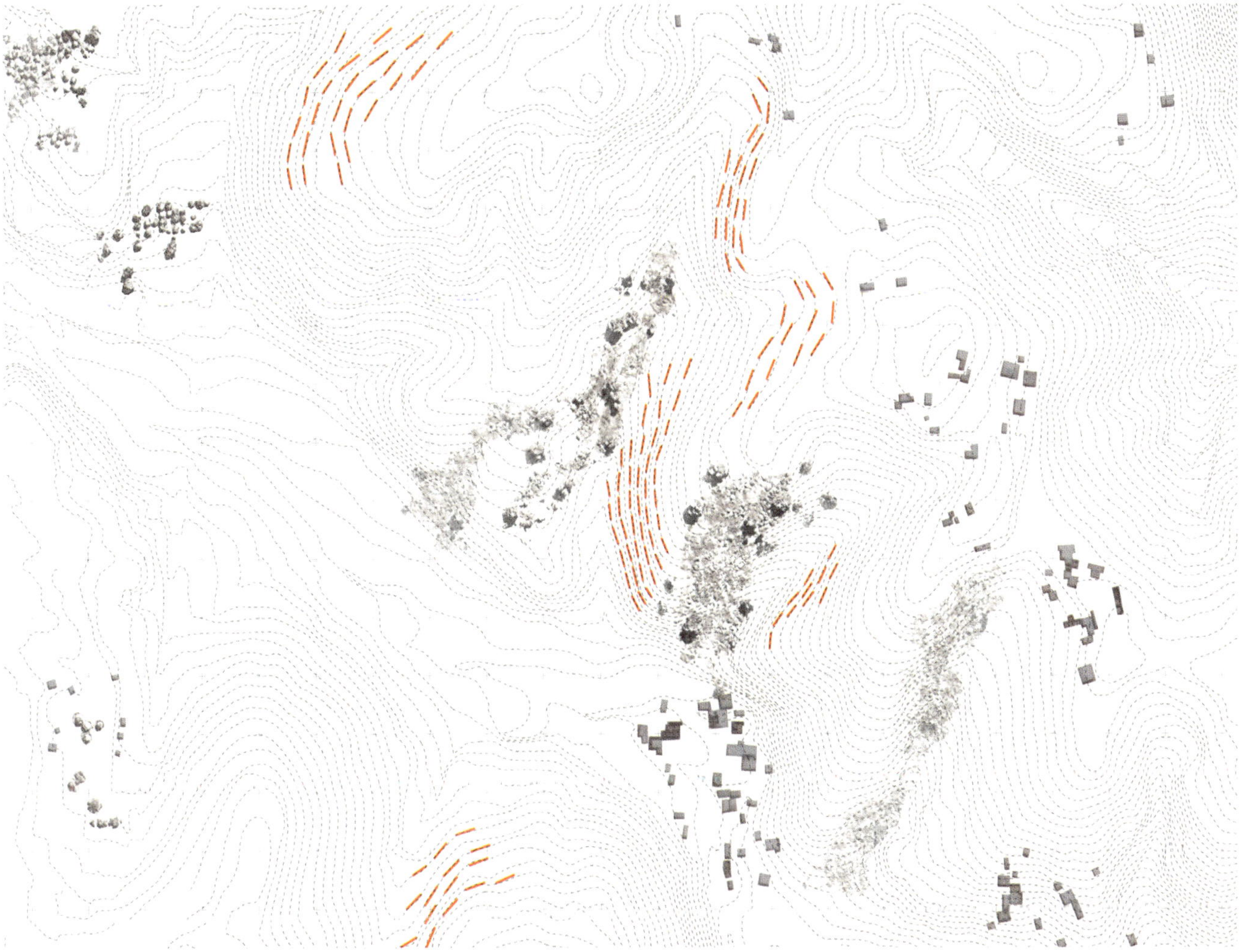

^ **Periphery A.**
At the periphery of the city, the government and NGOs can partner to set up networks that anticipate the growth of the city. Arranged on western slopes to catch the prevailing wind, specially-treated fog nets seep nutrients and water slowly into the soil, altering the soil profile to encourage the growth of vegetation and retard erosion.

characteristics that permit appropriation by inhabitants of the urban system. In other words, an infrastructure can withstand appropriation while remaining coherent as an intervention.

Fog farming, then, can only be fully understood as an urban intervention if it is understood as performing infrastructurally—as having effects on the urban system that extend far beyond the direct or spatial, to the alteration of streams and flows—liquid, capital, human, traffic, and so on. Thus we have sought not merely to provide a form of farming that meets a need for water, but also to understand how, in placing an infrastructure that would meet that need, we might also alter the streams and flows of the urban system for the better.

Consequently, the infrastructure we have developed is unified by its fundamental nature as a system for farming airborne moisture, but diversified in architecture, deployment, funding, distribution, and ownership, as the modification of each of those characteristics affects its interaction with the city.

Furthermore, we remain extremely sensitive to the fact that one of the difficulties of working as a designer in a city like Luanda is that the city government cannot always be relied upon to act in the interests of the poorest inhabitants. Indeed, even we—seeking as designers to act in ways that we perceive as beneficial those who live in the musseques—are likely (if not certain) to be mistaken at times about what will benefit the musseques. And so we have varied the degree of control embedded in the infrastructure, allowing it to be more rigid in the wealthier portions of the city where citizens have access to the levers of government and relinquishing nearly all control in the poorest regions. Some functions, which we hope will remain uncontroversially beneficial, are embedded in the nature of the infrastructure made available (most essentially, that it cultivates useable water from fog).

To these ends, we propose four basic forms for fog farming in Luanda (though the boundaries between forms exist more as gradients than hard delineations):

1. In the central districts of the city, particularly Ingombotas, a highly formal farming

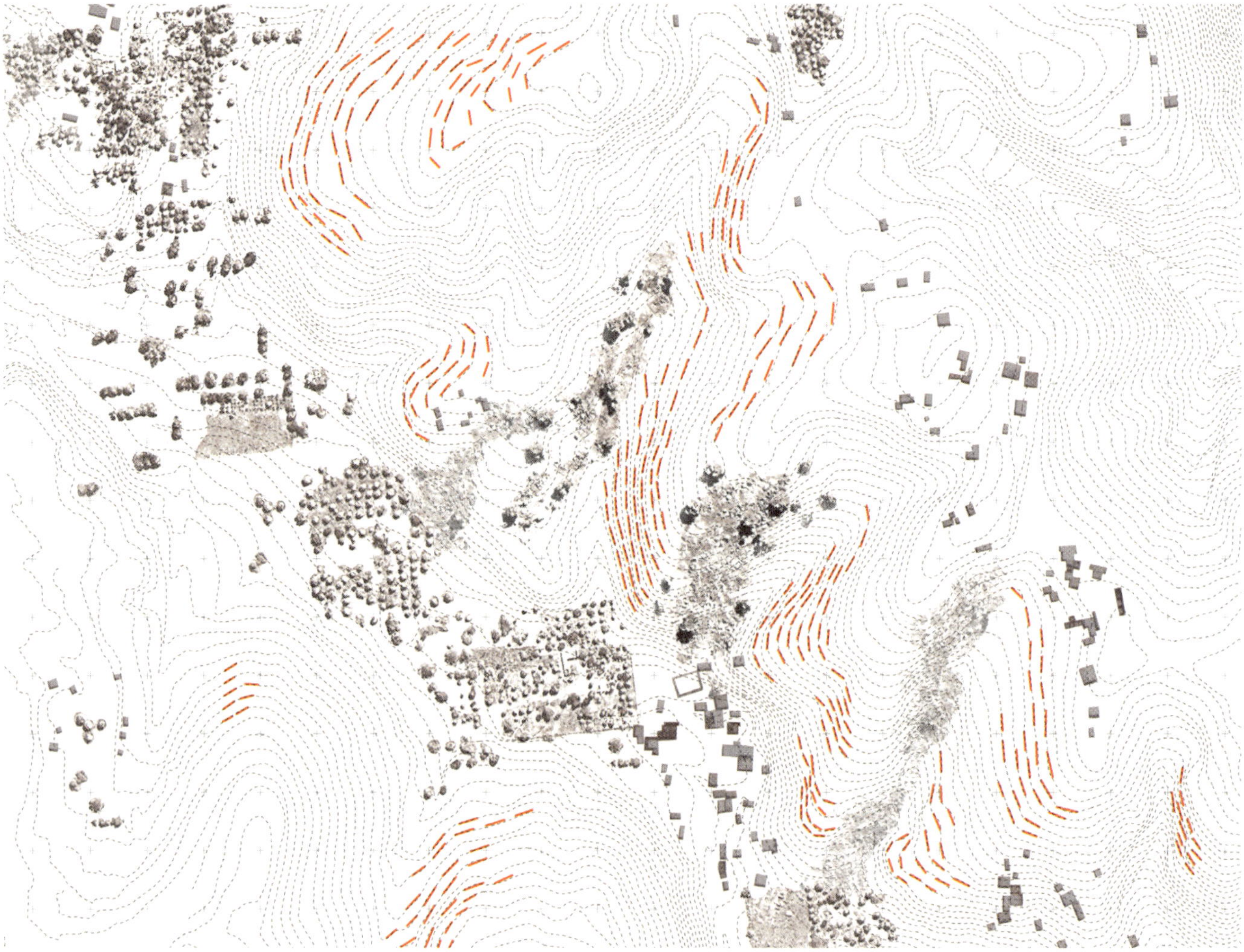

^ **Periphery B.**
As these networks are built up and proliferate, settlement creeps out from the city—driven both by immigration from the countryside as well as rising property values in the central districts (which have unfortunately led the government to bulldoze informal settlements in the past). The new settlements take advantage of the transformed soil profiles, building both homes and farms.

architecture appears, spanning the gaps between buildings and hanging off of scaffolding over streets, providing water treated with nutrients for the benefit of a system of tubes growing algae, which is in turn harvested at processing plants to make biodiesel. The profits from the oil industry with its fixed end-date can be invested to build an infrastructure that will meet the energy needs of Luanda far past that end-date. The distribution of this water is centralized; it is a striated, non-rhizomatic network. It would be government funded, with the intention that economies of scale (resulting from the erection of this network) would drive down the cost for building future fog networks in lower income areas of the city.

2. Middle class suburbs, which ring the central districts, would also receive investment from the government and see an interface arise between the municipal water system and the fog farming infrastructure, though some of the water supply generated there would be diverted to growing personal gardens and other vegetative uses.

3. In the musseques, a combination of NGOs and micro-lending practices would fund the development of an infrastructure, while the deployment would be at the whim of the citizens, who stand to benefit from the income generated by new sources of water. Appropriation by the community would be nearly total, while the benefits of the infrastructure would accrue entirely to the local communities. Distribution of the water would be completely local and decentralized, disconnected from the municipal supply and its filtration. Because of this, we propose nets treated with a slowly-released store of chlorine whenever affordable, with untreated nets supplementing the supply.

4. On the arid periphery of the city, fog farming would serve to seed the future of the city. Though the practice is deplorable, it is unlikely that the government will stop clearing informal settlements in the center of the city, so the periphery will likely continue to swell with a constant mixture of slum dwellers expelled from their homes and new arrivals from the hinterlands of Angola. Expanses farmed for fog would be gradually impregnated with nutrient-rich water, cutting

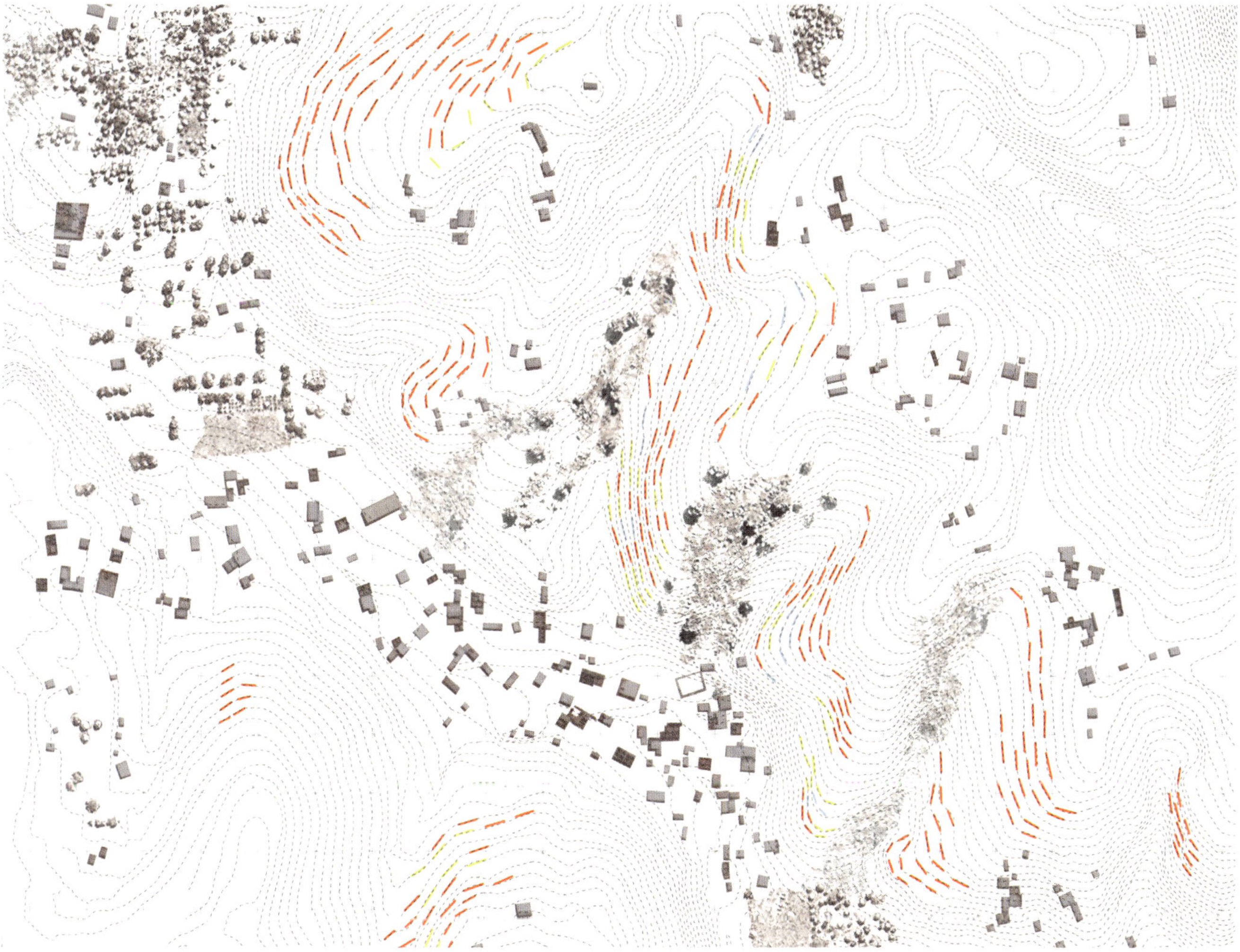

^ **Periphery C.**
Over time, the farms are replaced by homes, and the nutrient-treated nets replaced by chlorinated nets for drinking water. Still, the fog farms remain essential to the organization and growth of these new settlements, which lie too far removed from the city core to receive municipal water infrastructure.

down on dust storms, prompting the growth of farms and eventually settlements based around this infrastructure. Through farming water, we farm landscape, farm city, and farm future communities.

Key Sources: *Atmospheric Research*, *Environmental Engineering Science*, *IDRC Reports*, *Journal of the Chartered Institution of Water and Environmental Management*, *Journal of Water Supply: Research and Technology*, *New York Times*, NOAA, World Bank Group, University of Miami Rosenstiel School of Marine and Atmospheric Science, United Nations Department of Economic and Social Affairs.

>>> Stephen Becker and Rob Holmes are mammoth. Stephen is an architectural designer and project manager based outside of Boston. Rob practices as a landscape architect in Washington, DC.

A Statistical Index of Luandan Fog Farming

83%
the percentage of Luandan households without running water

12 cents
the price per gallon of water in the musseques of Luanda

0
the average monthly precipitation in Luanda from June to August

6%
the current yearly average growth of Luanda

4.8%
the current average growth of the next fastest growing major city in the world, Kinshasa

4,000,000
the population of Luanda in 2007

8,200,000
the project population of Luanda in 2025

12.9 liters
the projected average amount of water collected by a square meter of fog net in one day

244
the number of days it would take to recoup the cost, including both materials and installation, of a square meter of fog net at the current price per gallon of water in the musseques

1,200,000
the amount of water trucked into Luanda every day from the Bengo and Cuanza rivers

$35,152,969
the estimated cost of replacing the volume of water brought using trucks from the river with water from fog farming

$100
the estimated cost of a square meter of fog net, including materials and installation

2/3
the number of Angolans living on less than $2/day

450
the number of tanker trucks that travel from the rivers to Luanda every day

1,600
the number of Luandans killed by a major cholera epidemic, due to contaminated water, in 2006

hydroloops: mechanization and the command prompt

paul schuette

Modern hydroponic techniques were pioneered at the University of California Berkeley by William Frederick Gericke in the 1930s and continue to be used heavily in California both in industrial agriculture, as well as the legal gray zone of marijuana production. Hydroponic stores have multiplied across the California landscape in the past ten years selling scalable systems for both home gardeners and larger industrial productions. These systems served as a point of departure for the "HydroLoops" project, a series of hydroponics mechanisms that illustrate three typical forms of hydroponic cultivation. These hydro-mechanisms function as drawing machines, a type of automata originally described by André Masson and the Surrealists. Each hydro-mechanism has a simple drawing program associated with it, written in the Processing programming language and connected to the computer via an Arduino interface. The development of the plants and the operation of the hydro-mechanisms allow these drawing programs to run creating informal metrics of plants growth.

Hydromechanism 1-Tank is a dual hydroponic system. Plants in the tank rest in a nutrient solution. The roots of the plants sitting within the trays above are fed aeroponically via a spray mechanism. A webcam focuses on the tank and through a pixel counting program, conceptually similar to the type used with remote sensing in agribusiness, drawings are continually generated. *Hydromechanism 2-Slide* delivers a nutrient solution to plants on two inclined planes. The nutrient solution flows down planes into containers and is then pumped back up to the top. As with Hydromechanism 1, a pixel-counting program creates drawings based on plant growth. In *Hydromechanism 3-Tubes*, drip emitters deliver a nutrient solution via two tubes. Plants within the tubes grow in rockwool cubes, which soak up the solution then pass it onto the cube below when they are saturated. When the tray below the tubes fills to the top, a float switch is triggered, which allows the nutrient solution to be pumped back to the top where the process continues. The activation of the float switch also triggers a new phase in the drawing program.

The "HydroLoops" project is an experiment to see how a type of aesthetic production can be farmed out, or mapped to, the performance of another system. While "HydroLoops" was a starting point, in that it represented a self-enclosed system with measurable results, it's only a beginning in testing how growth, sensing and production can all be linked. As agricultural production continues to become increasingly industrialized, the means for visualizing agricultural data become increasingly complex; "HydroLoops" suggests alternate methods of feedback, visualization, and critical engagement with these new forms of visualization.

>>> **Paul Schuette is an architect and writer working in California.**

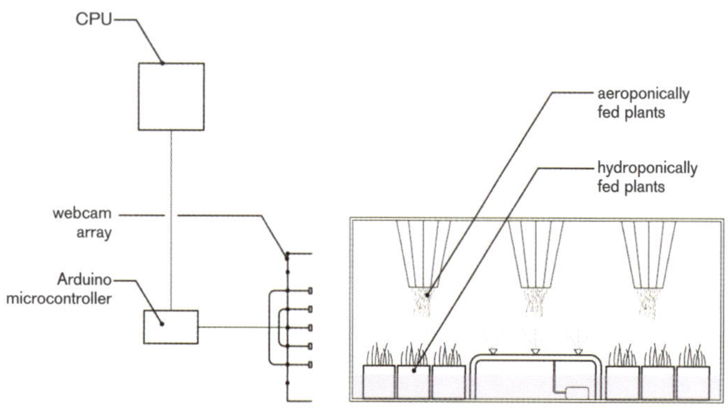

System Organization

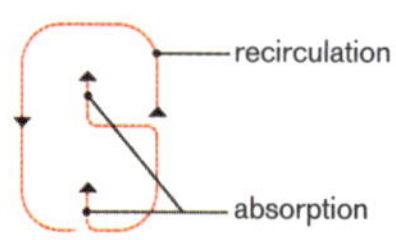

Nutrient Solution Cycle

^ ***Hydromechanism 1 – Tank.***
Photo by Kate Carr.

^ Stills from ***Hydromechanism 1 – Tank***
Processing program.

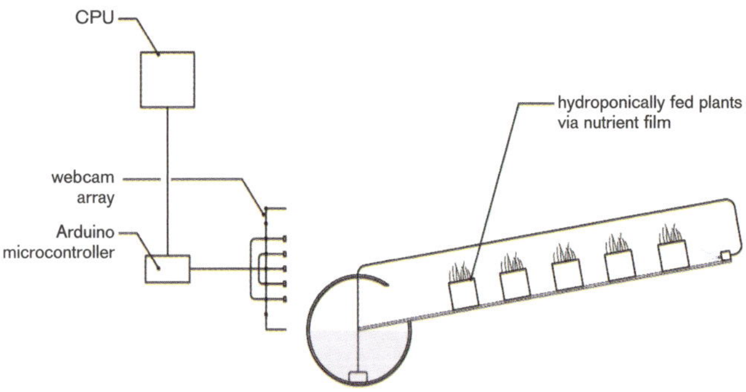

System Organization

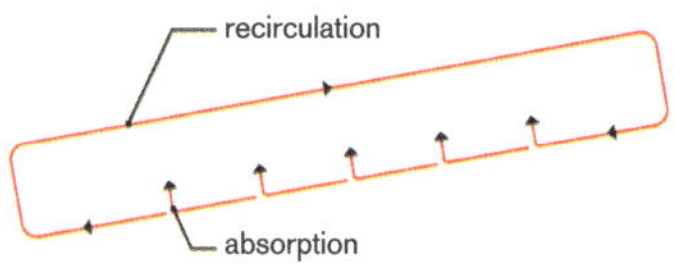

Nutrient Solution Cycle

^ ***Hydromechanism 2 – Slide.***
Photo by Kate Carr.

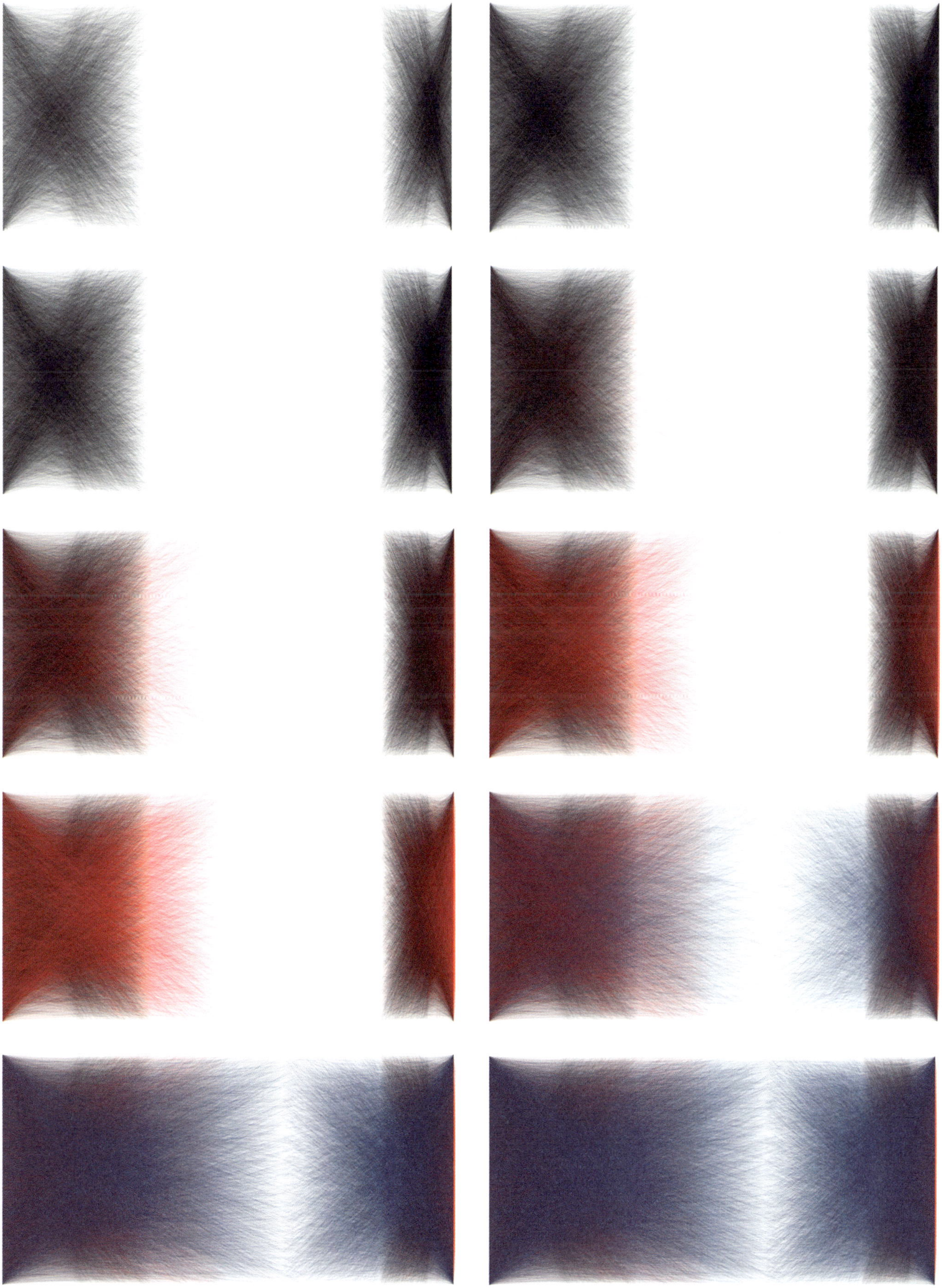

^ **Stills from *Hydromechanism 2 – Slide***
Processing program.

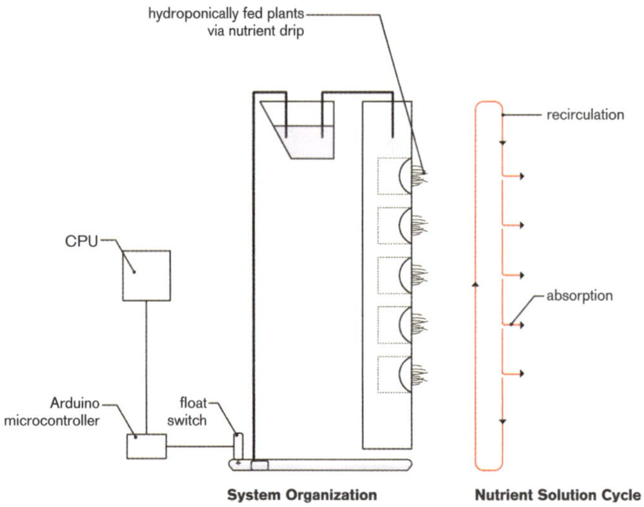

^ ***Hydromechanism 3 – Tubes.***
Photo by Kate Carr.

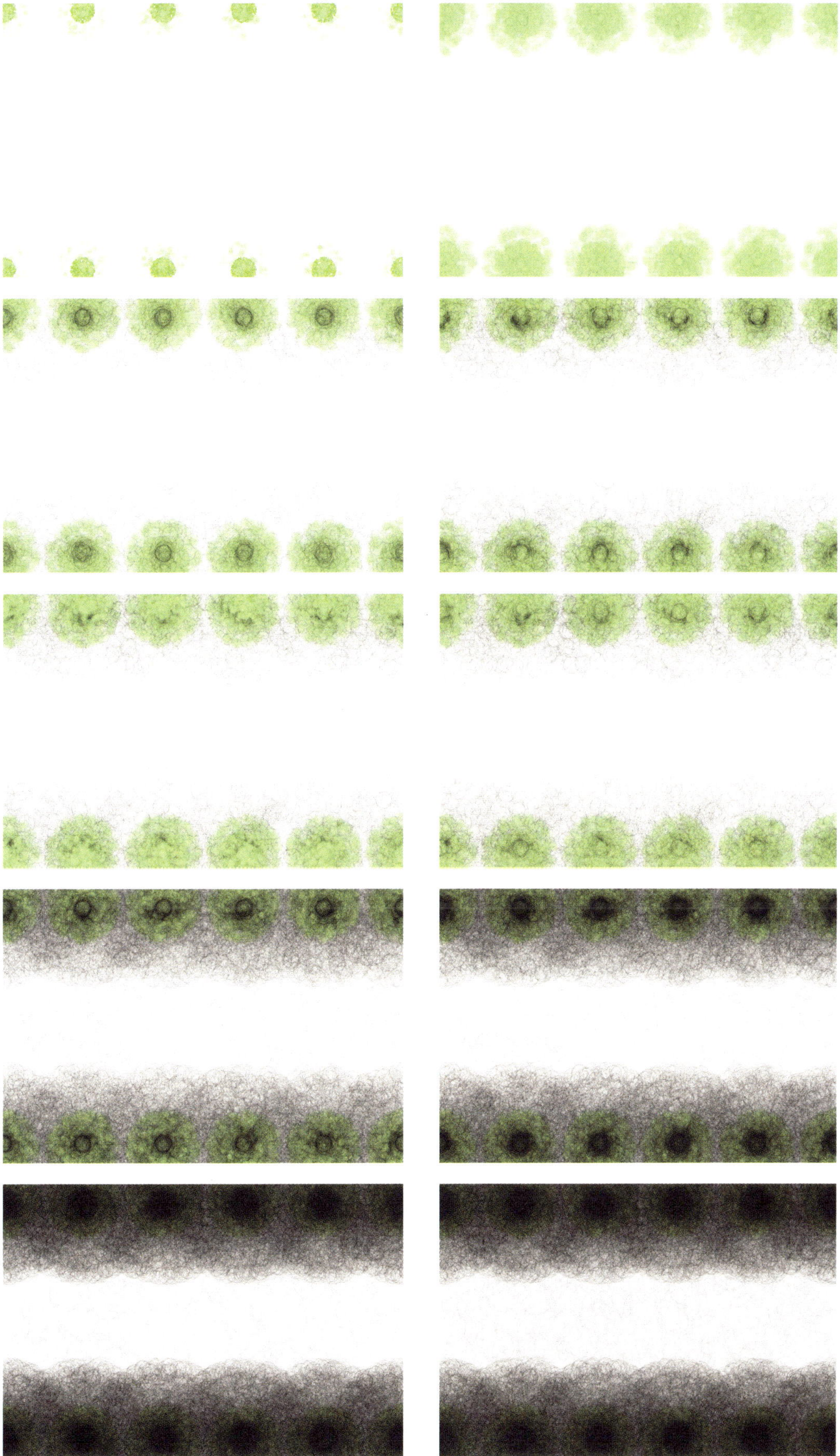

^ Stills from *Hydromechanism 3 – Tubes* Processing program.

what we are is what we eat

alexandros avlonitis

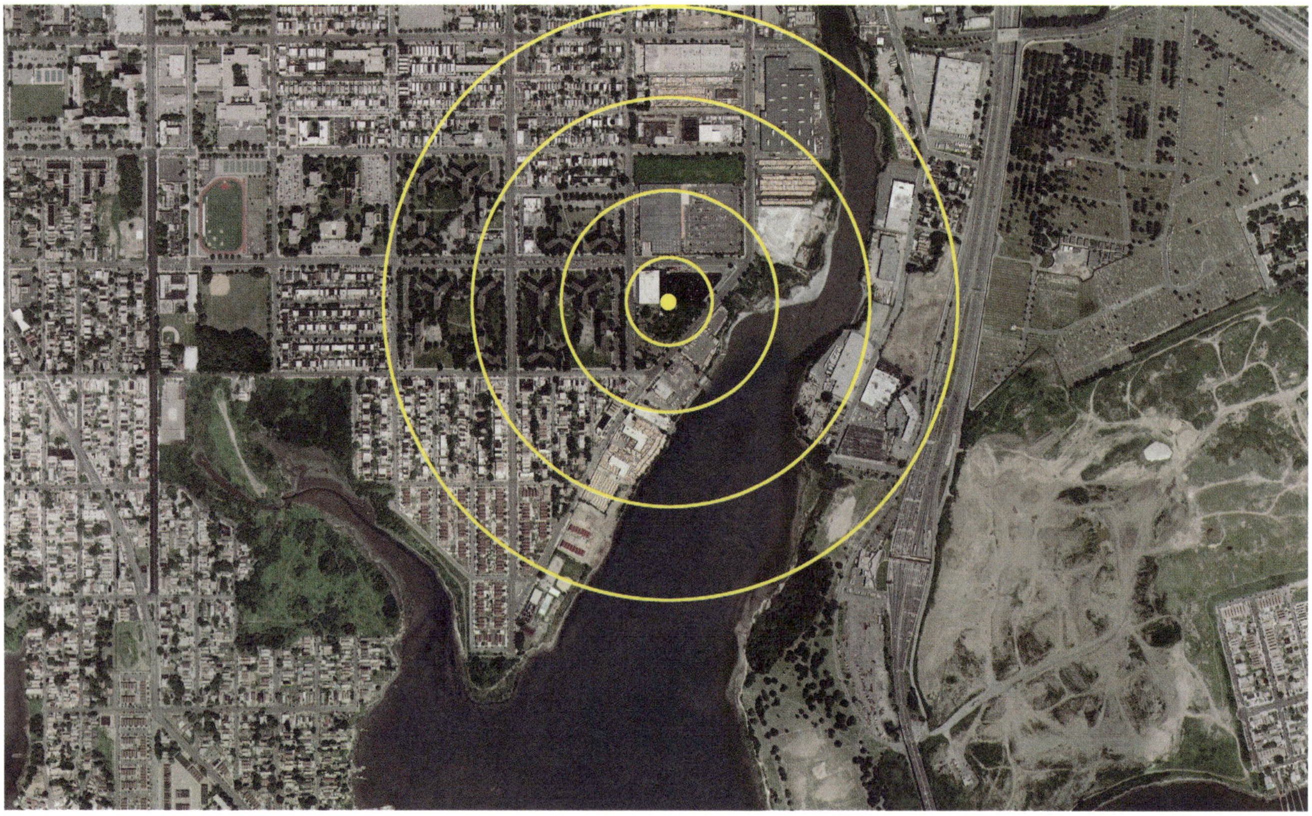

^ **Locavore: Castle Hill, Bronx, New York. Local zone of impact.**

It is estimated that in the next 50 years 80 percent of the world population will reside in cities, abandoning the rural and adopting new habits and ways of life. This cultural shift already has made a direct impact on agricultural production and food quality. Even today, urban consumers are surrounded by predominately low-quality food at low prices, while expensive organic products are accessible only to select social layers. Nutritious racism is becoming more and more evident in cities. Low social layers that cannot afford organic, nutritious goods are forced to consume second-class products of unknown origin and with great impact on their health. It is well-documented that food production has become increasingly industrialized with complex, unsustainable transportation practices. This project confronts these issues and offers an architecture that addresses current and emerging food culture challenges.

Primarily, the project serves as a public farm educating people on food culture, urban farming, and the benefits of consuming local produce in a collective, communal character. The site of this locavore project is an under-used community garden adjacent to Castle Hill Houses and the Bronx River in New York City. Castle Hill is one of the poorest areas in New York City with very limited access to nutritious food options. The public farm serves as a green magnet for the area, providing its residents with a range of seasonal products at reasonable prices. Residents would be provided a place where they can grow their own produce, socialize, and participate in a range of other food-related activities. On the site, the farm is divided into three distinct zones. Zone 1 houses the entire farm infrastructure, such as the market, distribution, and other commercial functions. Zone 2 is the tree park, or public orchard, where visitors gather and socialize. Zone 3 is the productive farm zone, where plots are defined and distributed to locals to grow and manage their own crops. The farm zone is organized through a system of strips that alternate between raised ground and pedestrian access.

This proposal establishes a landscape that respects the existing circumstances, resisting the trends of big vertical farming structures. It is about building food communities and spatializing a movement to combat junk food, malnutrition, and the loss of food culture. Furthermore, the project serves as a catalyst and prototype for similar sites. The public farm demonstrates the possibility to exploit more and more abandoned urban voids in the city to offer affordable, locally-produced organic and nutritious food. The intention here is not that of a community garden, but a community farm-park.

>>> **Alexandros Avlonitis is an Athens-based architect. He studied architecture at Aristotle University of Thessaloniki and obtained his MScAAd from Columbia University.**

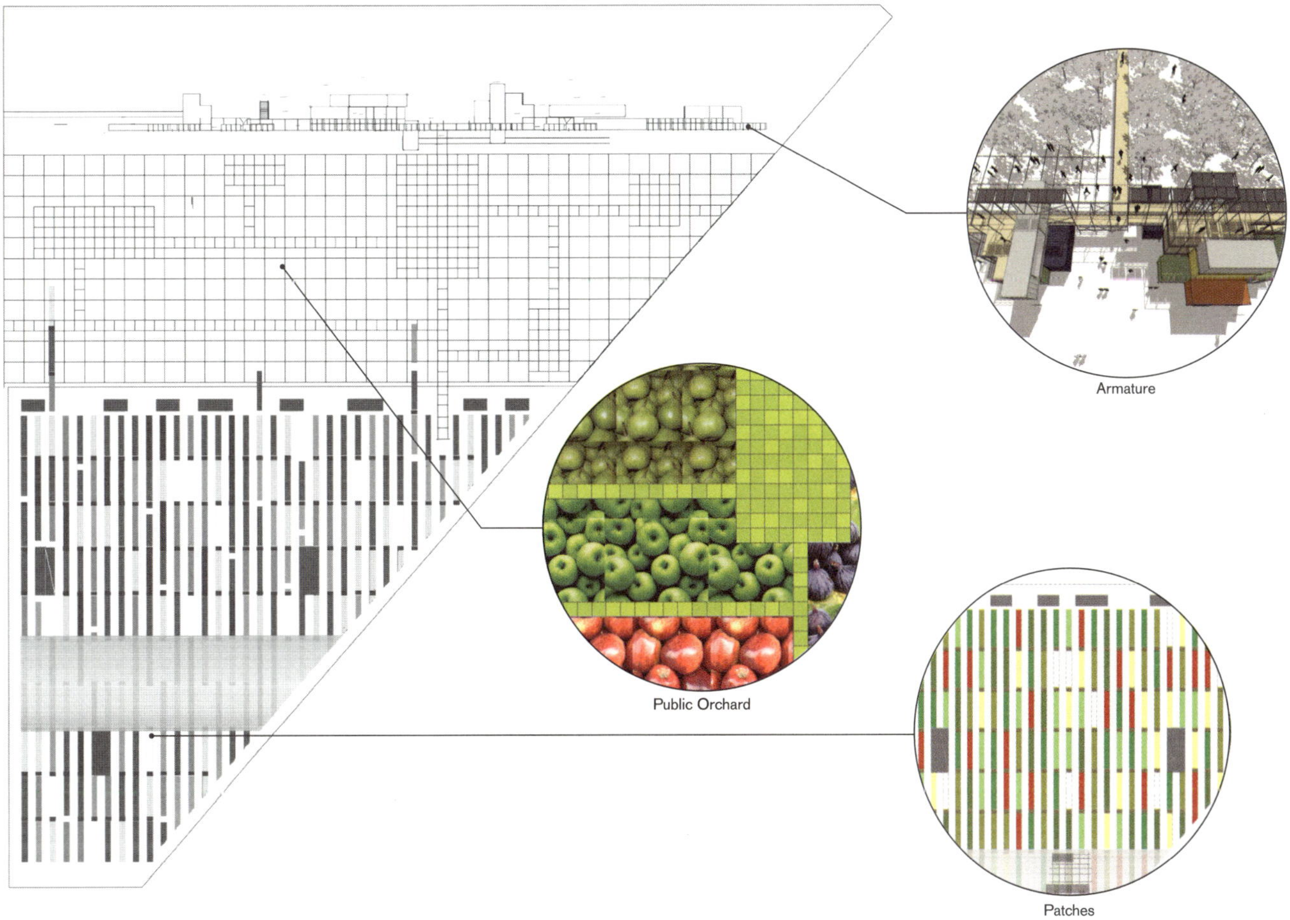

^ **Spatial organization of the site showing the Armature, the Public Orchard, and the Patches.**

^ **General site section. Slopes exploit best radiation of the sun. The shallow slopes function as paths and the higher ones as garden plots. The orchard mediates between urban and rural conditions. The armature zone functions as the reception to the site. It contains all the facilities, infrastructure, and markets.**

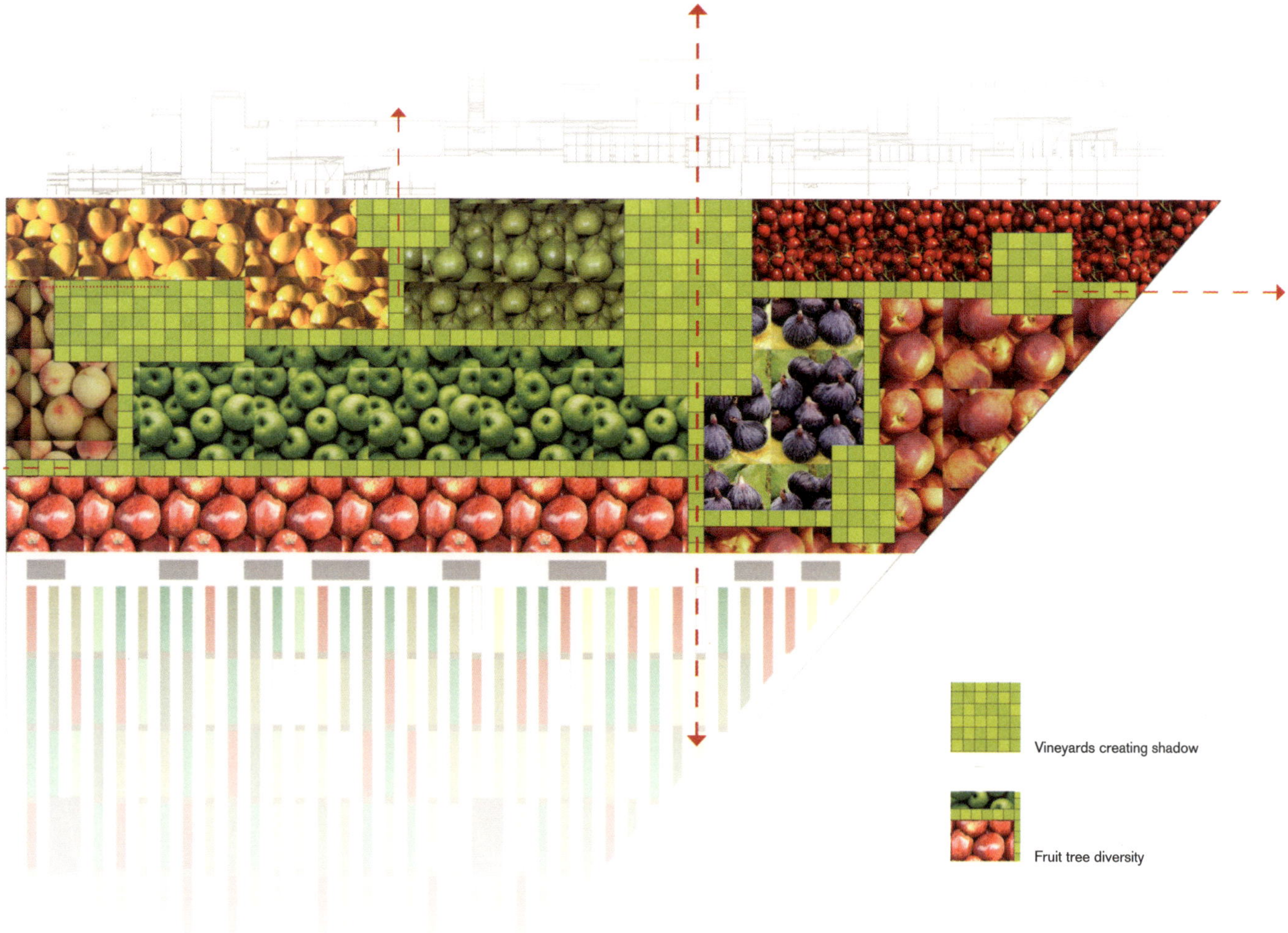

^ **The Public Orchard. Organization plan.**

^ **The Public Orchard. Managing and preparing gathered crops.**

^ The Patches. View of the patches and their distribution.

^ The Patches. Beans, lettuce, wheat, watermelon, basil, spinach.

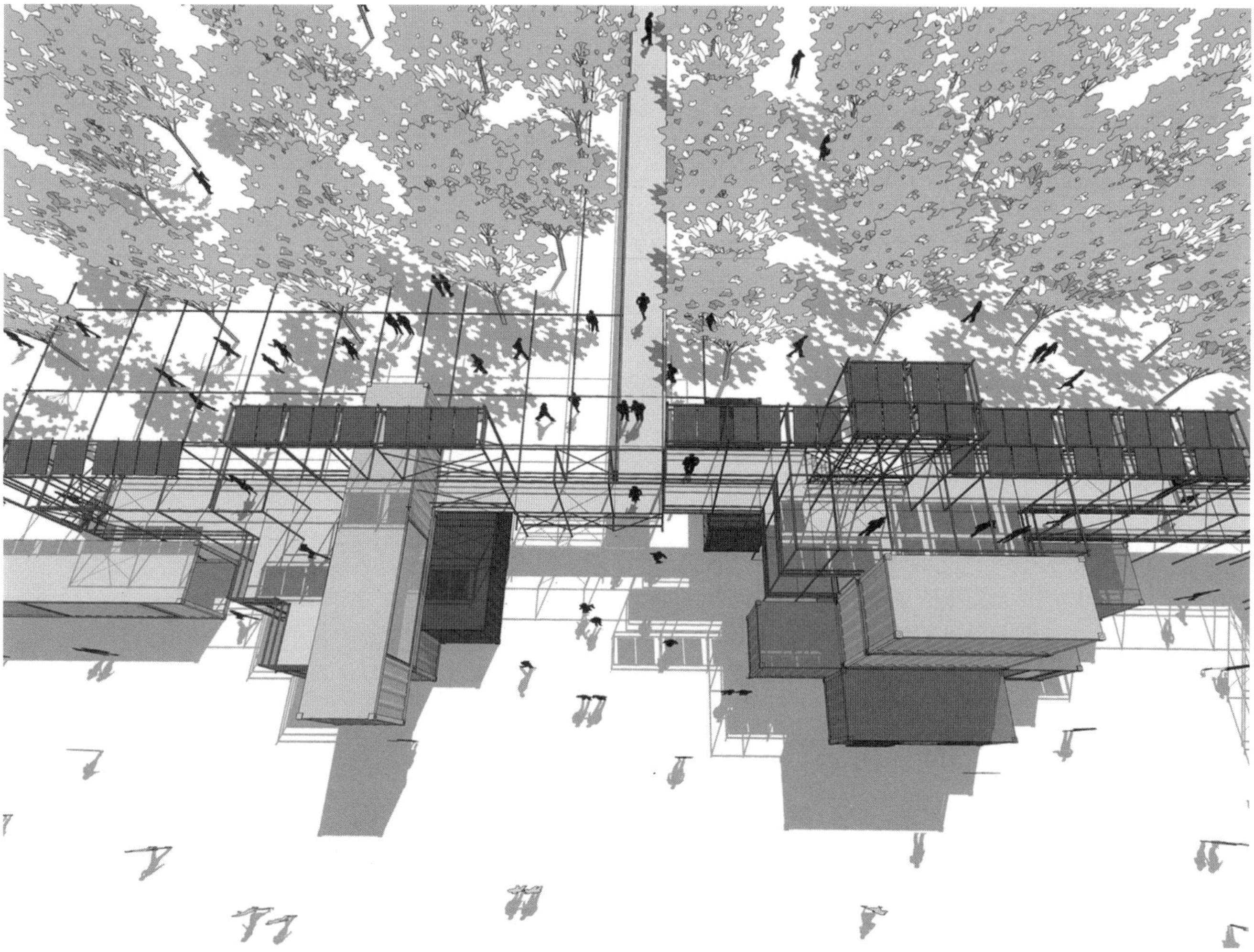

^ The Armature. Construction elements: used shipping containers, scaffoldings, wooden pallettes, wind turbines, solar panels.

^ The Armature. Structured and natural shading.

^ **The Armature. Diversity of functions: Local open market, farming educational programs, food festivals, schools tourism, flower market, cooking seminars, vineculture.**

vertical farming in las vegas? beyond pragmatism, toward desire

joyce hwang

^ **The Fremont Experience, Las Vegas.**
Las Vegas News Bureau / LVCVA.

In January 2008, a rumor spread quickly through internet sources claiming that Las Vegas would be the first city to construct a 30-story vertical farm, modeled after a concept propagated by Dr. Dickson Despommier, a professor of Environmental Health Science at Columbia University.[1] While this pairing may initially seem to be impossibly incongruous, an examination of the logics of vertical farming will reveal that it fits opportunistically into the context of Las Vegas. The city's abundance of consumer activity—particularly food-driven activity (think: mega-buffets)—could certainly stand to benefit from the environmental, economic, and social consequences promised by vertical farm concepts. The aim of localizing the food industry could not be more appropriate for a city like Las Vegas, given its' location in the middle of a drought-ridden desert landscape. The promise of insulated, environmentally-controlled interiors would enable farmers to more securely manage depleting natural resources. Ideas of water harvesting and reclamation, for example, strike a particularly poignant chord for Las Vegas, a city which is currently combating symptoms of a chronic water shortage.

The introduction of vertical farming into the City of Las Vegas could indeed begin to address an array of environmental and economic problems. But beyond pragmatism, its logics and sensibilities embody a collection of desires. Vertical farming strategies for containing and controlling environmental conditions resonate beyond the economic and environmental needs for ensuring sustainability in Las Vegas. The pragmatic concerns of urban food accessibility, fuel conservation, energy efficiency, pesticide elimination, water reclamation, and mixed-use urban zoning are nested within a swirl of wild desires permeating throughout Las Vegas' world of resort development. Already, the city thrives on the premise that hyper-controlled environments create experiences of touristic pleasure. Located on the Strip, the Mirage casino and hotel, for example, houses a functioning rain forest, complete

^ **Venetian Hotel and Casino, Las Vegas. Interior canals, gondola rides, and faked skies.**

with 60-foot high palm trees, tropical flora, waterfalls, and lagoons;[2] an environment for white tigers, white lions, panthers, leopards, and dolphins;[3] a 20,000 gallon saltwater aquarium;[4] and an artificial volcano, scheduled to erupt on a regular interval and enhanced with a soundtrack by musicians Mickey Hart (former Grateful Dead drummer) and Zakir Hussain.[5] Mandalay Bay, another casino-hotel located on the Strip, is home to the Shark Reef Aquarium, a 1.6 million gallon tank of seawater containing over 2,000 animals including sharks, sawfish, giant rays, sea turtles, piranha, and crocodiles.[6] Other Las Vegas establishments brazenly attempt to recreate the auras of complex urban metropolies—such as New York, Paris, and Venice—through mechanisms of environmental and social control, ranging from the creation of artificial waterways and faked skies to the regulation of employee-hiring criteria. Considering these situations, it is not a far stretch to imagine the sight of fruits, vegetables, and animals towering above the Las Vegas Strip and sealed within tightly controlled biosphere-inspired enclosures. In many ways, the concept of a vertical farm already subscribes to Las Vegas' obsession with controlling nature.

A survey of Las Vegas' resorts will confirm that the city's entertainment industry today relies on the production of exuberance, grandness, and perhaps even a touch of

^ **MGM Grand Hotel Las Vegas prospectus. Exterior rendering by Martin Stern. Martin Stern Collection, Special Collections, University of Nevada, Las Vegas Libraries.**

^ **Xanadu Hotel and Casino. Exterior rendering by Martin Stern. Martin Stern Collection, Special Collections, University of Nevada, Las Vegas Libraries.**

imperialism. Caesar's Palace and Imperial Palace are relics from the 1950s to 1970s which have (so far) survived the city's recent trend of demolitions. Meanwhile, hotel-casinos such as Dunes and Sands were imploded in 1993 and 1996, respectively. New mega-resorts which have been built since 1989 possess identifiers such as Excalibur, Paris, Monte Carlo, New York New York, Luxor (built in the form of an Ancient Egyptian pyramid, complete with a recreation of the Great Sphinx of Giza above the porte cochere), and Stratosphere (the tallest observation tower in

1. As it turned out, the project images which had so quickly propagated on the web were actually lifted from www.verticalfarm.com, the website of Dr. Dickson Despommier, a professor of Environmental Health Science at Columbia University who has been developing vertical farming concepts over the past decade. The visualizations themselves were of a speculative farm proposal by Chris Jacobs at United Future. Neither Despommier nor Jacobs were aware of any plans to build a 30-story farm in Vegas. James Murray, businessGreen.com, http://blog.businessgreen.com/2008/01/so-is-las-vegas.html, accessed January 17, 2008.

2. http://www.mirage.com/attractions/atrium.aspx, accessed January 29, 2009.

3. The Mirage Hotel and Casino, http://www.mirage.com/attractions/secret-garden.aspx, accessed January 29, 2009.

4. The Mirage Hotel and Casino, http://www.mirage.com/attractions/aquarium.aspx, accessed January 29, 2009.

5. In the words of Hart: "To put percussive sound to the image of fire and have it dance around a lagoon is something I've never done; I've never birthed a volcano… Zakir and I immersed ourselves in the legends and myths and studied what really makes a volcano work. The result, a true union of the senses, will rattle your bones, just like a real volcano." The Mirage Hotel, http://www.mirage.com/attractions/volcano.aspx, accessed January 29, 2009.

6. The Mandalay Bay Hotel and Casino, http://www.mandalaybay.com/entertainment/sharkreef.aspx, accessed January 29, 2009.

the United States, at a height of 1149 feet).[7] Imagined exuberances can already be seen in the initial design phases of many of the city's resorts. Original color renderings situate various resort and casino designs within atmospherically extreme environments, complete with artificial lighting effects and dramatic weather conditions. In an architectural rendering of the Stardust resort's future expansion, for example, an intensely starlit sky is depicted, only to be outdone by the illuminated "Stardust" signage. In the MGM Grand Hotel rendering, the sky is shown as an array of color and sweeping cloud formations. Renderings of the Xanadu resort, a scheme by Martin Stern that was never realized in Las Vegas, boasts not only an theatrical lighting scheme in its atrium design, but also a grand pyramidal form, shown as a giant hulk among a sea of low-lying cars and people. Like these exuberantly illustrated resorts, the imagined presences of vertical farms almost always promise a vision of techno-environmental grandeur. The Vertical Farm prototype tower by Chris Jacobs, for example, appears at first to resemble a typical high-rise office building, with the only anomaly being the gigantic solar panel on the roof. The night rendering emphasizes the tower's verticality with a light beam rendered to perpetually extend its height. Combining the rooftop-searchlight with the warm yellow-orange glow of interior lighting, the overall effect of the tower is that of a solitary light beacon, perhaps an island of hope with the rest of the city off on the horizon, literally. This sentiment is also reflected in SOA Architects' Living Tower Project. It is also portrayed as a lonely beacon surrounded by low-lying darkness, while its urban context is far off in the distance. The role of interior lighting also plays a significant role in defining the projected attitude of these projects. In Jacobs' tower, interior lighting emphasizes the blurry silhouettes of indoor trees, which are all cleanly encapsulated within each floor. At first glance, it seems uncanny to produce an image of a farm with large, bushy trees instead of low, slender crops. The effect of this portrayal, though, demonstrates two significant desires. First, the building exerts its ability to control nature. Unlike Le Corbusier's Esprit Nouveau Pavilion, where a circular opening in the building's roof plane accommodates the growth of a multi-story tree, the Vertical Farm image suggests that trees are always contained on a single level, without room to grow beyond a set height limit. Second, the rendering hints at a grandness of scale. The presence of volumetric trees in the rendering suggests that perhaps the floor-to-floor height is much taller than one might imagine when looking at a building that bears resemblance to an office tower. It does not come as a surprise, then, that *New York* Magazine notes Jacob's Farm as having been "inspired by the Capitol Records Building in Hollywood."[8] The idea of "grandness" is also present in other vertical farm projects. Gordon Graff's proposal for Toronto towers over its neighboring buildings. A speculative rendering of the new Toronto skyline shows that the tower, although aloofly set apart from a cluster of downtown skyscrapers, also rises as one of the tallest buildings in the city, bested only by the CN Tower. The Pyramid Farm by Eric Ellingsen and Dickson Despommier, a volume of growing vegetation visible within a giant transparent pyramid, is a symbolic reference and blatant allusion to power. MVRDV's Pig City, a proposal that envisions pigs living and breeding within a series of towers, strikes a resemblance to hyper-commercial zones in touristic resort cities such as Benidorm, Spain. Grandeur in this case is not about the elegance of a light beacon, but the sheer density and scale of the development and the surprising luxury provided to its inhabitants compared with industrial pig farms.

Given the enormous size, scale, and mechanization of the vertical farms discussed so far, it could be inferred that the design of these farms have been considered with industrial agriculture in mind. Unlike the community garden or family farm models, vertical farm prototypes propagate imaginations laden with hyper-efficiency and "current technology."[9] In Jacobs' model, an automated crop picker distinguishes between ripe and unripe fruit by means of a reflectometer, technology used for color-detection. Crop cultivation is managed through a device called "The Feeder." "Like an ink-jet printer, this dual-purpose mechanism directs programmed amounts of water and light to individual crops," reports *New York* Magazine.[10] Jacob's model even includes a "control room," a location from which all operations are regulated by means of monitoring through multiple CCTV-type screens, "allowing for year-round, 24-hour crop cultivation." Yet, despite the desired rhetoric of high-efficiency and latest-technologies, vertical farm proposals are almost always portrayed as lush, leisurely settings. Most rendered images project a mood of calm, not at all one of industrial efficiency. In "Towers of Food," a graphic slideshow by the *New York Times*,[11] one projected scenario depicts a generously proportioned outdoor terrace which includes several people picnicking on a lawn and two children playing near the base of a tree. A rendering of this project exhibits a large atrium-like space showing several children leisurely picking fruit while a woman strolls down a spiral ramp.

The desire for high-tech, controlled efficiency stands in contrast to an equally strong desire for idyllic, pastoral environments. Indeed, industrial agriculture, as we know it today, is anything but bucolic.

> Expectations for experiences with nature in farming country are very low these days in the context of modern agriculture... In the dying small towns, clerks apologize to tourists attracted to their Main Street's country craft shops in lovely old Victorian houses as they catch the strong stench of hog manure between car and shop. Agriculture as ecological sacrifice is generally accepted (though not acknowledged out loud) in U.S. Department of Agriculture offices where farmers go to fill out farm program papers..."[12]

But perhaps it is the promise of future technology that might liberate industrial agri-environments from this conundrum. In vertical farm models, the clash between control and leisure optimistically plays out in a utopian vision, where machines flamboyantly perform all the visible work, while humans control and monitor from a hidden and distanced center. In the visionary renderings, an occasional farm worker can be detected within the luscious vegetation.

7. The Stratosphere Hotel and Casino, http://www.stratospherehotel.com/thrills/, accessed January 29, 2009.

8. Lisa Chamberlain,"Skyfarming," *New York* Magazine (April 2, 2007).

9. Ibid.

10. Ibid.

11. Bina Venkataraman, "Country, the City Version: Farms in the Sky Gain New Interest," *The New York Times*, July 15, 2008. Image by Blake Kurasek of University of Illinois Champaign's Graduate School of Architecture, featured in "Towers of Food" slideshow, a series of accompanying images. This image is also featured on the Vertical Farm Project website: www. verticalfarm.com.

12. Dana Jackson and Laura Jackson, "Agriculture as Ecological Sacrifice" in *The Farm as a Natural Habitat: Reconnecting Food Systems with Ecosystems*, ed. Dana Jackson and Laura Jackson, (Washington D.C.: Island Press, 2002), 11-12.

^ Vertical farm proposal by Chris Jacobs. Daytime exterior rendering Image courtesy of Chris Jacobs.

^ Vertical farm proposal by Chris Jacobs. Nighttime exterior rendering. Image courtesy of Chris Jacobs.

^ Skyfarm. Vertical farm proposal for Toronto by Gordon Graff.

^ Living Tower by SOA Architects. Night-time exterior rendering. The Vertical Farm Project (www.verticalfarm.com).

^ Pyramid Farm by Eric Ellingsen and Dickson Despommier. The Vertical Farm Project (www. verticalfarm.com).

But could it really be so easy? The cultivation of high-tech agriculture is already in full swing; yet its circumstances tell us that the utopian vision is far from reality. The Southern Spanish town of El Ejido, for example, houses over 177 square miles of highly-controlled greenhouses, a phenomenon which has been famously dissected by Keller Easterling as a "comedy of intense control and endless research" which accompanies "the quest for the perfect tomato."[13] She writes:

> In the most high-tech greenhouses the plants grow in linear bags of growth medium punctured with tubes that deliver fluids and nutrients. The entire enclosure is computer-controlled, buzzing, burping, and opening flaps in response to a shift in wind or a moment when the sun temporarily slips behind a cloud...

The increasing ubiquity of automated control mechanisms in greenhouse farming, however, still does not eliminate the need for intensive manual labor. "However perfectly engineered they may be," continues Easterling, "Spain's greenhouse agripoles need something extra to stay alive. They need to tighten the boundaries of their world while also cheating selectively outside of those boundaries." "Labor," she defines, is "the secret requirement for perfection and control in Almeria."[14] In order to maintain the greenhouses in El Ejido, approximately 20,000 workers are required. Immigrating primarily from North Africa (often illegally), many of these laborers reside and work in tentative and sub-standard conditions. As reported by Gregory White: "...most immigrants live in wretched conditions, an 'apartheid' at the core of economic success. Dwellings are often no more than sheds or old motor vehicles, lacking running water, electricity and privacy."[15] Although not quite as drastic, a somewhat parallel universe of labor exploitation exists in California's world of farming. In 2001, "unauthorized" individuals comprised approximately 50% of the state's farming workforce. Since 1986, when Congress passed the Immigration Reform and Control Act (IRCA) act which made it illegal to hire illegal immigrants, the farming industry has been propagating a merry-go-round of short-term labor stints among undocumented workers.[16]

^ **Vertical farm proposal by Chris Jacobs. An interior rendering of crop fields imagines the use of technology in farming practice. Image courtesy of Chris Jacobs.**

Yet, despite the bleak diagnosis, there is indication that some tides have been turning. A 2007 study by the Center for Immigration Studies shows that farm worker wages are increasing and that the introduction of mechanization has drastically decreased harvesting labor needs. In 1960, 45,000 "peak harvesters" were employed in California (80% of whom were Braceros, or Mexican guestworkers). By 1970, there were less than 5,000 workers who rode on machines during harvesting. But while these processes of mechanized, bulk harvesting reduced labor in one sector, they increased labor in another, that of science. Tomatoes, for example, have been "redesigned" to ripen uniformly, ensuring a more efficient harvesting. Their water content and shape have also been altered to "reduce cracking when carried in bulk."[17]

The automation of farming has slowly shifted labor from hand-harvesting to a more diversified set of activities such as driving machinery and genetically modifying fruits and vegetables. So, one might say that labor has not altogether disappeared, but has instead been dispersed. The pastoral desires projected by the dream of vertical farming will inevitably be tempered with the labor-intensive reality of industry—if not that of 'traditional' farming, then that of scientific research. Even if nearly all manual labor were to be eliminated (in an impossibly extreme scenario), the fact of human labor—however hidden—will still be in existence and integral to the success of any general farming scheme.

In thinking about the role of humans in an increasingly automated world, the imagery of CCTV and other monitoring techniques come to mind. But despite the number of cameras, recording devices, sensors and biometric readers scattered within any given spatial boundary, it is still the human mediator who is patiently and painstakingly managing, interpreting, and acting upon collected data or information. To this end, Las Vegas, once again, may serve as a model of inspiration for vertical farming. It is difficult to think of a city which better encapsulates the conflicting desires of both fantasy and security. For a tourist in Las Vegas today, idyllic leisure is a sensation which is practically guaranteed. Unlike the old days when the city was run by the mob, today's family-friendly Las Vegas strives to create a hypnotizing concoction of both soothing and exciting experiences. Visitors can easily meander on the strip (once defined by

13. Keller Easterling, "El Ejido," in *Enduring Innocence* (Cambridge: MIT Press, 2005), 44.
14. Ibid, 45.
15. Gregory White, "Risking the Strait: Moroccan Labor Migration to Spain," *Middle East Report*, no. 218 (Spring 2001): 27.
16. Phillip Martin, "Farm Labor in California: Then and Now," (working paper 37, The University of California San Diego Center for Comparative Immigration Studies, April 2001).
17. Phillip Martin, "Farm Labor Shortages: How Real? What Response?" Center for Immigration Studies, Washington D.C. (November 2007): 10.

^ "The Living Skyscraper: Farming the Urban Skyline" by Blake Kurasek. Exterior rendering. The Vertical Farm Project (www.verticalfarm.com).

^ "The Living Skyscraper: Farming the Urban Skyline" by Blake Kurasek. Interior rendering. The Vertical Farm Project (www.verticalfarm.com).

the automobile), often drifting above traffic on the numerous pedestrian bridges and stopping every so often to see an erupting volcano or a water light show. Within the resorts, visitors over 21 are able to wander comfortably from one fantasy to another, while occasionally being drawn into the excitement of blackjack or roulette. Walking hand in hand with this kind of fabricated leisure, yet mostly hidden from plain view, is one of the most complex and comprehensive security networks in existence, with parallels to those of the CIA and the U.S. Department of Homeland Security.[18] Yes, visitors can see the seemingly endless field of "eye in the sky" surveillance cameras, as well as the army of security guards infiltrated within every casino floor. But their presence does little to prevent tourists from enjoying themselves with careless abandon. For the most part, tourists are relatively unaware of—or at least oblivious to—the invisible forces of checking and monitoring that occur pervasively throughout these fun-filled environments. Gaming tables, playing chips, cards are coded with tracking technologies from computer chips to invisible inks and barcodes; human behavior is monitored not only by guards and observers but also by computer software that can "mathematically calculate whether somebody is a good player or a cheat;"[19] more software pieces together disparate data to detect "non-obvious relationships."[20] But ultimately, human labor is still the driving force of resort security. At a lecture for a recent casino surveillance conference, Richard Marcus, a man who claims to have netted millions of dollars by cheating casinos for over 25 years and is the author of the book *World's Greatest Gambling Scams*, told his audience "Even as technology progresses, my thing was still to beat the people on the floor so they didn't bring the cameras into play."[21] His 'winning' strategy was to "psychologically manipulate the people his teams dealt with," by capitalizing on his understanding that dealers, security guards, and pit bosses become complacent and "robotic" over time. It could be said that it is Las Vegas' human work force that is instrumental in nearly guaranteeing idyllic leisure for the city's paying guests. From waiters and bartenders to dancers and singers, front-of-house employees are all performers in some sense. The controlled invisibility of back-of-house employees even further contribute to the tourist's sense of fantasy; almost as if by magic, beds are made, towels are cleaned, food is delivered, flowers are always fresh, glass is always sparkling, and dirt disappears.

Like farm laborers, back-of-house hotel laborers in the U.S. have historically dwelled in a borderline world of low wages, exploitation, job instability, and public disaffection. The hospitality industry in Las Vegas, however, exemplifies one of very few exceptions. Thanks to Culinary Local 226, the city's largest labor union (covering over 90% of hotel workers on the Las Vegas Strip), waiters, bartenders, dishwashers, maids, and other "unskilled" workers are guaranteed relatively decent wages (on a national level, Las Vegas waiters net the most money per hour before tips), 40 hour workweeks, three weeks of vacation per year, health care benefits, a proper pension plan, and even access to a high-quality job-training school.[22] While willpower, patience, and struggle

18. Will Sturgeon, "Las Vegas Cheat-Busting: The Truth About Tech Detection," *Silicon.com*, August 19, 2005, http://www.silicon.com/research/specialreports/gambling/0,3800010160,39151555,00.htm (accessed January 29, 2009).

19. Tim Stanley (CIO of Harrah's Casino Group) quoted in Will Sturgeon. "Las Vegas Cheat-Busting: The Truth About Tech Detection," *Silicon.com*, August 19, 2005, http://www.silicon.com/research/specialreports/gambling/0,3800010160,39151555,00.htm (accessed January 29, 2009).

20. Will Sturgeon, "Las Vegas Cheat-Busting: The Truth About Tech Detection," *Silicon.com*, August 19, 2005, http://www.silicon.com/research/specialreports/gambling/0,3800010160,39151555,00.htm (accessed January 29, 2009).

21. Arnold Knightly, "Blink and You'll Miss Him," *Las Vegas Review-Journal, McClatchy-Tribune Business News*, February 25, 2007.

22. Steven Greenhouse. "Local 226, 'the Culinary,' Makes Las Vegas the Land of the Living Wage," *New York Times*, June 3, 2004.

among laborers were certainly some of the factors that pushed this condition into being, "The Culinary's success," reports the *New York Times*, "cannot be separated from the industry's wealth. With profits rolling in, the casinos have decided to be relatively magnanimous to their workers to ensure labor peace and a happy work force."[23]

Vertical farming, as a concept, has been widely accepted as an exciting new frontier for agriculture, architecture, and urbanism. But as a potential agricultural industry, it can stand to look toward the unlikely model of Las Vegas, particularly its uncanny ability to mix extreme leisure with extreme security. Further, in advocating for environmental ethics and social consciousness, vertical farms must begin to incorporate the agency of human labor into their schemes and play out its impact through design. After all, as in other businesses of leisure and food production, these farms will simply not be able to manage themselves, regardless of the number of control mechanisms dispersed within its systems. It would be ludicrous to propose a new model for farming while maintaining the same model of human exploitation. Dickson Despommier lamented that the primary factors which have prevented the realization of a vertical farm are construction and energy costs.[24]

It is time that vertical farm advocates take advantage of a city (and its developers) that regularly sees astronomical sums in creating ever-new worlds of fantasy. Operating on logics of grandeur, spectacle, leisure, and control—not unlike those of resorts in Las Vegas—a vertical farm would add a fantastic and pragmatic new dimension to the city's food and entertainment industries. If one considers Las Vegas as a hub of super-scaled cultural experiments, which rides the waves of trends and simultaneously creates new ones, the vertical farm vision indeed fits snugly within this sentiment. Perhaps it is time for Dr. Despommier to contact Steve Wynn.

>>> **Joyce Hwang is an architect and Assistant Professor at the University at Buffalo, State University of New York.**

^ Wynn Resort, Las Vegas. Buffet, view from entry area.

23. Ibid.

24. Bryan Walsh, "Vertical Farming," *Time*, December 11, 2008, http://www.time.com/time/magazine/article/0,9171,1865974,00.html (accessed February 1, 2009).

the catalog: from ploughs to clouds

maya przybylski

Since the late nineteenth century, the catalog served an important role as a disseminator of knowledge and access to tools, becoming through this role an emblematic symbol of geographically dispersed communities. Often considered a means of selling products, the catalog has always performed a more nuanced social role. In its earlier forms, the catalog's role was limited to a one-way, single-voiced, broadcaster of information. The reimagining of the catalog's potential in the 1960s, exemplified by Stewart Brand's *Whole Earth Catalog*, extended the catalog's role into a participatory forum of information exchange. Through this evolution, the catalog developed into a prototypical information service for decentralized, geographically distributed communities. Further, the catalog operated as a testing ground for exploring the potentials of leveraging mass collaboration as a way to add value—a concept now commonly referred to as crowdsourcing. Coupled with the global reach of the internet, an inversion of sorts has now taken place. At their inception, catalogs served farms; now, catalogs are themselves the farmers.

Fueled by the Homestead Act of 1862, American populations headed west to claim their tenure in the new frontier. This westward expansion marked the 1890s as the period with the largest increase in land under cultivation in American history where more than 200 million acres were added to the country's farming stock, an increase of 25 percent.[1] The notion that the frontier would one day close, when all land was claimed, provided the incentive to move quickly. As a result, the farmstead, not the village, became the basic organizational unit of farmers across the landscape. While living directly on the farm benefited production levels and encouraged the family-farming model, the farmstead created conditions of geographic dispersion and social isolation. At the beginning of the twentieth century a farm "was in a separate world with infrequent communications with the outside."[2] This isolation was problematic in that these upstart operations needed to achieve relatively efficient levels of production and this, in turn, depended on access to knowledge, tools, and techniques. By the beginning of the nineteenth century, the American landscape was dotted with a geographically distributed, decentralized collection of non-expert farmers in need of an effective communication mechanism in order to stay current with respect to changing market conditions, improving tools and techniques, and shifting policies.

Taking advantage of frontier-pushing infrastructure, such as the cross-country railway, improved shipping and storage strategies and favorable government policies, mail order businesses began distributing catalogs in hopes of tapping into this isolated market. Catalogs distributed on a local or regional scale can be found as early as the eighteenth century; but it wasn't until the advent of Rural Free Delivery in 1896, extending timely and accessible postal services to rural communities, that the catalog became a forum for information exchange and community connectivity. Remarkably, the US Postal Service recognized and fostered this potential for catalogs by classifying mail order publications not as commercial articles but as aids in the dissemination of knowledge thereby entitling them to a reduced postage rate of one cent per pound. Recognizing this opportunity, pioneers of the mail order industry such as Montgomery Ward and Sears Roebuck & Company issued as many as six catalogs each year. Every issue featured a wide selection of products ranging from agricultural equipment, implements, and even buildings to the latest, urban, fashions, toys and domestic appliances.[3] Upon each catalog's arrival, farmsteads across the country were immediately connected by a shared document and collectively presented with new opportunities. As an example, when the gasoline tractor became a practical reality after 1915, they immediately began to appear on catalog pages and farmers saw the opportunity to increase productivity by expediting work and reclaiming land previously used to grow feed for plough-pulling horses and mules. These catalogs extended beyond their obvious mandate of selling goods to include a variety of articles describing new techniques, changing policies, and shifting market information. With this, the catalog began to deliver, with higher frequency, some of the knowledge traditionally found in the annually issued *Farmer's Almanac*.

The catalog's central place in the household was demonstrated by its nickname, the "Homesteader's Bible", as it was

1. Gardner, Bruce L., *American Agriculture in the Twentieth Century: How it Flourished and What it Cost* (Cambridge, MA: Harvard University Press, 2002), 3.

2. United States. Bureau of the Census. *American Agriculture: Its Structure and Place in the Economy* (New York: John Wiley & Sons, Inc., 1955), 152; and Gardner, Bruce L. *American Agriculture in the Twentieth Century: How it Flourished and What it Cost* (Cambridge, MA: Harvard University Press, 2002), 54.

In 1910 there were over 6 million farms cultivating more than 850 million acres of land. A quick calculation returns 132 acres as the average farm size. Beyond farmsteads, trading centers provided another order of organization. With their locations initially calibrated to the rail lines, these centers acted as economic hubs where farmers sold and traded their goods. While they offered the notion of centralized centers of activity, the distance between the farms and trading centers was prohibitively far in relying on the centers for daily contact and support. Average distance from farm to nearest trading centre was 6 miles. By 1950 two thirds of farms were within 0.2 miles of an all weather road, making travel from farm to center take 20 minutes.

3. Cherry, Robin. *Catalog: An Illustrated History of Mail-Order Shopping* (New York: Princeton Architectural Press, 2008).

^ **Rural Free Delivery. Postal carrier with horse and wagon, 1905. Smithsonian National Postal Museum.**

^ ***Our Grocery List.* Catalog Cover, Sears, Roebuck & Co., 1907.**

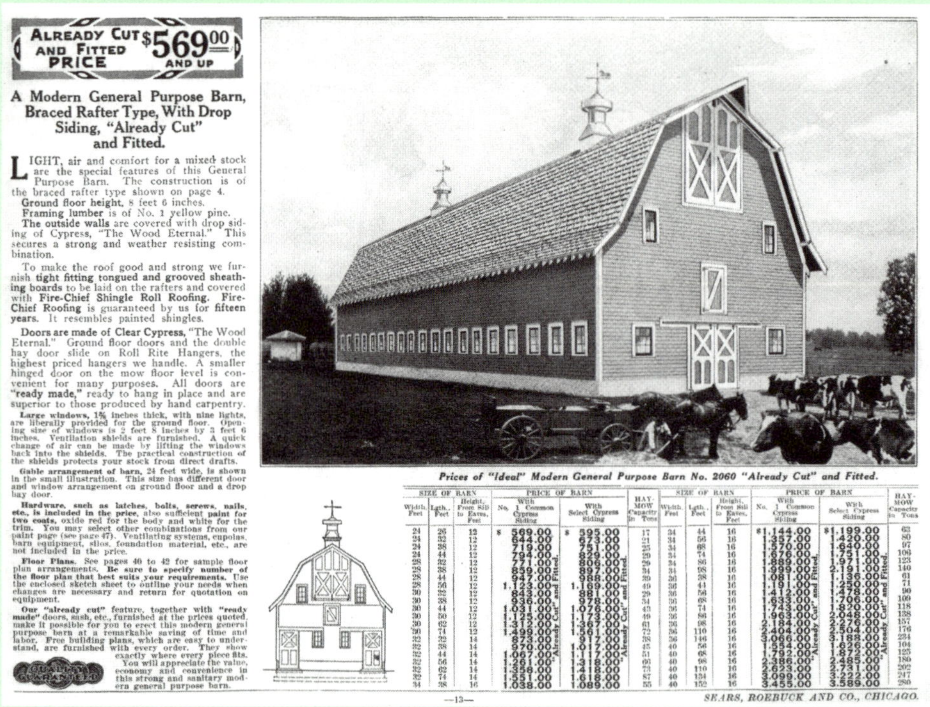

ALREADY CUT AND FITTED PRICE $569.00 AND UP

A Modern General Purpose Barn, Braced Rafter Type, With Drop Siding, "Already Cut" and Fitted.

LIGHT, air and comfort for a mixed stock are the special features of this General Purpose Barn. The construction is of the braced rafter type shown on page 4.

Ground floor height, 8 feet 6 inches.

Framing lumber is of No. 1 yellow pine.

The outside walls are covered with drop siding of Cypress, "The Wood Eternal." This secures a strong and weather resisting combination.

To make the roof good and strong we furnish **tight fitting tongued and grooved sheathing boards** to be laid on the rafters and covered with **Fire-Chief Shingle Roll Roofing. Fire-Chief Roofing** is guaranteed by us for **fifteen years.** It resembles painted shingles.

Doors are made of Clear Cypress, "The Wood Eternal." Ground floor doors and the double hay door slide on Roll Rite Hangers, the highest priced hangers we handle. A smaller hinged door on the mow floor level is convenient for many purposes. All doors are **"ready made,"** ready to hang in place and are superior to those produced by hand carpentry.

Large windows, 1⅜ inches thick, with nine lights, are liberally provided for the ground floor. Opening size of windows is 2 feet 8 inches by 3 feet 6 inches. Ventilation shields are furnished. A quick change of air can be made by lifting the windows back into the shields. The practical construction of the shields protects your stock from direct drafts.

Gable arrangement of barn, 24 feet wide, is shown in the small illustration. This size has different door and window arrangement on ground floor and a drop hay door.

Hardware, such as latches, bolts, screws, nails, etc., is included in the price, also sufficient **paint for two coats,** oxide red for the body and white for the trim. You may select other combinations from our paint page (see page 47). Ventilating systems, cupolas, barn equipment, silos, foundation material, etc., are not included in the price.

Floor Plans. See pages 40 to 42 for sample floor plan arrangements. **Be sure to specify number of the floor plan that best suits your requirements.** Use the enclosed sketch sheet to outline your needs when changes are necessary and return for quotation on equipment.

Our "already cut" feature, together with "ready made" doors, sash, etc., furnished at the prices quoted, make it possible for you to erect this modern general purpose barn at a remarkable saving of time and labor. Free building plans, which are easy to understand, are furnished with every order. They show exactly where every piece fits. You will appreciate the value, economy and convenience in this strong and sanitary modern general purpose barn.

QUALITY GUARANTEED

Prices of "Ideal" Modern General Purpose Barn No. 2060 "Already Cut" and Fitted.

Width, Feet	Lgth., Feet	Height, From Sill to Eaves, Feet	With No. 1 Common Cypress Siding	With Select Cypress Siding	HAY-MOW Capacity in Tons
24	26	12	$ 569.00	$ 595.00	17
24	32	12	644.00	673.00	21
24	38	12	719.00	751.00	25
24	44	12	794.00	829.00	29
28	32	12	771.00	806.00	29
28	38	12	859.00	897.00	34
28	44	12	947.00	988.00	39
28	56	12	1,123.00	1,169.00	49
30	32	12	843.00	881.00	29
30	38	12	936.00	978.00	34
30	44	12	1,031.00	1,076.00	43
30	50	12	1,125.00	1,173.00	49
30	62	12	1,312.00	1,367.00	61
30	74	12	1,499.00	1,561.00	72
32	32	14	873.00	917.00	38
32	38	14	970.00	1,017.00	45
32	44	14	1,067.00	1,117.00	51
32	56	14	1,261.00	1,318.00	66
32	62	14	1,358.00	1,418.00	73
32	74	14	1,551.00	1,618.00	87
34	38	16	1,038.00	1,089.00	55

Width, Feet	Lgth., Feet	Height, From Sill to Eaves, Feet	With No. 1 Common Cypress Siding	With Select Cypress Siding	HAY-MOW Capacity in Tons
34	44	16	$1,144.00	$1,199.00	63
34	56	16	1,357.00	1,420.00	80
34	68	16	1,570.00	1,640.00	97
34	74	16	1,676.00	1,751.00	106
34	86	16	1,889.00	1,971.00	123
34	98	16	1,999.00	2,191.00	140
36	38	16	1,081.00	1,136.00	61
36	44	16	1,191.00	1,250.00	71
36	56	16	1,412.00	1,478.00	90
36	68	16	1,633.00	1,706.00	109
36	74	16	1,743.00	1,820.00	118
36	86	16	1,963.00	2,048.00	138
36	98	16	2,184.00	2,276.00	157
36	110	16	2,404.00	2,504.00	176
36	146	16	3,066.00	3,188.00	234
40	56	16	1,554.00	1,626.00	104
40	68	16	1,792.00	1,872.00	125
40	98	16	2,386.00	2,485.00	180
40	110	16	2,623.00	2,731.00	202
40	134	16	3,099.00	3,222.00	247
40	152	16	3,455.00	3,589.00	280

"Already Cut" and Fitted.

—13—

SEARS, ROEBUCK AND CO., CHICAGO.

^ **Catalog Listing for a Modern All-purpose Barn. Sears, Roebuck & Co., 1918.**

often the only other book in the farmhouse. In short time, the catalog became a unifying emblematic symbol of the geographically dispersed rural populations it served. The catalog became so identified with its social milieu that wounded World War I soldiers recovering in France were provided with the most recent versions of the Sears Catalog as a reminder of their lives back home.

While the years leading up to World War I were marked by the rise of the American family farm and the "golden age" of agriculture, after 1915 the social conditions of agriculture changed sharply. As technological innovations further mechanized agricultural processes, fewer farmers were needed to cultivate more and more land. This increased mechanization also encouraged the restructuring of family farms into larger scale, increasingly specialized farm enterprises. As farmer refugees headed back to urban neighborhoods, the nation's attention turned back to cities.

The role of the catalog as a social network would not resurface until its reconceptualization in the 1960s. In "Greening of America," Charles A. Reich describes a series of two shifts in the socio-economic consciousness of the country between the 1920s and 1960s.[4] First, as a result of the nuclear reality of World War II and the Cold War, the values associated with rural farmers and small business were replaced by values associated with people management and complex organizations, and technologically directed research and development. The next shift was marked by a paradoxical duality. On the one hand, the world was in danger of nuclear war, and on the other hand the individual was being destroyed by the very bureaucracies trying to protect them. This double-sided threat began to inform and motivate a young generation of counter cultural revolutionaries seeking an alternative. Interestingly the catalog was once again leveraged for its ability to connect and disseminate. This time, however, the catalog was not designed to serve an expansive population of farmers but instead to empower a new population heading back to the land.

Rejecting hierarchical political and social structures, the back-to-the-landers of the late 1960s left their urban lives to rediscover the potentials of flattened and localized societies within rural contexts. Among the tens of thousands of young people who fled to America's remote landscapes, there is a subset of particular interest: while they shared the common intention of building self-sufficient communes valuing small-scale egalitarian structures, the New Communalists, also took with them a trust in knowledge, technology and an understanding of the value of interdisciplinary collaborations established in the research projects of the Cold War era.[5]

The western American landscape was once again dotted with a geographically distributed set of communities thirsty for a way to communicate, collaborate and connect. It was for these populations that Stewart Brand published the *Whole Earth Catalog*. Brand, aligning himself with earlier catalogers in recognizing the potential of the format as tool for information dissemination, began to assemble a collection of outdoor equipment, building plans, mechanical blueprints, books, periodicals and how-to guides ranging from bee culture to glass blowing that would make rural relocation for the New Communalists easier. Considered an empowering device, the purpose of the catalog was described as follows:

> *We are as gods and might as well get used to it. So far, remotely done power and glory—as via government, big business, formal education, church—has succeeded to the point where gross obscure actual gains. In response to this dilemma and to these gains a realm of intimate, personal power is developing—power of the individual to conduct his own education, find his own inspiration, shape his own environment, and share his adventure with whoever is interested. Tools that aid this process are sought and promoted by the WHOLE EARTH CATALOG.*[6]

On the one hand, the *Whole Earth Catalog* adopted the underlying principle of traditional consumer catalogs in that both served geographically disconnected yet functionally connected populations by delivering curated sets of useful items and making clear the channels by which these items could be acquired. Effectively both versions served as vehicles of critical information delivery. On the other hand, the *Whole Earth Catalog* both expanded and entirely re-envisioned the potentials of its predecessors' functions and logics. Embracing, extending and reimagining the catalogs' potential with respect to community and connectivity, the *Whole Earth Catalog* clarified and defined a contextual framework fundamental to our current digital age.

Brand's ambitions as a Network Entrepreneur allowed him to imagine a single catalog that served a wide-ranging readership.[7] Whereas earlier catalog readership was understood as one group—the consumers—Brand saw value in weaving together often disparate groups as a way of strengthening the potentials of a loosely defined social network. For this reason, while the New Communalists were its first audience, the *Whole Earth Catalog* quickly expanded its readership to include scientific research communities, specifically those interested in technology, systems theory and ecology, and artist communities, specifically those in San Francisco and New York City. This broadening of readership led to the sale of 2.5 million copies of the *Whole Earth Catalog* between 1971 and 1972. This interdisciplinary audience further amplified the value of the publication in the eyes of the New Communalists. Through this interwoven integration of disparate audiences, the publication achieved a flattening between experts and communards where all shared and were equally empowered by the same information.

Picking up on earlier catalogs' nascent roles with respect to offering modes of exchange between readers, the *Whole Earth Catalog* sought to establish itself as a forum where New Communalists could meet not only with each other but also with the catalog's expanded readership of academics, scientists and artists. Establishing reader-publisher and reader-reader relationships required a means by which readers could become more actively involved. One basic mechanism used to foster this involvement ensured that tools listed in the catalog reflected the broad readership rather than the single voice of the publisher. The editorial team's continuous review of the catalog's content based on readers' insights, suggestions and referrals reinforced this diversification. A second, more instrumental, mechanism with which to achieve reader-reader connections was the *Supplement*. Published as a quarterly update to the Catalog, the *Supplement* privileged reader-generated content. The Catalog offered access to tools, the

4. Reich, Carles A. *The Greening of America* (New York: Random House, 1970).

5. Fred Turner coins the term "New Communalist" in his work *From Counterculture to Cyberculture: Stewart Brand, the Whole Earth Network, and the Rise of Digital Utopianism.* For Turner the term was necessary to differentiate the New Communalists from the New Left (page 33). While rejecting many of the effects of the rising status of the military-industrial complex, New Communalists did value the potentials of knowledge work associated with the interdisciplinary collaborations favored and instigated by the very institutions they rejected.

6. *Updated Last Whole Earth Catalog: Access to Tools.* Point, 1974. *Whole Earth Catalog*'s Purpose Statement. Found on page 2 of each issue.

7. Turner, Fred. *From Counterculture to Cyberculture: Stewart Brand, the Whole Earth Network, and the Rise of Digital Utopianism* (Chicago: University of Chicago Press, 2006), 5, 251-55.

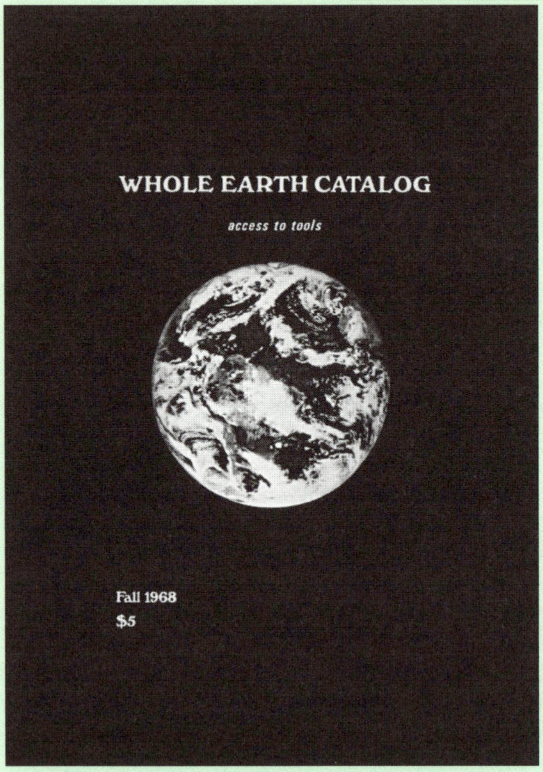

^ **Fall 1968 cover, *Whole Earth Catalog: Access to Tools.***

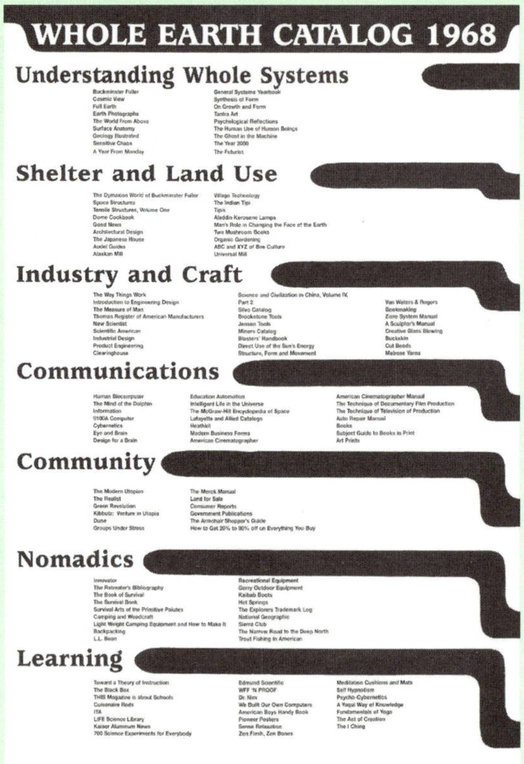

WHOLE EARTH CATALOG 1968

Understanding Whole Systems

Buckminster Fuller
Cosmic View
Full Earth
Earth Photographs
The World From Above
Surface Anatomy
Geology Illustrated
Sensitive Chaos
A Year From Monday
General Systems Yearbook
Synthesis of Form
On Growth and Form
Tantra Art
Psychological Reflections
The Human Use of Human Beings
The Ghost in the Machine
The Year 2000
The Futurist

Shelter and Land Use

The Dymaxion World of Buckminster Fuller
Space Structures
Tensile Structures, Volume One
Dome Cookbook
Good News
Architectural Design
The Japanese House
Audel Guides
Alaskan Mill
Village Technology
The Indian Tipi
Tipis
Aladdin Kerosene Lamps
Man's Role in Changing the Face of the Earth
Two Mushroom Books
Organic Gardening
ABC and XYZ of Bee Culture
Universal Mill

Industry and Craft

The Way Things Work
Introduction to Engineering Design
The Measure of Man
Thomas Register of American Manufacturers
New Scientist
Scientific American
Industrial Design
Product Engineering
Clearinghouse
Science and Civilization in China, Volume IV, Part 2
Silvo Catalog
Brookstone Tools
Jensen Tools
Miners Catalog
Blasters' Handbook
Direct Use of the Sun's Energy
Structure, Form and Movement
Van Waters & Rogers
Bookmaking
Zone System Manual
A Sculptor's Manual
Creative Glass Blowing
Buckskin
Cut Beads
Melrose Yarns

Communications

Human Biocomputer
The Mind of the Dolphin
Information
9100A Computer
Cybernetics
Eye and Brain
Design for a Brain
Education Automation
Intelligent Life in the Universe
The McGraw-Hill Encyclopedia of Space
Lafayette and Allied Catalogs
Heathkit
Modern Business Forms
American Cinematographer
American Cinematographer Manual
The Technique of Documentary Film Production
The Technique of Television of Production
Auto Repair Manual
Books
Subject Guide to Books in Print
Art Prints

Community

The Modern Utopian
The Realist
Green Revolution
Kibbutz: Venture in Utopia
Dune
Groups Under Stress
The Merck Manual
Land for Sale
Consumer Reports
Government Publications
The Armchair Shopper's Guide
How to Get 20% to 90% off on Everything You Buy

Nomadics

Innovator
The Retreater's Bibliography
The Book of Survival
The Survival Book
Survival Arts of the Primitive Paiutes
Camping and Woodcraft
Light Weight Camping Equipment and How to Make It
Backpacking
L.L. Bean
Recreational Equipment
Gerry Outdoor Equipment
Kaibab Boots
Hot Springs
The Explorers Trademark Log
National Geographic
Sierra Club
The Narrow Road to the Deep North
Trout Fishing in American

Learning

Toward a Theory of Instruction
The Black Box
THIS Magazine is about Schools
Cuisenaire Rods
ITA
LIFE Science Library
Kaiser Aluminum News
700 Science Experiments for Everybody
Edmund Scientific
WFF 'N PROOF
Dr. Nim
We Built Our Own Computers
American Boys Handy Book
Pioneer Posters
Sense Relaxation
Zen Flesh, Zen Bones
Meditation Cushions and Mats
Self Hypnotism
Psycho-Cybernetics
A Yaqui Way of Knowledge
Fundamentals of Yoga
The Act of Creation
The I Ching

^ **Fall 1968 table of contents, *Whole Earth Catalog: Access to Tools.***

Supplement, through articles, product news and reviews, how-to guides and event listings, offered a dynamic view into how the tools were used and, perhaps more importantly, a way for readers to connect with one another.

Another extension carried out by Brand and his team included a broadening of the notion of tools. While earlier catalogs focused on physical objects such as utensils, equipment and clothing, *Whole Earth Catalog* injected informative tools into its pages. Some of these informative tools were also physical, such as maps, books and journals while others were more intangible, such as educational classes and recorded lectures. This broadened notion of tools coincides with the New Communalist's interest in tools not only as objects but as processes of transformation through knowledge.

Perhaps the boldest, and most consequential, departure from traditional catalogs was *Whole Earth Catalog*'s underlying organization. Whereas traditional catalogs organized themselves according to rigid hierarchical categories, *Whole Earth Catalog* was loosely organized around highly abstract themes. Organizational hierarchies such as Appliances, Fitness and Sport, Automotive, and Outdoor Living were replaced by themes such as Understanding Whole Systems, Shelter and Land Use, Community, and Nomadics. Instead of seeing twelve different shovels or four different sewing machines on a single page, readers of the Catalog were confronted with a wide array of objects, concepts, manuals, and other resources on each spread. As an example, one spread presents readers with techniques of television production, an auto repair manual, and a book-industry trade journal. Further, in the *Whole Earth Catalog* there is no such thing as a feature item. Each entry is weighted equally and appears just as valuable and necessary as the rest. Admittedly, this cloud-like organization made finding things more difficult; but as a result of the absence of hierarchy, the reader was compelled to draw connections and to explore the overlapping potentials of the multi-dimensional systems distributed throughout the document. With this reorganization, the Catalog was further amplified as an informational tool. Not only did it show what could be done, it encouraged innovation, reinterpretation, criticality and other intellectually focused processes.

Brand's reconceptualization of the catalog as a multilateral forum focused on both information exchange and community building established a paradigm for the nascent digital culture emerging in the early 1970s.[8] It has now been over forty years since the first *Whole Earth Catalog*, a period which has seen the continuous expansion and reimagination of the potentials of web-based platforms in their support of geographically scattered knowledge-based communities. Similar to the catalog's evolution from a unidirectional, singularly voiced disseminator of information to a multidirectional knowledge forum, the Internet

8. It is beyond the scope of this essay to recount in more detail Brand's relationship to digitally-based ideologies, movements, and innovations. For an in-depth examination see: Turner, Fred. *From Counterculture to Cyberculture: Stewart Brand, the Whole Earth Network, and the Rise of Digital Utopianism* (Chicago: University of Chicago Press, 2006).

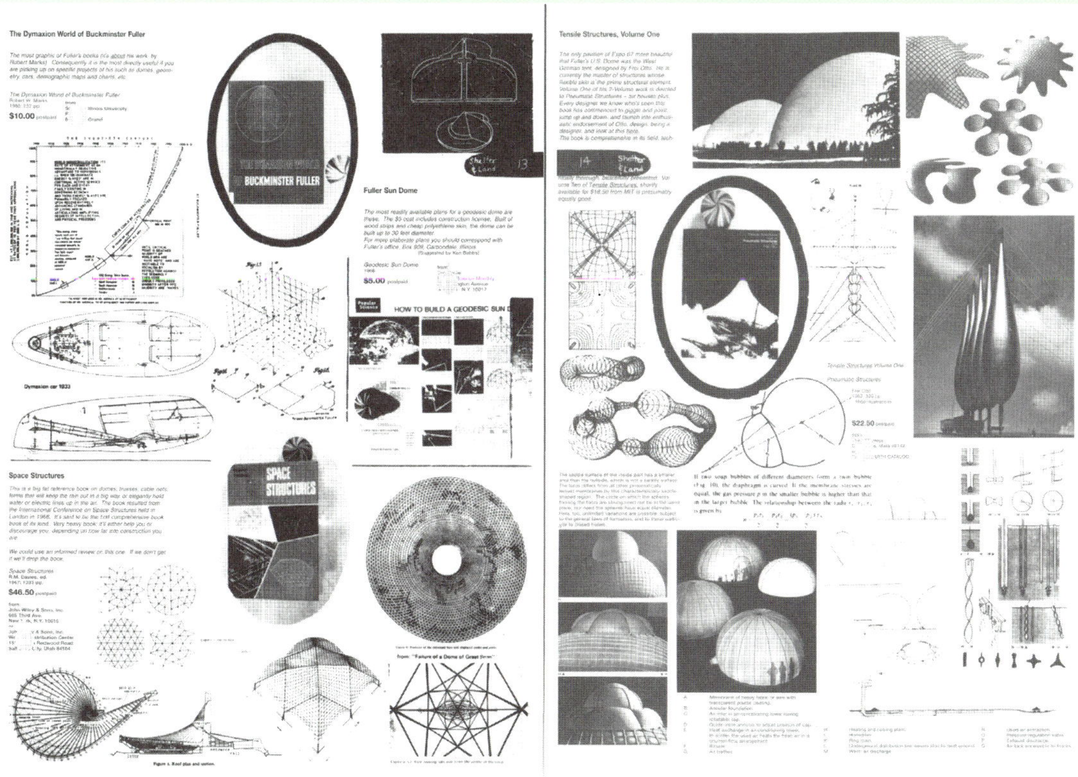

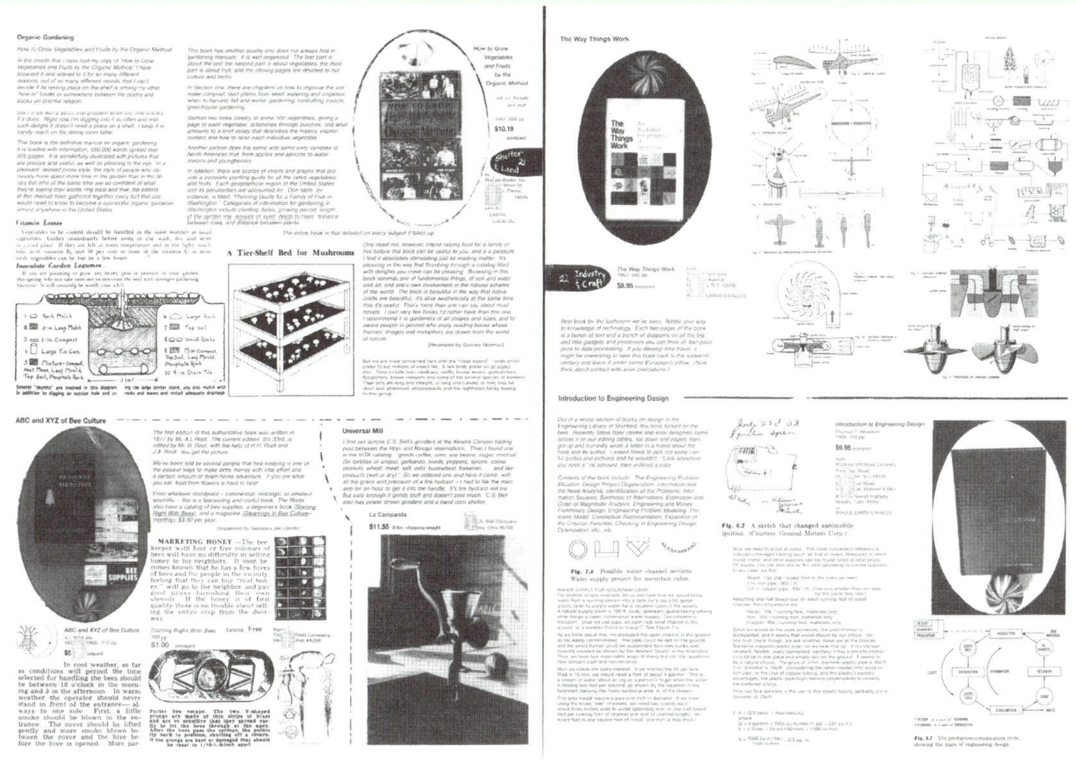

^ **Fall 1968 spreads, pages 13-14 (top), pages 21-22 (bottom), *Whole Earth Catalog: Access to Tools.***

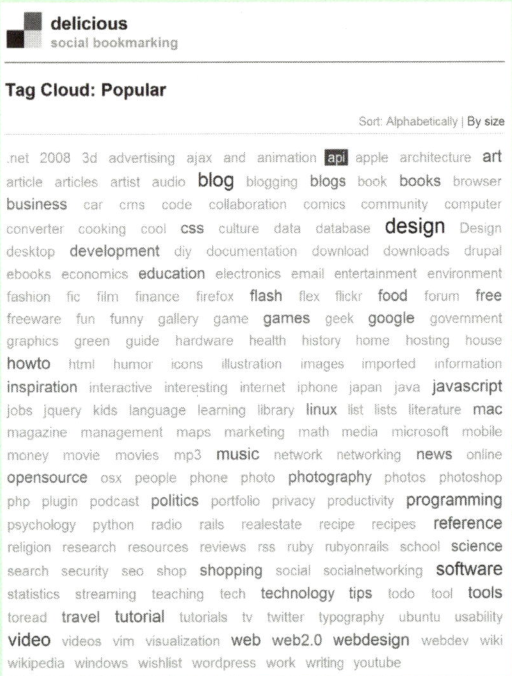

^ Tag Cloud. Screenshot of Delicous' *Popular Tag Cloud* on April 04, 2010 at 3:51PM. http://delicious.com/tag.

has seen a steady growth of the value associated with the participatory creation of content. Amazon.com is a prototypical example of this transition. On the one hand, Amazon.com is a screen-based equivalent to a traditional print catalog in that at its highest level, the site organizes books, and now other product types, in normative categories and offers them for sale. On the other hand, Amazon.com has embedded within itself concepts of user agency to move away from a static catalog towards a dynamic marketplace complete with consumer feedback and vendor competition. Utilizing crowdsourcing in order to build up a collection of reader reviews, Amazon.com leverages site users to create content that is useful to potential consumers and thereby further secures its position as a leading retailer. Further, taking advantage of its own sale data, Amazon.com continually re-shuffles its catalog by offering context-specific listings showing complimentary books that are frequently bought together. At the same time, the site provides a forum for readers to become participants within a community rather than simply consumers of a product. Users' roles on Amazon.com have expanded to allow the creation of virtual bookshelves, curate collections wherein books take on additional layers of meaning through association. In a similar vein, Amazon.com added another layer of agency and potential collaboration when it offered many of their computer code libraries to outside developers, allowing for new functionality through the creation of independent tools. Through the extension of user agency to include the role of curator and developer, Amazon.com advances *Whole Earth Catalog*'s underlying concepts of open systems, dynamic organization and mass-collaboration.

Multilateral communication on the Internet is now omni-present. From book reviews to hotel reviews, from Facebook News Feeds to blogs and their comments, conversations between geographically dispersed populations are ubiquitous throughout cyberspace. Equally prevalent are web-based services whose value is created fundamentally by means of mass collaboration and exchange. Crowdsourcing has emerged as the common idea behind contemporary web-based concepts such as social bookmarking, volunteer computing, citizen science, and other participatory-centered web-based services and sites.[9] The amplitude of participation in such forums has led to the emergence of the "tag cloud" as a new organizational logic proliferating in cyberspace. Free from predetermined hierarchies and anticipated results these clouds create dynamically generated indices based on the real-time aggregation and harvesting of user-generated input across an entire network. Thus the invention of the catalog as an informational service for spatially disconnected communities has ultimately inverted the notion of "farming": initially conceived of as a tool for farmers, and later expanded to seed ideas of mass collaboration and multilateral communication, the catalog has now led to the cultivation and farming of the readership.

9. The concept of "social bookmarking," or "collaborative tagging," is rooted in the pairing of crowdsourcing and emergent, non-hierarchical organization. Social bookmarking services allow users to store, share, and organize links to Internet content. In addition to sharing the address of specific web content, users develop their own categories, tags, and descriptions for organizing the content being indexed—the collections of which are called tag clouds. Without relying on a predetermined system of classification, these lists of bookmarks get infused with another layer of value through dynamic associations. Delicious (www.delicious.com), one of the web's leading social bookmarking services boasting as many as 5.8 million users, organizing 180 million unique URLs in 2008, further leverages its users by generating real-time snapshots of activity, thus projecting a collective awareness.

Volunteer computing, a type of distributed computing where individual computer owners offer their computing resources such as processing power and storage, to a project that has exorbitant computational requirements. Instead of taking advantage of humans' cognitive ability to organize, like in the case of social bookmarking, volunteer computing takes advantage of the enormous amounts of unused processing power of interconnected computers around the world. SETI@home is a volunteer computing project with over 3 million users. SETI, an acronym for Search for Extra-Terrestrial Intelligence, uses distributed processing power to analyze observational data in the form of radio signals generated by the Arecibo radio telescope in search of signs of extraterrestrial intelligence. Enormous sets of data are broken down into smaller chunks and sent for processing to participating computers and then, in turn, the volunteer computers return their findings. Simply, a project that would normally involve an insurmountable amount of data analysis is made manageable by using a large, Internet-based community. The SETI@home project is also an example of citizen science in that the project includes networks of volunteers, many of whom may have no specific scientific training, perform or manage research-related tasks such as observation, measurement or computation.

ploughs/ combines

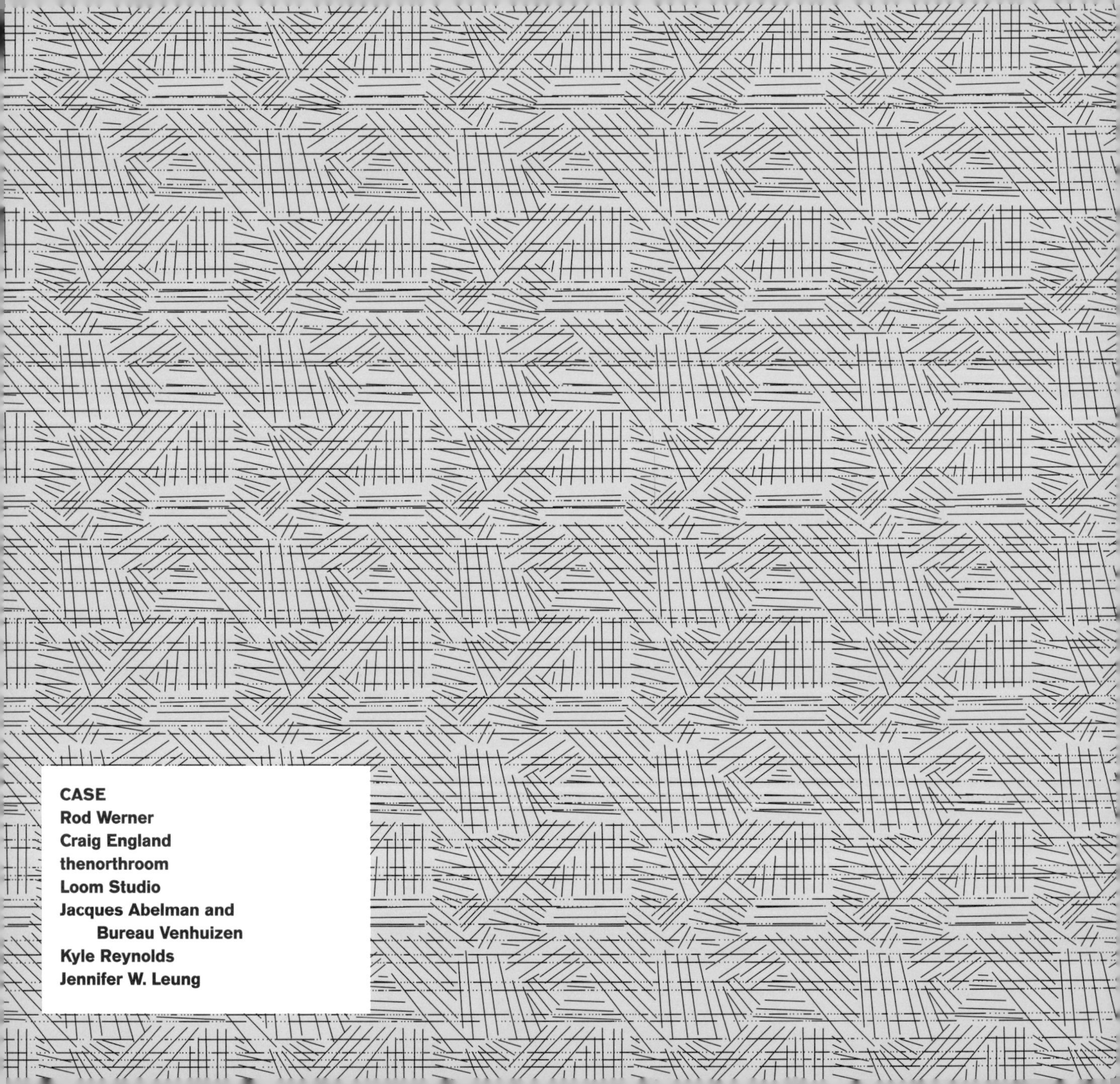

CASE
Rod Werner
Craig England
thenorthroom
Loom Studio
Jacques Abelman and Bureau Venhuizen
Kyle Reynolds
Jennifer W. Leung

bldg 2.0: crowd-sourcing building energy performance

case (steve sanderson & josh emig)

Buildings are the largest contributor to greenhouse gas emissions, and three-quarters of U.S. energy goes toward building operations. Yet, there is no way to accurately predict building performance. There is a need for simple tools to provide meaningful and actionable feedback on building performance to designers, owners, and occupants. BLDG 2.0 seeks to fill this void between design intent and building performance. Built upon principles of mass collaboration and collective intelligence, BLDG 2.0 is an open-source interface to building performance databases, a collaborative community of experts, and an open innovation marketplace for building performance solutions.

BLDG 2.0 addresses these two, somewhat paradoxical, barriers to informed sustainable design: the lack of publicly accessible performance data and the difficulty of extracting meaningful, consistent information from the unstandardized data when it is available. BLDG 2.0 reconciles this by creating a knowledge resource for building performance. The steps to follow are simple to enumerate.

Make data accessible

Numerous organizations currently collect and report building performance data, but there is little consistency in the way this data is shared. BLDG 2.0 provides an open platform where data from numerous sources can be accessed, visualized and meaningfully compared. The platform utilizes open data whenever possible and encourages the development and adoption of open standards for collecting and reporting building performance data.

Information interface

Once the datasets have been identified, an intuitive and open interface is used to view building performance data. While there have been several successful proprietary tools designed for this purpose, none of them can be extended by end-users. BLDG 2.0 will be designed with an Application Programming Interface (API) that will allow members of the BLDG 2.0 community to enhance the functionality of the system for their own purposes and for use by others.

Open innovation network

The BLDG 2.0 platform transforms users from passive consumers of information into active contributors, creating a network of 'citizen-scientists' in which new knowledge, technical information, insights and experiences are shared, debated, and refined. This network is leveraged to build an 'innovation marketplace' for building performance solutions. Organizations are able to submit 'design challenges' for the community to solve, with the best solutions receiving awards.

As with the Human Genome Project, decentralized collaboration has the potential to yield new knowledge by harnessing the analytical power of the collective. This new knowledge enables planners and designers to make more informed decisions, engineers to design more effective

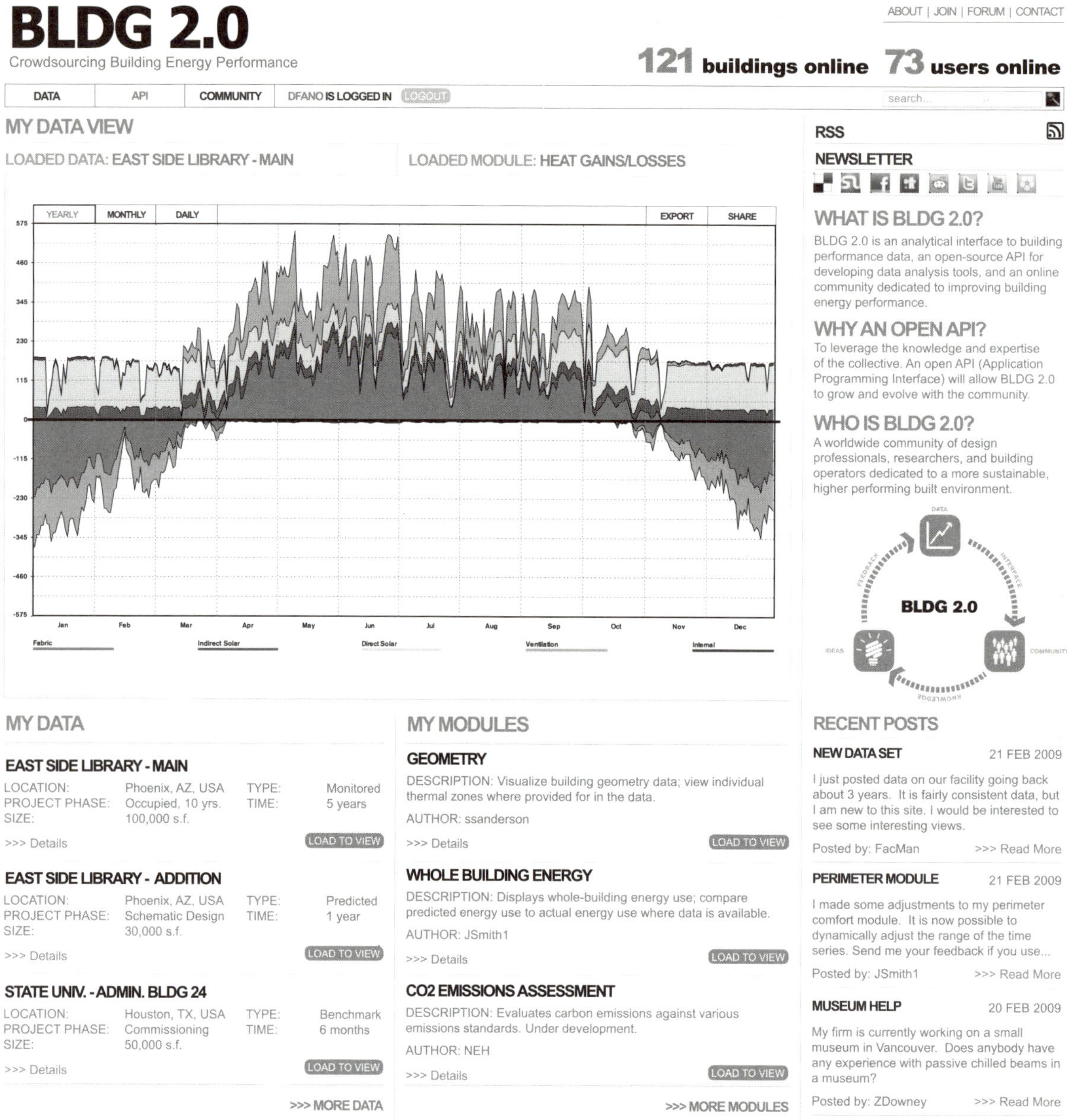

∧ Sketch of BLDG 2.0 platform.

systems, manufacturers to develop better products, communities to adopt more ambitious policies, software companies to build more useful tools, owners to set higher standards and schools to better educate our future leaders. As an ever-evolving and improving collaborative infrastructure, BLDG 2.0 provides a foundation for a more effective and adaptive community of design professionals dedicated to a sustainable built environment.

BLDG 2.0 is a collaboration between architects, technologists, sustainability experts. Led by CASE, a building technology and project integration consultant based in New York, the team includes: Arup Sustainability, Columbia University, IDEO and SHoP Architects, with the New Buildings Institute, an authority on building energy performance.

>>> Steve Sanderson is a partner at CASE, a virtual design and construction (VDC) and integrated-practice consultancy based in New York City.

>>> Josh Emig is a visiting assistant professor in Auburn University's Master of Design Build program, focusing on the practice and pedagogy of integrated design and construction.

down on the body farm

rod werner

Adjacent to The University of Tennessee-Knoxville's Medical Center is 2.5 acres of land known as the "Body Farm." The Body Farm belongs to the University's Forensic Anthropology Center and was established in 1981 by Dr. William Bass. Bass who had arrived ten years earlier to lead the Anthropology Department founded the Body Farm to study the postmortem changes that occur to a human body. Since its inception Bass' brainchild has become an invaluable source for forensic scientists, crime scene investigators, and law enforcement. Bodies, usually numbering around fifty at a time, are placed in various locations around the site. Some rest in water, while others are buried under poured-concrete slabs. Some are placed in the trunks of cars, and others are left exposed to the elements. Throughout the decomposition process, the changes of each corpse are painstakingly documented by the researchers and added to the lexicon of forensic science.

In his book *Death's Acre*, Dr. Bass, with coauthor Jon Jefferson, writes of his desire to make a scientific study of human decomposition. "The idea of making a scientific study of decomposition had been germinating in some recess of my mind. The seed had been planted back in 1964." It took the doctor nearly twenty years for the seed to sprout. Surprisingly, no extensive study of a corpse's postmortem transformation had been compiled since the Chinese text *Washing Away of Wrongs*, which was written by Sung Tz'u in 1247. Sung Tz'u was a senior criminal judge in the Southern Song Dynasty, and during his long tenure amassed an impressive amount of case experience and forensic data. Sung Tz'u recorded a wide variety of types of death. From blunt force trauma to poisoning, from the characteristics of hanging to drowning, this great atlas was a significant contribution to forensic science.

Although the 700-year-old handbook was extensive, Dr. Bass saw the need for an update. In addition to leading the Anthropology Department, Bass was also named Tennessee's state forensic anthropologist and later became a special consultant for the Tennessee Bureau of Investigation. In this capacity, he was involved in many fascinating and often puzzling cases and as his academic program and reputation grew, Bass eventually developed a team of graduate students to assist him in his investigations. Through his work with the Tennessee Bureau of Investigation, Bass realized the need to be able to answer what he calls "laughably elementary questions," such as "at what point does the arm fall off?" and "what causes that greasy black stain under the body?" These were questions that could only be answered through observation. And to make these observations Bass needed bodies. After sending scores of letters to funeral directors and medical examiners throughout the state of Tennessee, Bass received his first donated corpse in May 1981. In a small clearing that had been used for years as site for hospital trash burning, Bass poured a sixteen square-foot concrete slab and surrounded and roofed it with chain link to deter larger pests. The donated body, for anonymity's sake known simply as "1-81"—referring to the first corpse received in 1981—belonged to a seventy-three year old white male who had suffered from chronic alcoholism, emphysema, and heart disease. "1-81" was placed on the slab and the Body Farm was born.

Research at the Body Farm began paying immediate dividends. One of the first and most important forensic studies to originate at the Body Farm was "Insect Activity and Its Relationship to Decay Rate of Human Cadavers in East Tennessee" by Bill Rodriguez. Rodriguez was a graduate student in Anthropology and the abstract of his research which was published in the *Journal of Forensic Sciences* reads:

> This is the first report on an ongoing study conducted to collect data on the specific insects that are found in association with decaying human cadavers. Four nude unembalmed human cadavers were each placed, at various times of the year, within a decay research facility located in open wooded area. Data were collected daily throughout the entire decay cycle on the various insect populations that frequented each cadaver. Analysis of the data shows that there is a direct correlation between the rate of decay and the succession of insect families and species found in association with a decaying cadaver. Application of this entomological information can contribute to a more accurate estimation of "time since death" of an individual.

The forensic database at the Body Farm is being built by people who spend their days documenting the arrival time and egg-laying habits of blowflies and the development of maggots. Some study adipocere (the soapy substance which develops when fatty tissue decomposes in a moist environment), while others document putrefaction (the decomposition of soft tissue by bacteria). Beginning with Rodriguez's studies, research gleaned from the Body Farm has provided an essential "time since death" codex, which enables any murder victim's time of death to be pinpointed. Recently emergent studies in DNA degradation and facial recognition programming have utilized the contributions from the Body Farm. Over the years this research has served to convict murderers and to acquit the innocent.

While research at the Body Farm has been going on quietly for almost thirty years, the last decade has witnessed an explosion of interest in forensic science. Thanks to the success of the hit television franchise *CSI*, which has a weekly audience of sixty million, applications to university forensic science programs across the United States have skyrocketed. This influence, known as the "CSI effect," was written about by Stefan Lovgren in his 2004 *National Geographic* article, "CSI Effect Is Mixed Blessing For Real Crime Labs." Lovgren reported that in 2003, 180 potential students applied for just 20 vacancies at Michigan State's Forensic Science Masters Program. The proliferation of forensic anthropology programs has also raised eyebrows among some forensic anthropology educators. On its website, the Council of Forensic Science Education, a professional group that was established over two decades ago to develop excellence in academic standards, warns of hurriedly organized and underfunded programs that do not educate or prepare their students for the rigors of the profession.

The interest in forensic sciences is not always viewed as a positive development, however. Prosecutors and the scientists who provide expert testimony in America's courtrooms are seeing and feeling the influence of the "CSI effect" as jurors increasingly anticipate the same forensic flash witnessed in prime time. In 2007, The San Antonio Express reported on the new expectations appearing in America's halls of justice. "With crime-show scenes swimming in their heads, many jurors nowadays bring unrealistic or heightened expectations into the courtroom, said Maricopa County Attorney Andrew Thomas of Phoenix. Lawyers are increasingly forced to deal with how to adapt their cases accordingly." In accordance, law journals are publishing a flurry of academic papers with titles such as "A Study of Juror Expectations and Demands Concerning Scientific Evidence: Does the CSI Effect Exist?"[1] and "CSI and its Effects: Media, Juries, and the Burden of Proof."[2] Even the real investigators have been influenced by the "CSI effect." When Anthony Zuiker, the creator of *CSI: Crime Scene Investigation*, first convinced the Las Vegas crime scene unit to allow him to shadow them for his research, they were known as Las Vegas Field Services. It was only after the success of the show that Field Services changed its name to CSI.

The first bodies that arrived at the Body Farm in 1981 are still being studied today. These were the corpses of the indigent and the forgotten. But, in the quarter century since the Body Farm was founded, the corpses studied have—with more frequency—come from the Body Donation Program. According to the University of Tennessee's Forensic Anthropology Center website, in the years since the Body Farm was established the Center has amassed a Forensic Skeletal Collection of 100 sets of remains. These belong to mostly unidentified victims of perimortem trauma. The Donated Skeletal Collection houses over 650 sets of remains. This collection is of identified persons and potential research includes a range of pursuits from skeletal, biological, and forensic to dental, biomedical, and genetics. The skeletons range from fetal to 101 years of age. 1300 future dead souls, from 35 different states and four different countries, have pledged their remains to the Body Farm. And the Body Farm is growing. From its humble beginnings, body donations have continued to increase and an expansion of over ten acres is planned.

New facilities have also been established. Western Carolina University operates a research station through its Human Identification Laboratory where decomposition standards are being developed for the western North Carolina mountain habitat. Texas State University also studies human decomposition rates. Each of these facilities relies on body donation for their research.

So what is the motivation to donate one's body? Many donations come from people who had intended to sign their bodies over to science but did not make adequate plans with a medical school. A number of bodies are donated because funeral arrangements were not made. The AARP (American Association of Retired People) estimates the combined costs of a traditional funeral to average $10,000 dollars. And although costs are significantly less than a traditional burial, cremation still averages about $2,000 dollars. A growing trend in the funerary profession is the "green burial." A "green burial" takes place on a site certified by the National Green Burial Council. Bodies are placed in biodegradable caskets or shrouds and are not preserved with any formaldehyde-based embalming fluid. Even the most economic of eco-friendly burials, however, will cost several hundred dollars. Donation to the Body Farm is free and they will even pick your body up if you are within a two hundred mile radius. Beyond that distance, a funeral home

1. ***Vanderbilt Journal of Entertainment & Technology Law*, Vol. 9 p.330, 2006.**

2. ***New England Law Review*, vol. 41, no. 3, 2007.**

is needed to set up transportation. So for the cash-strapped survivors of the dearly departed, body donation is an attractive option.

With the increased interest in forensic study, a number of donations likely come from those intrigued with the science, and the Body Farm allows one to participate directly in the process. In life these people may have theorized on the identity of Jack the Ripper. They may have written books about who killed JonBenét Ramsey. Or perhaps they worked in the field and want to keep contributing after they have moved on. Though, most donations come from average people who realize the importance of the information their bodies can provide.

Donating one's body is simple and straightforward. The form for donation is downloadable from the Forensic Anthropology Center's website and requires only the donor's name, address, signature, date and the signatures of two witnesses. The form need not be notarized. It's just that easy to give oneself to the Body Farm.

For most, however, body donation is the only way one can access the Body Farm. While a number of outside forensic anthropologists, students, law enforcement officers and coroners do research at the facility, no public tours or visits are allowed. A spokesman for University of Tennessee at Knoxville estimated that the facility receives over one hundred requests per year for visits. Some are made by legitimate sources, such as documentary filmmakers and journalists but, the spokesman also noted that after an extensive vetting process is done to keep the morbidly curious away, only a fraction of those requests are granted. Beyond the simple wooden fence that runs the perimeter of the facility important work is being done. Work that cannot be disturbed. Work that will solve crimes. Work that will heal wounds: the work of the Body Farm.

>>> **Rod Werner is a Freelance Design Thinker based in Indianapolis, Indiana.**

food matrix

craig england

Food Matrix is a quick-reference guide for the small-scale producer. It is intended to be utilized at the initial planning stages for a sustainable farming development. At a glance, one can compare energy input and output ratios, average yields, and water or land requirements. It assists with the development of a yearly plan spanning the seasons responding to typical farming cycles and practices. Food Matrix provides a fast and convenient method to contrast the benefits or disadvantages of many common crops and livestock.

Comparing energy efficiencies (input and output ratios) allows users to strategize which crops are suitable for their needs and capabilities. Energy ratios listed in Food Matrix are calibrated for conventional farming practices with typically high fossil-fuel externalities such as fertilizers, pesticides, and herbicides. The energy input ratios will be different under sustainable farming practices with no externalities but there will be an increase in energy expended through labor .

To increase productivity per acre, the user may quickly compare average yields of crops to determine an initial planting strategy. During harvest, the user may initially encounter lower yields than those in the reference guide, as they stem from conventional farming practices. However, yields will increase within a few years of sustainable practice to match, if not exceed, conventional yields.

Another method to increase productivity per acre while reducing externalities is to utilize companion plantings. Companion plants have beneficial relationships that strengthen the health of each individual plant through its association with its companion. This relationship typically causes an increase in combined yields versus individual yields from monocultural plantings. It also offers a natural alternative to pest management when plants are combined effectively. The Food Matrix provides typical companion combinations, but requires further research by the user to specifically develop an integrated system approach.

The Food Matrix provides data on typical cycles within farm practice. Crop cycles allow the user to pre-plan their year by referencing sowing, development, and harvest periods. This also allows for general tallying of yields throughout the year, and thus economic projections of farm income. Livestock cycles are used to strategize when to breed animals, their gestation periods, and time until slaughter maturity. Depending upon Food Matrix calculations, the byproducts provided by certain livestock could provide either the primary source of income derived from livestock or a secondary income.

Food Matrix is a farm planning tool beneficial to a wide spectrum of farmer experience levels. By supplying the ability to develop an initial farming strategy, Food Matrix allows anyone to begin small-scale food production in urban, peri-urban, or rural areas.

>>> **Craig England is a Toronto-based designer interested in the development of sustainable food systems for urban/peri-urban areas.**

Name	Latin name	Symbol	Season	Duration	Ratio	Yield
paragus	agus officinale	P		1yr 8wks	ND	7.3 lbs/100 sq.ft
Aubergine	Solanum melongena	F	SU	10wks 13wks	ND	55.1 lbs/100 sq.ft
Beans, bush	Phaseolus spp	L	SP SU	8wks 12wks	0.345:1	17.6 lbs/100 sq.ft
Beets	Beta vulgaris spp	R	SP SU FA	8wks 4wks	ND	34 lbs/100 sq.ft
Garlic	Allium sativum	O	SP FA	1½yr	ND	40.9 lbs/100 sq.ft
Kale	Brassica sabellica	G	SP FA	8wks	ND	16 lbs/100 sq.ft
occoli	ca oleracea	C		8wks 5wks	0.33:1	33.9 lbs/100 sq.ft
Brussels Sprouts	Brassica oleracea gemmifera	C	SP FA	11wks 12wks	0.69:1	36.7 lbs/100 sq.ft
Cabbage	Brassica	C	SP FA	12wks 4wks	0.76:1	69.4 lbs/100 sq.ft
Carrots	Daucus carota	R	SP SU FA	10wks 4wks	ND	72.5 lbs/100 sq.ft
Peas	Pisum sativum	L	SP FA	10wks 12wks	1.7:1	9.2 lbs/100 sq.ft
Potatoes	Solanum tuberosum	R	SP SU	12wks	1.23:1	84.2 lbs/100 sq.ft
uliflower	ca botrytis	C		10wks	ND	38.5 lbs/100 sq.ft
Celery	Apium graveolens	S	SP FA	14wks 4wks	0.599:1	160.7 lbs/100 sq.ft
Chard, Swiss	Beta vulgaris	G	SP SU FA	8wks	ND	34 lbs/100 sq.ft
Cucumber	Cucumis sativus	V	SU	8wks 12wks	0.35:1	39.3 lbs/100 sq.ft
Sweet Corn	Zea mays	M	SU	11wks	2.5:1	27 lbs/100 sq.ft
Sweet Pepper	Capsicum annuum	F	SU	10wks 17wks	0.14:1	68.7 lbs/100 sq.ft
alfa	cago sativa	F		12wks	6.17:1	14.9 lbs/100 sq.ft
Barley	Hordeum distichon	Ce	SP FA	12wks	4.14:1	6.5 lbs/100 sq.ft
Buckwheat	Fagopyrum esculentum	O	SP SU	11wks	ND	2.6 lbs/100 sq.ft
Lentils	Ervum lens	L	SP SU	12wks 8wks	ND	2.8 lbs/100 sq.ft
Cattle, Beef	Bos taurus	R_L	FA	284 days 12 mo.	0.029:1	34-56 L/day; 110 lbs/day
Cattle, Dairy	Bos taurus	R_L	FA	284 days 12 mo.	0.053:1	80-160 L/day; 72 lbs/day
ts	sativa	Ce		15wks	5.1:1	4.8 lbs/100 sq.ft
Rice	Oryza sativa	G_r	SU	17wks	2.1:1	15.3 lbs/100 sq.ft
Rye	Secale cereale	Ce	FA	17wks	2.75:1	3.5 lbs/100 sq.ft
Sorghum	Sorghum	O	SU	13wks	1.96:1	6.8 lbs/100 sq.ft
Geese	Anserinae anser	W	SP SU	28-30 days 5-6 mo.	ND	0.6 L/day; 0.5-0.7 lbs/day
Goat	Capra aegagrus hircus	R_S	SU FA	150 days 10 mo.	ND	3.8 L/day; 2-3 lbs/day
bean	e max	L		9wks 4wks	4.15:1	4.6 lbs/100 sq.ft
Sunflower	Helianthus annuus	O	SP	17wks	ND	4.1 lbs/100 sq.ft
Wheat, durum	Triticum spp	Ce	FA	17wks	2.2:1	4.6 lbs/100 sq.ft
Wheat, white	Triticum aestivum	Ce	FA	17wks	3.69:1	3.7 lbs/100 sq.ft
Rabbit	Lagomorpha leporidae	A	Year round	32 days 11-13 wks.	ND	0.3-0.5 L/day; 1 lb/day
Sheep	Ovis Aries	R_S	FA	145 days 12 mos.	0.063:1	3.8 L/day; 5 lbs/day

Lettuce, head *Latuca spp* — S — SP FA 12wks — 0.14:1 — 85.8 lbs/100 sq.ft

Onion *Allium spp* — O — SP FA 15wks — ND — 101.4 lbs/100 sq.ft

Apple *Malus domestica* — T — SP 3yr — 1.1:1 — 51.4 lbs/100 sq.ft

Blueberry *Vaccinium* — P — SP 3½yr — 0.075:1 — 8.8 lbs/100 sq.ft

Cherry, sour *Prunus cerasus* — T — SP 4yr — 0.5:1 — 14 lbs/100 sq.ft

Cherry, sweet *Prunus avium* — T — SP 4yr — 1.527 — 15 lbs/100 s

Spinach *Spinacea oleracea* — G — SP FA 6wks — 0.23:1 — 34.6 lbs/100 sq.ft

Squash, zucchini *Cucurbita pepo* — V — SU 8wks 25 — ND — 27 lbs/100 sq.ft

Grapes, raisin *Vitis spp* — V — SP 3yr — 0.986:1 — 45.4 lbs/100 sq.ft

Grapes, wine *Vitis spp* — V — SP 3yr — 0.174:1 — 31.6 lbs/100 sq.ft

Peach *Prunus persica* — T — SP 3½yr 10 — 0.38:1 — 53.4 lbs/100 sq.ft

Pear *Pyrus communis* — T — SP 4yr 75 — 0.51 — 66.6 lbs/100 s

Tomato *Lycopersicon esculentum* — F — SU 10wks 17 — 0.60:1 — 67 lbs/100 sq.ft

Turnip *Brassica napus* — R — SP FA 8wks 4 — ND — 47 lbs/100 sq.ft

Plum *Prunus domestica* — T — SP 4yr 25 — 0.765:1 — 26.7 lbs/100 sq.ft

Raspberry *Rubus idaeus* — P — SP 2yr 8 — 0.34:1 — 12.3 lbs/100 sq.ft

Strawberry *Fragaria virginiana* — P — SP 2yr 4 — 0.21:1 — 102.4 lbs/100 sq.ft

Watermelon *Citrullus edulis* — V — SU 12wks 13 — 0.07 — 58.7 lbs/100 s

Chicken *G. gallus domesticus* — P — SP SU 21 days 5 mo. — 0.063:1 — 0.5 L/day — 0.3-0.8 lbs/day

Duck *Anatinae anas* — W — SP SU 28 days 12 mo. — ND — 0.5 L/day — 0.4-0.5 lbs/day

Lamb *Ovis aries* — Rs — n/a n/a 10 wks. — 0.005:1 — 2.0 L/day — 3.5 lbs/day

Pigs *Sus scrofa domestica* — S — FA 113 days 5 mo. — 0.015:1 — 11-26 L/day — 8 lbs/day

Turkey *Meleagris gallopavo* — P — WI 28 days 10 mo. — ND — 0.6-1.2 L/day — 1.2 lbs/day

Veal *Bos taurus* — RL — n/a n/a 3-14 wks. — ND — 34-56 L/day — 3.7 lbs/day

FRUITS, VEGETABLES, & FIELD CROPS

Common Name / Botanical Name — Class: C

- Class
- Seeding Season
- Time to Maturity
- Harvest Period
- Water requirements: low, medium, high
- X.xx:1 — kcal output/kcal input
- Companion plants
- XX.x lbs/100 sq.ft — Conventional yield

Class Notations

Fruits & Vegetables
- C - Cole/Cabbage
- F - Fleshy-fruited
- G - Greens
- O - Onion
- P - Perennial
- S - Salad
- V - Vine
- L - Legume
- T - Tree
- R - Root
- M - Miscellaneous

Field Crops
- Ce - Cereal
- F - Forage
- L - Legume
- Gr - Grain
- O - Other

Livestock
- RL - Ruminant, la
- RL - Ruminant, sr
- P - Poultry
- W - Waterfowl
- S - Swine
- A - Alternative

LIVESTOCK

Common Name / Scientific Name — Class: C

- Class
- Breeding Season
- Gestation Period
- Time to Maturity
- Secondary Products
- Land requirements: low, medium, high
- X.xx:1 — kcal output/kcal input
- X.x L/day — Water Requirements (Avg.)
- XX.x lbs/day — Feed Requirements (Avg.)

Letter Codes

Seasons
- SP - Spring
- SU - Summer
- FA - Fall
- WI - Winter

Other
- ND - Not Determined

project :: farm

thenorthroom

Project :: Farm is an ongoing investigation into the procedural and cultural economies of farming the high plains of West Texas. The Texas Panhandle is a region whose settlement pattern, infrastructure, and industry reflect the influence of large-scale agriculture. Ranging from tactical maneuvers to playful experiments, Project :: Farm is as much a study of weeds, drought, and dust storms as it is of crop yields and efficiency. Embedded in the everyday practices of farming, these projects exploit the temporalities of cultivation, the contingencies of weather, and the astonishing beauty of productive landscapes.

In addition to operating as a demonstration and production arena for organic agriculture, the 250-acre farmstead is site of thenorthroom and its investigations.

Contributors to Project :: Farm include: Jesse Vogler, Szu-Han Ho, John Houser, Louis Schalk, Emily Vogler, Tina Larkin.

thenorthroom is a research and production platform that explores the material practices of institutions and landscapes through collaborative, project-based work.

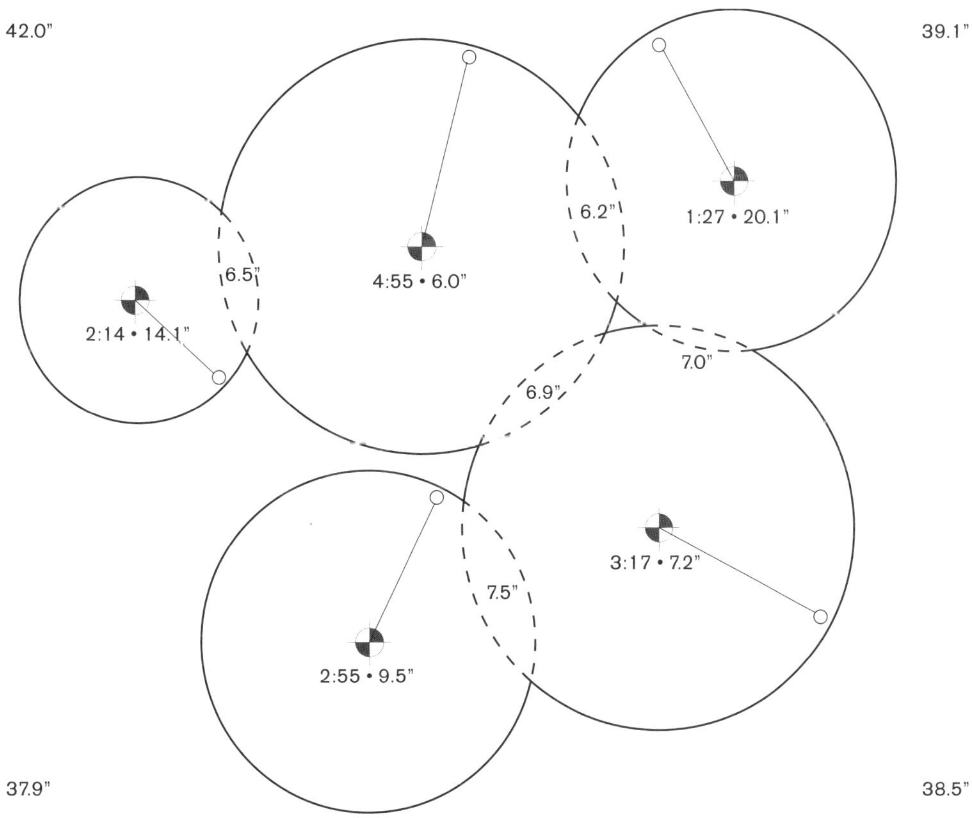

^ Goat Circles

Five goats staked in position graze on weeds in overlapping areas, creating circular markings of various heights based on duration and digestion.

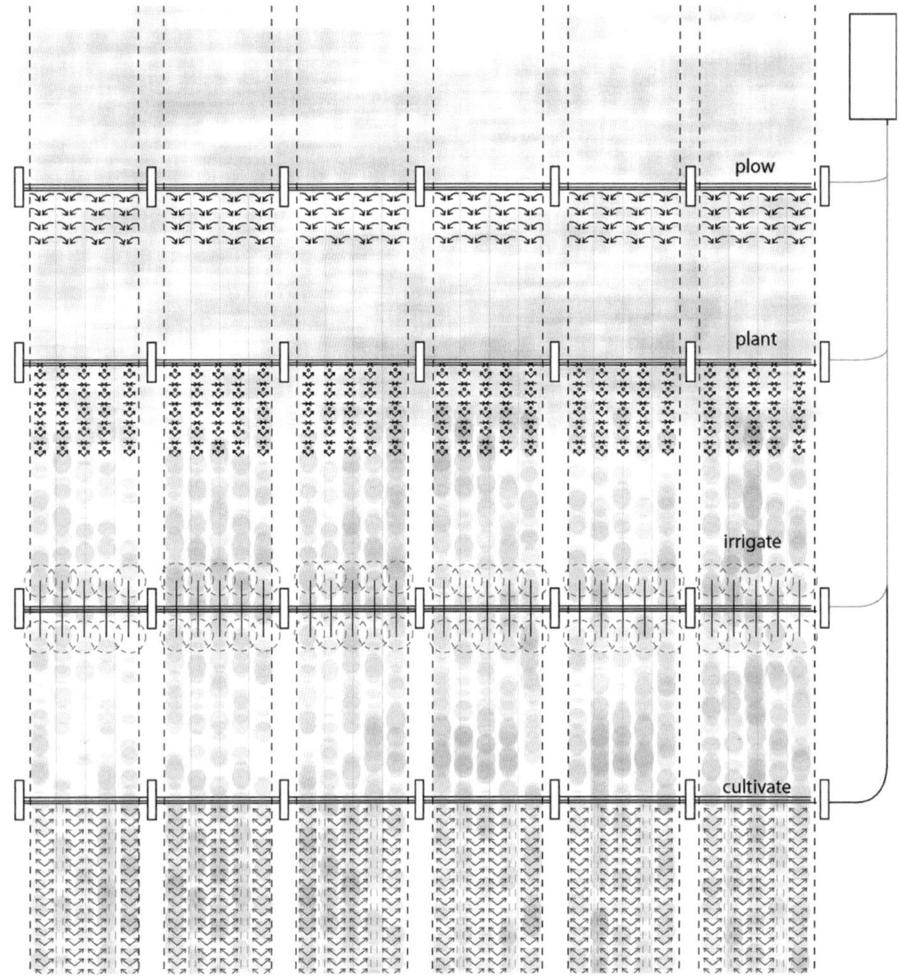

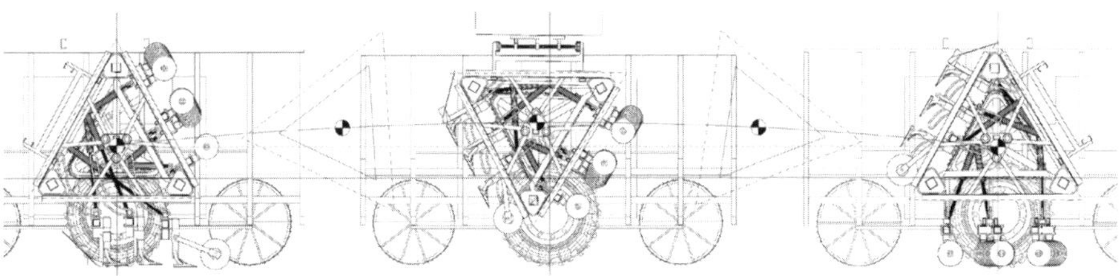

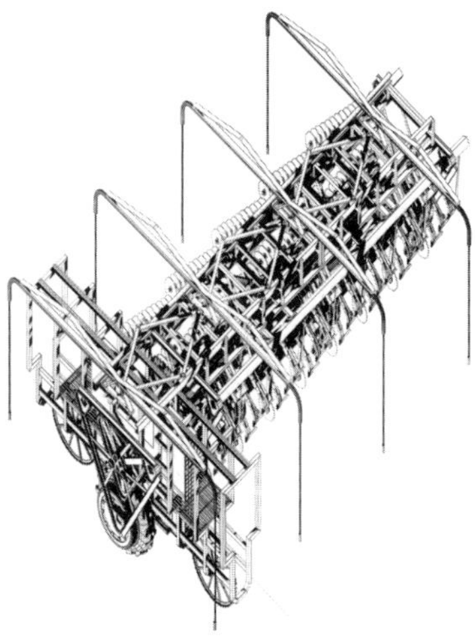

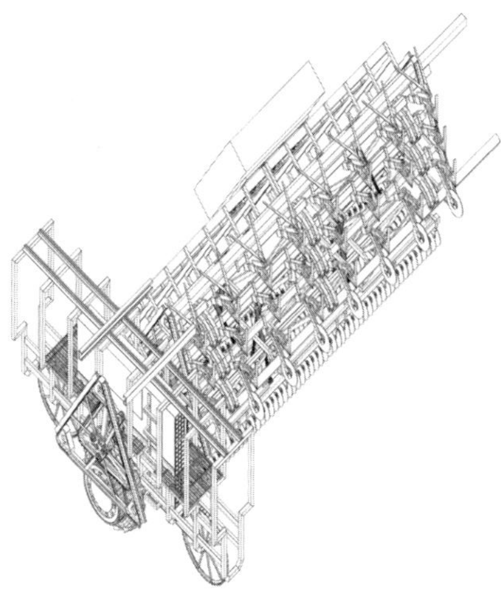

^ Farm Scanner

A proposed all-in-one system for plowing, planting, irrigating, and cultivating enacts a diagram for total agricultural implementation: the center pivot meets the tractor.

thenorthroom

Project :: Farm

^ **Implements**
Livestock and John Deere 1790 split row 16/31 planter (as seen from the soil).

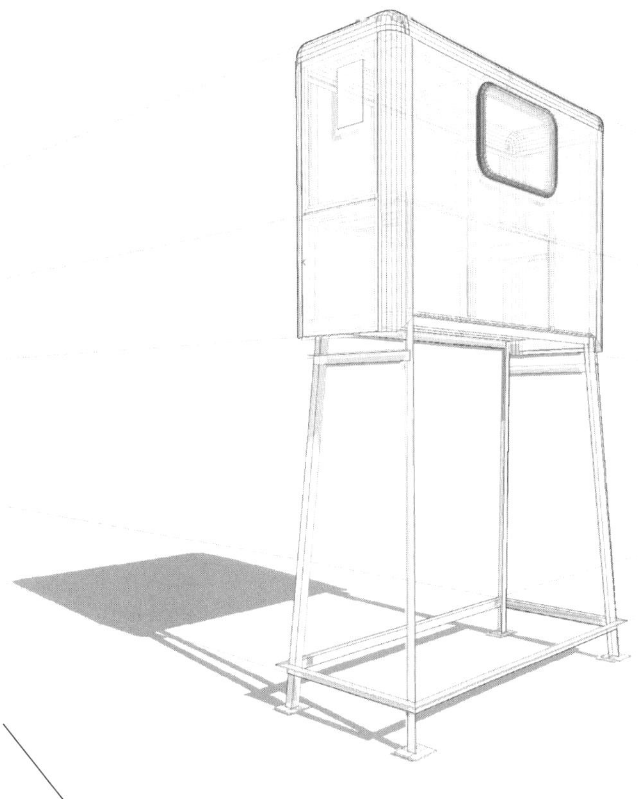

^ **Shepherd's Quarters**
A one-person shelter provides vantage and all-weather accommodation for tending livestock and landscape.

loom studio

Site Context

Shepherd's Way Farm is located on a windswept plateau of farmland in the glaciated landscape of the Cannon River watershed, flowing to the Mississippi River. The property is an eighty-acre plot surrounded by farm fields, sloping fifty feet down from the south road to the north fence. The site includes a forty-acre field for crops, thirty-seven acres of woods and sheep pasture, and three acres devoted to buildings and gravel cover. Agricultural buildings are near the south dirt road and a farmhouse sits further north in a grove of oak trees.

On the night of January 24, 2005 arsonists set fire to the former sheep barn, killing 273 ewes and 190 lambs. This senseless act could have destroyed the farming family, yet they persevered, assisted by literally thousands of people who came to their aid from around the country.

^ Farm logic.

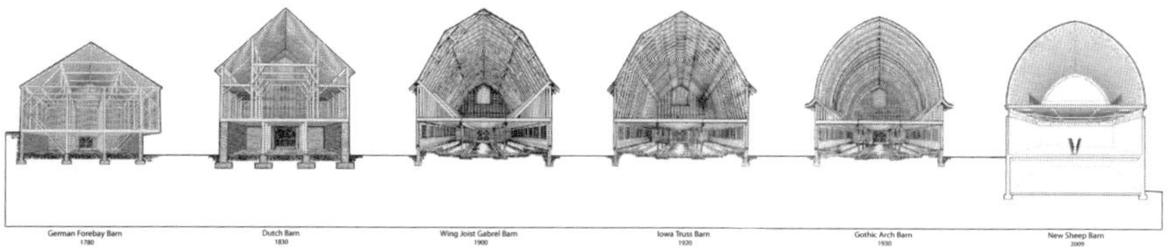

^ Pragmatic evolution.

Farm Logic

As we began design of the new sheep barn, we asked ourselves "What would a farmer do?" in order to filter all decisions made from a purely architectural perspective. Farm logic is characterized by sober creativity, directness, and integrity. In response to pragmatic necessity, small opportunities for maximum architectural effect were embraced and elaborated upon. Farm logic meets architectural invention to create something simple and durable, wholesome and remarkable. A desire to minimize disturbance guided the research process to identify what works well on the farm. At the scale of the larger landscape, the land is fertile with clean air, water, and pastures of native plants. At the scale of buildings and morphology, the farm vernacular has been refined for generations and retains several virtues in this climate and pragmatic context. No agricultural paradigms were harmed or destroyed in this process.

Client Ethic

The clients structure their life around an ethic based on "companionship." This is centered on a belief that people, plants, animals and rocks are all in the same place working together and inter-dependent. For example, the sheep depend directly on the plants, water, topography, and people for their life. The client's long-term desire is that five families will live together on the site in direct relationship to each other and sheep. The clients practice a life filled with profoundly hard work and an ethical ideal. They moved to this land in 1999 and began with twelve sheep and four young sons. They began cheese-making in 2002 and have transformed their farmstead into a local and regional institution, winning national honors for their cheese. They were early members of the Slow Food movement and have practiced sustainable farming and animal husbandry. They have a vision to create a social and educational institution within their community and hope to have their sons and friends carry the institution into the future.

Renew Existing

The farm master plan proposes a new industrial model that renews both the family farm and current corporate models, creating a positive hybrid. Industrial values of earnest effort, constant diligence, and productive enterprise directed the design. The sheep barn design is based on an existing barn roof (moved to the site from a nearby abandoned farmstead) that is transformed through an integrated renewal process. Cast-off building materials from two dismantled buildings on site are utilized, as well as new ones for the new construction. As more farmsteads in the community are abandoned, their dismantling, re-use, and renewal reveals architectural opportunities beyond mere cost savings and moral gratification.

Immediate Need: Sheep Barn

The clients articulated immediate need of a new sheep barn to replace the old sheep barn that burned to the ground. The new barn is sited near an existing barn that currently houses a milking parlor, cheese-making rooms, and cheese caves. The clients conduct public tours of their operation and requested that the site design provide clear entry and a legible tour route.

A new entry courtyard defined by the existing grain and feed silos provides transition from the parking area to a reception room and on to the tour route. Public tours enter the cheese production rooms first, then the milking parlor, followed by the sheep barn. Following a tour of the cheese production and milking operation, visitors can enter the new lamb courtyard to pet the young sheep, then move into the new sheep barn.

The new barn requires open space on the main level to shelter 500 sheep, including an area devoted to lambing. The clients requested a column-free space for ease of maintenance and daylighting for the main level with even control of brightness. The new barn includes an open hay loft above (defined by a reused barn roof moved from an abandoned farmstead) that will accommodate social events such as a dance or wedding, and a lower level for housing machinery and farm vehicles. A new corral on the north side of the barns allows controlled herding of sheep to and from the milking parlor and is animated by watering troughs lining the north face of the barn.

Pragmatic Evolution

Farm logic has dictated barn design for centuries, leading to creative and subtle evolutionary changes driven by pragmatic necessity as well as the desire to achieve a maximum effect through minimum means. While rarely inventing new technology, barn design embraces current technology and pares it down to its minimum. The new sheep barn continues this lineage, utilizing minimum material and energy to provide column-free space, a performative enclosure, and functional apertures for daylight and ventilation.

To support the reused barn roof, new steel Queenpost trusses weave through the existing floor beams. The new trusses use a minimum of steel for tension and compression members. The members are fabricated off-site and assembled on-site around each beam. The shepherd will assemble them himself with the help of his sons. To guide and direct rainwater from the barn roof, a series of water passages are incorporated. Precipitation falling on the translucent corrugated roof panels direct water to deep eave gutters leading to downspouts that empty into cisterns and watering troughs for the sheep.

Small gaps of daylight pass through chinks in the old wood-slat roof sheathing to animate the hayloft/event hall. Continuous clerestory glazing around the barn allows daylight to softly reflect off the rain gutter and truss sheathing down to the sheep. Modified aircraft-hanger bi-fold doors allow the ends of the hay loft to open to light, air, and access for the hay conveyor while keeping out rain. Sliding barn doors at grade glide open for sheep and machinery without being affected by high winds.

Future Desire: Five-Family Farm

The client's long-term desire is that five families will live together on the 80-acre site in direct relationship to each other and sheep. Sheep are the core around which five families will be linked. The design is calibrated around the maximum number of sheep and people who can live together comfortably on the site and sustain it.

The typical family farm in America is transforming rapidly, challenged by the dominance of domestic industrial farming and international free trade. This project finds "cottage industry" as a sustainable position, striving to maximize profit from the land and market while maintaining the social virtues of close neighbors and small community.

As single-family farms are abandoned, thousands of barns lay dormant in America. Around Nerstrand, Minnesota only half of the farmsteads are currently occupied and over 20 unused barns in excellent condition sit within a five-mile radius of town. As Shepherd's Way Farm grows to include five families, we propose moving more barn roofs to the farm to be transformed and repurposed. A single 150-foot long roof can be segmented into five studios, each dedicated

^ April 24, 2010, just before the storm.

^ June 10, 2010, hayloft cleared for upcoming wedding.

^ May 1, 2010, just after Spring barn cleaning.

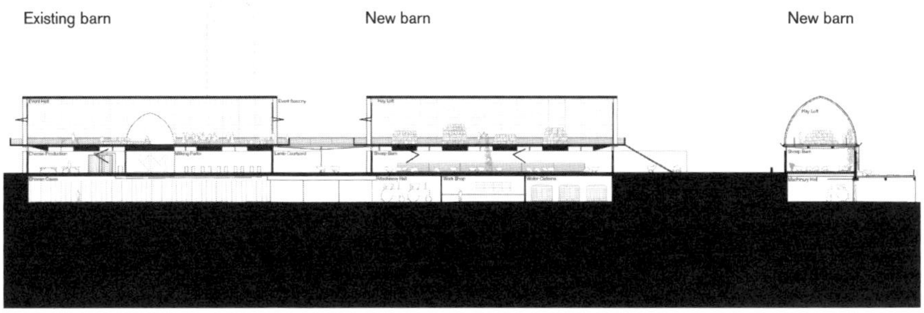

^ Old and new barn sections.

to the five disciplines working on the site. The studios will act as retail outlets and a hybrid between agriculture building, artisan studio, and strip mall.

Wind

The farmstead is located on a high plateau, in a region where the average wind speed is 12 miles per hour. The state of Minnesota, along with the Federal Government offers generous grants to develop wind-generation electricity in this region. Taking strong, steady wind as a virtue, the design responds to this phenomenal fact at the scale of the region, farm, and building.

Joining with an existing regional network and master plan for wind-generation infrastructure, the twenty-two vertical wind turbines proposed for the western edge of the site will serve the needs of the farm and supply electricity back to the grid. Each wind tree is composed of three to five caged plane rotors with soft mesh surrounds to prevent bird injury. They will be installed over a period of five years.

Following the farmstead tradition, windbreaks of existing and new tree groves are orientated to buffer the harsh northwest winter winds. Summer breezes direct the malodorous perfume from barns and manure piles away from the houses and studios based on appropriate building placement and orientation.

Sliding doors at the main level of barns, sheds, and studios allow easy opening in strong breezes. Bi-fold doors for the hay loft and upper studios allow each upper story to be vented naturally in warm weather and sealed tight in the winter. Courtyards between buildings buffer the end walls and provide transitional spaces to the broader landscape.

Water

Three watershed zones on the property drain to a seasonal creek on the north side of the site. The creek joins a stream which flows two miles to the Cannon River before reaching the Mississippi. To assure clean water enters the creek, all dirty site water goes through several filtering stages.

Located just north of the sheep barn, the manure pile is surrounded by vegetative riparian strips on the downside slope, where selected plants filter the runoff. Manure is spread seasonally on the grain field to the east.

Extending downhill from the barns to the north creek, the sheep run is also capped by vegetative riparian strips, forming stepped lagoons where selected plants filter the water. Extending along parallel lines of topography in easterly and westerly

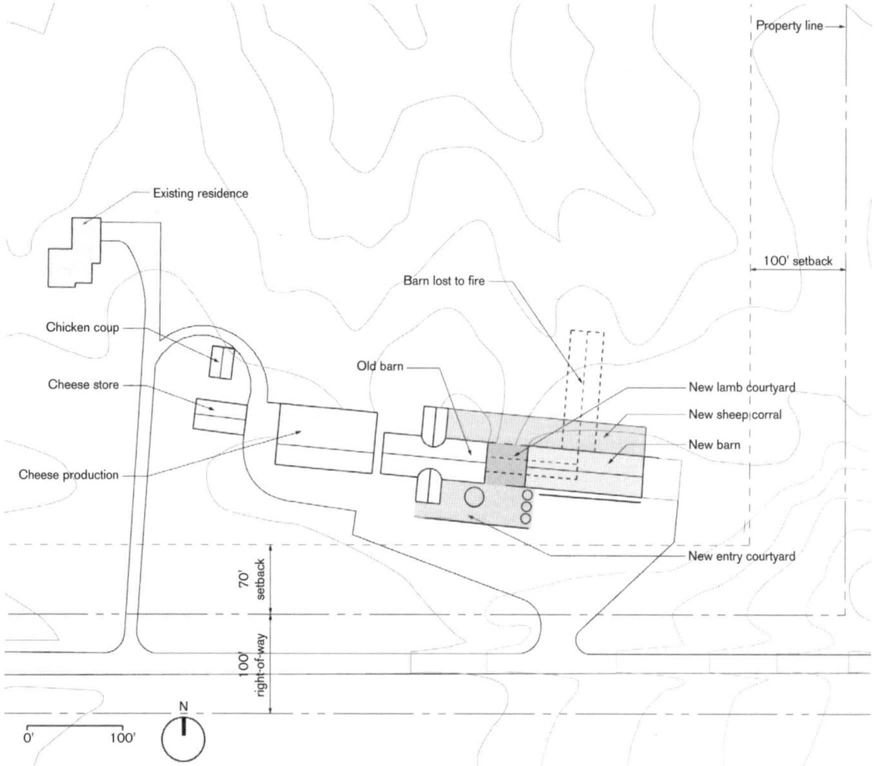

^ **Site plan showing existing and new constructions.**

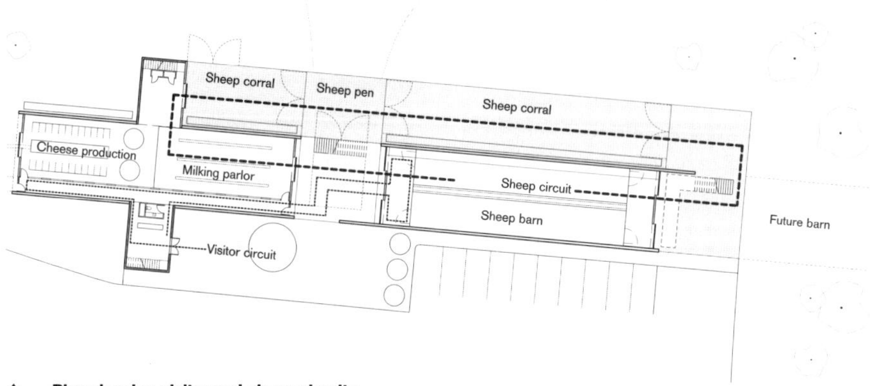

^ **Plan showing visitor and sheep circuits.**

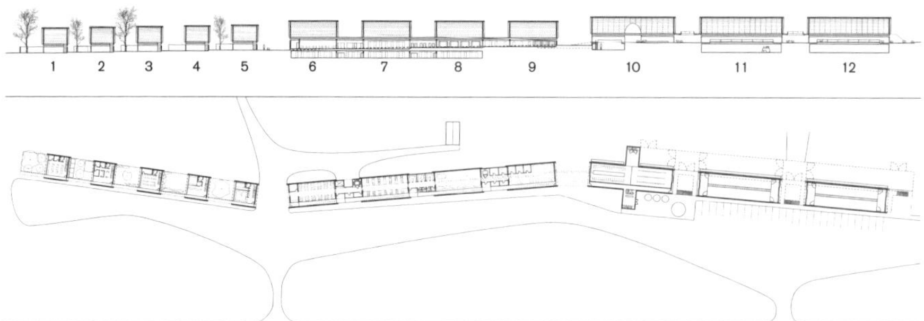

1. Weaver's studio and shearing court
2. Veterinarian's clinic and animal court
3. Cheese shop and tasting court
4. Butcher's shop and slaughter yard
5. Farm office and flower garden
6. Raw milk cheese aging barn—cheese caves below, loft above
7. Cheese packing barn—cheese caves below, loft above
8. Cheese processing barn—loft above
9. Milk pasteurization barn—hay loft above
10. Milking and feeding barn—hay loft above
11. Lambing barn and nursery—hay loft above, workshop below
12. Sheep barn—hay loft above, tractor garage below

^ **Five-family farm, plan and section.**

directions, the sheep paths connect the sheep pastures to the sheep run. Each path is enclosed by low earth berms lined with hedges to retain water and soil runoff. Driving surfacs drain to low ditches lined with riparian vegetation.

Sheep Pastures

There are fourteen pastures in four sizes, each enclosed with low earth berms, hedges with fenced gates, and trees. The largest pastures to the east allow sheep to graze on grain stubble. The five smallest pastures to the west are adjacent to each family house; the sheep mow the fescue grass for a well-trimmed lawn.

Farm Logic was designed by Loom Studio and assisted by Tamer Azzazi, Dan Clark, Krista Toperzer and Federico Garcia-Lammers.

Ralph Nelson is founding principal of Loom Studio and Adjunct Associate Professor at the University of Minnesota. Don Vu is a partner at Loom Studio.

^ January 24, 2020.

^ September 22, 2010, early morning frost on riparian lagoons.

^ Moss-covered ripple block.

^ Water-channeling tornado block.

ecologically emergent leisure landscapes [eells]

jacques abelman & bureau venhuizen

One of the remarkable characteristics of The Netherlands, especially from a foreigner's perspective, is the amount of protected open green space surrounding densely populated urban centers. The Dutch are excited to have access to verdant fields with placidly grazing cows and sheep always a bike ride away from the city, and this is actually true for most Dutch cites. Urban sprawl is kept to a minimum. However, as land becomes ever more precious, the preserved status of these green zones is called into question. In many cases these peri-urban areas are carefully managed by several partners in order to preserve their rural appearance, yet they no longer function as agricultural spaces for a variety of reasons. In some areas, soil has been too contaminated by dioxins, PCBs, and other pollutants to support food production. In other areas it is no longer economically viable. An enormous amount of energy and coordination is necessary for the maintenance of these lands that appear to be agricultural but are in fact park landscape reminding inhabitants of their agricultural origins. As urban populations increase and diversify what future role could these once vital farmlands play?

The Krabbeplas project deals specifically with this challenge on a site in Vlaardingen, near Rotterdam. The municipality, accompanied by SKOR (Foundation for Art and Public Space in The Netherlands) and Bureau Venhuizen (cultural project management) set out to investigate if these green zones could be reimagined and put back to work. The task of the designers was to investigate the meaningful re-purposing of place and the exploration of new forms of recreation. Experiments were carried out by selected designers and citizens on designated parcels of land and the results recorded.

The Ecologically Emergent Leisure Landscapes' departure point is the materiality of agricultural processes as a catalyst for recreation. The forms emerge from the potential contained in the ecological processes found in situ. Straw and hay are freely available in this context as a basic product of the landscape. Gathered and stuffed into long tubes of pliable bio-plastic, they form an organic matrix offering a new formal vocabulary for building. Wrapped, interwoven, folded, and piled, the EELLs create spontaneous outdoor shelter, which invite visitors to the fields, into the softness and refuge of an unexpected lounging spot. The golden straw within the slick, yet soft, bioplastic creates a sensual

^ **Making tubes: Harvest of straw from site and making of flexible building system in biodegradable plastic tubing.**

^ **Tubelounge: Lightweight tubing system generates on-site shelter.**

combination. The form of the EELL lounges can be endlessly recombined to suit specific needs.

Well-suited for events in the fields, the EELLs also have another purpose. Once their use as outdoor furniture is complete, they can be left on site to begin another process. The straw-filled tubes are soaked in water and inoculated with mushroom spores. Over the course of several weeks, the mushroom spores spread through the straw while the bioplastics break down, bringing the growing fungi into contact with the earth. The fungi are then able to deeply colonize the soil of the site. Studies have shown that the enzymes present in fungal mycelial networks break down complex molecules such as dioxin and PCB, metabolizing them into harmless substances. This form of bioremediation is called mycoremediation. The mycelial net, which can expand to cover the entire ground of a forest in some species, does the work of purifying the polluted ground. The fruiting bodies it creates, which we call mushrooms, remain safe to eat.

The EELLs project seeks to address new ways of enjoying agricultural space, actively connecting users to ecological cycles and bioremediation. From hay harvest to lounging and through to mushroom production and soil purification, pleasure and utility are combined in a new leisure landscape.

Designer: Jacques Abelman; Project Initiator: SKOR (Stichting Kunst en Openbare Ruimte); Concept Management Office and Commissioner: Bureau Venhuizen, Hans Venhuizen, Francien van Westrenen.

>>> Jacques Abelman researches designing with ecological systems and works in landscape architecture. He attended Amherst College and is currently at the Academy of Architecture in Amsterdam, The Netherlands.

>>> Bureau Venhuizen is an office for cultural planning in Rotterdam, The Netherlands. Led by Hans Venhuizen, the bureau's methodology is based on the re-use of existing landscape and cultural qualities in spatial planning.

^ Party: Lounging participants and gourmet mushroom harvest party.

^ Mushroom: Inoculation of tubes, breakdown of biodegradable plastic, composting and on-site mushroom production.

^ **Tubelounge: Once their use as outdoor furniture is complete, tubes can be left on site to begin another process.**

recycling takes command

kyle reynolds

This project examines the urban farm as a byproduct within the larger ecology of waste management. Contemporary models of waste management use a conduit system—through trucking or piping—to set trash in motion, moving it far from the original point of disposal. In an attempt to re-appropriate a model of collection and percolation of waste, this project represents a new 'green logo' for the city of Chicago. A network of five facilities, shaped by the processes of waste management, are strategically distributed across the city and linked by an existing light-rail infrastructure. The shape of each facility is unique, yet together they establish a collective identity for a green Chicago. The project focuses on three processes underlying waste management, which all yield productive byproducts: composting (yields soil), recycling (yields raw materials), and waste-to-energy incineration (yields power). Each process has certain characteristics and formal requirements that promote effective translation.

Every day, Chicago generates a volume of trash equivalent to one city block filled one foot high. In order to process all of this trash the city would need to create a distributed network of five equal facilities. This project proposes the main facility to be located on the formerly razed Block 37 in the downtown Loop area, which is a prime location, both symbolically and functionally. Block 37 has access to Lower Wacker Drive, a service drive used for the delivery of goods and removal of waste for towers in the loop area. This allows all of the waste from the loop to be delivered to the Block 37 site underground without disrupting traffic on the streets above. Once delivered, trash is sorted into the three essential categories, which correspond to the three waste management processes. Compostable waste is sent to the top of the building where it is composted, cured, and used in agriculture. Recyclable waste is sent to the middle of the building where it is resorted by type and distributed by gravity into large bins on the outer face of the building. Each subsequent floor processes one material type. Any waste that passes through those two processes is sent to the waste-to-energy incinerator where it is vaporized to produce ethanol. The byproducts (soil, recycled materials, ethanol) are sent back down below the Lower Wacker level where they are picked up nightly by the 'trash train,' which uses existing red- and blue-line light-rail infrastructure to redistribute trash and byproducts to any of the four other facilities.

The physical characteristics of this facility align it with a family of similar projects that fit in the overarching aesthetic category of 'shape' projects. Shape projects distinguish themselves from formal projects (as in the work of Peter Eisenman) in that formal projects attempt to create difference out of sameness whereas this facility creates sameness out of difference. The project is further distinguished from mass projects (as in the work of Frank Gehry) in that mass projects are expressionistic and excessive whereas this facility is calculated and expedient. Each of the waste management facilities read as singular objects, yet their final shape is not indexical of their processes of creation. The project methodology employs a series of operations broken into three categories used to create shapes: generating a primitive, operations on a primitive,

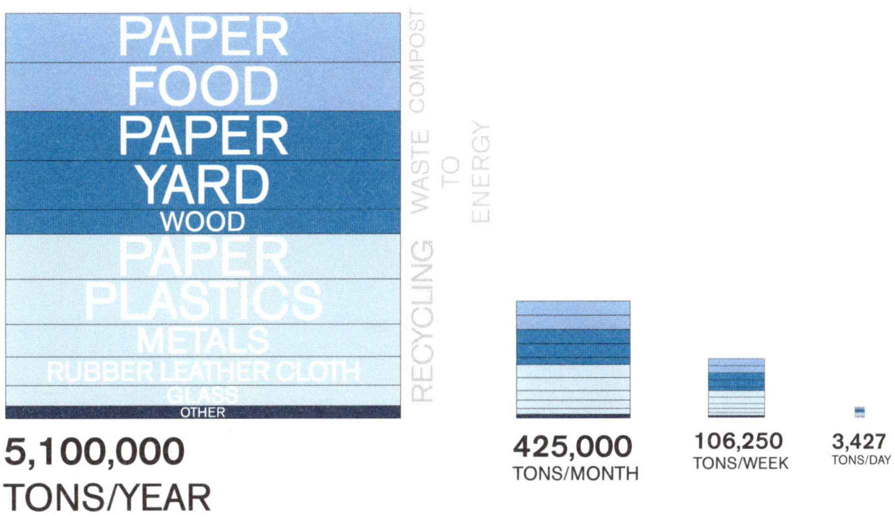

^ **Chicago produces 3,427 tons of trash or roughly the equivalent of one city block filled 1'-0" high every day.**

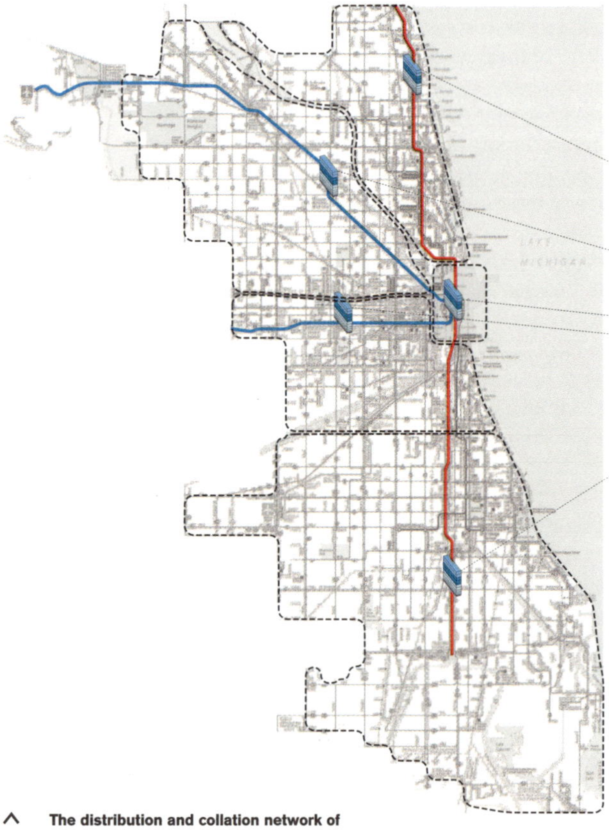

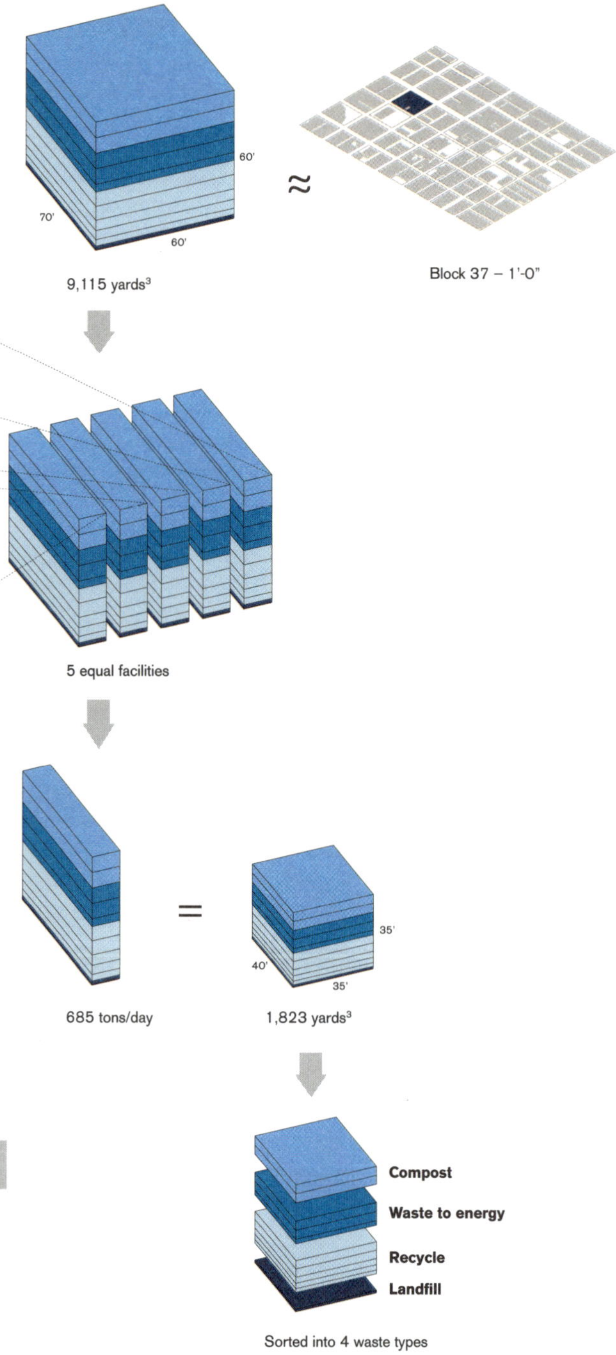

^ **The distribution and collation network of waste is sorted into compost, waste-to-energy, recycling, and landfill.**

Compost

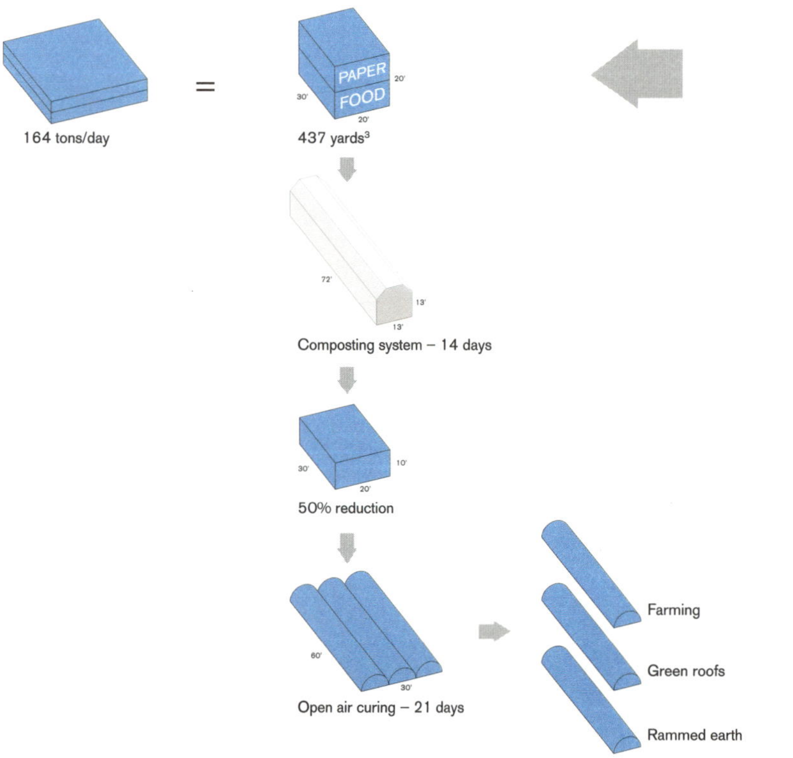

and accumulations or collections of primitives. These three shape operations are completely separate from the functions or processes they contain yet maintain an association to the programmatic requirements. The overall facility shapes may be constructed in innumerable ways through purely operational terms. This eliminates an indexical relationship to process, ultimately defining the intrinsic qualities of a 'shape' project.

Not only are the new waste management facilities singular collectors of different shapes, different processes, and different types of trash, but they are also collectors of different identities within the city. The network of buildings is at once public art, institution, infrastructure, architectural monument, public park, green building, and urban farm. Through accommodating these different roles the network of waste management facilities establishes a new collective identity and infrastructural logo for a green Chicago.

>>> **Kyle Reynolds is a Chicago-based designer, urbanist, educator, and daydreamer.**

^ This facility sits in a commercial district and has the largest farm and open-air curing surface.

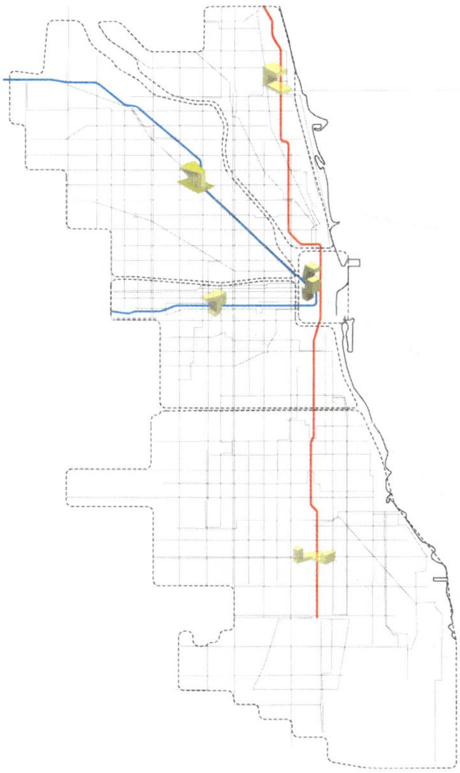

^ Five waste management facilities serve the city of Chicago. While distinct in their character, collectively they form the new green logo for the city of Chicago.

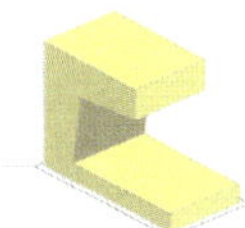

Facility 1
Commercial district

Facility 2
Residential district

Facility 3
Downtown loop

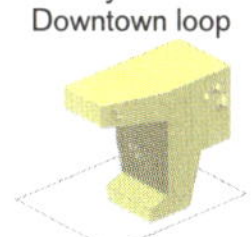

Facility 4
Housing projects

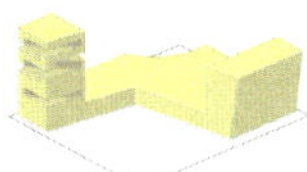

Facility 5
Brownfield

^ This facility looms over a residential district and has three waste-to-energy incinerators.

^ This facility maintains a small footprint near housing projects.

^ Atypically horizontal, this facility occupies a brownfield site.

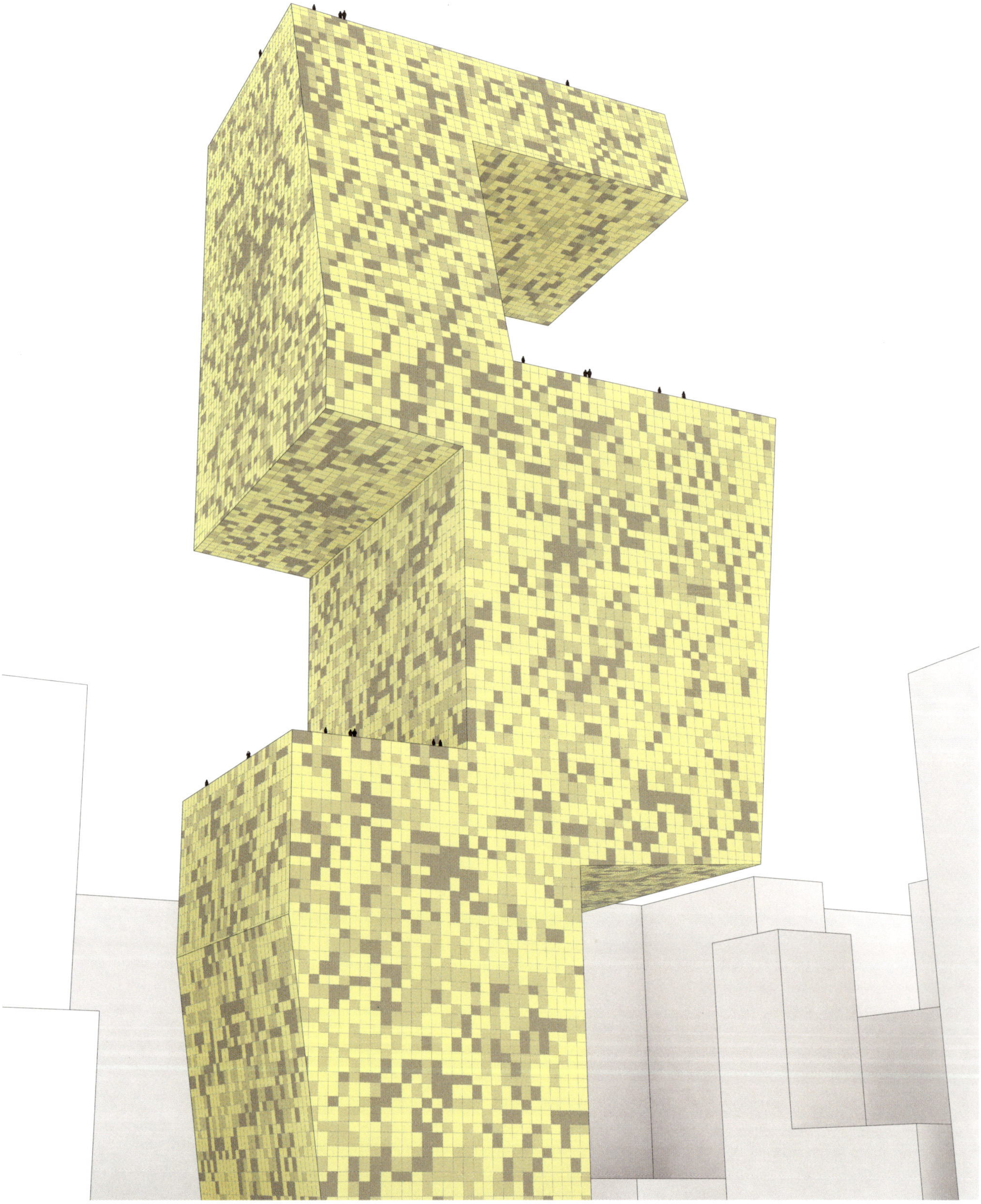

^ **Bordered by Lower Wacker, trash is delivered to Facility 3 on the lower level where it is sorted into four types of waste.**

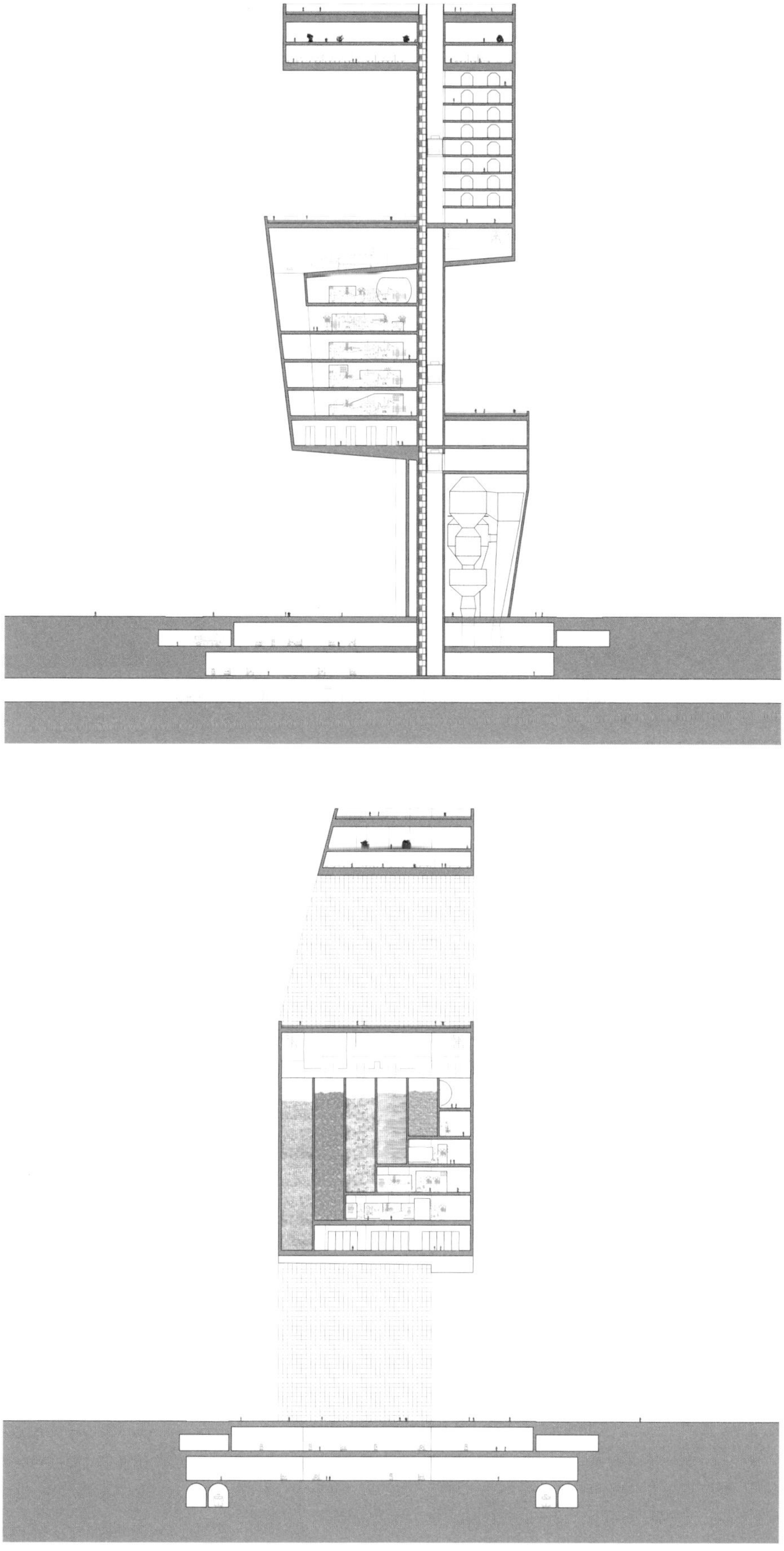

∧ **Sections through Facility 3: Downtown loop showing composting, recycling, and waste-to-energy operations.**

fructus vegetabilis: growing profit in the war on error

jennifer w. leung

The 2006 Crops of the Americas stamp series depicts a cornucopia of imagined pure native vegetables[1]—fruits of a pre-industrial landscape in "Five Different Designs." This Edenic fiction resurrects the rhetoric of bounty that helped legitimize the founding of the nation. A self-adhesive image of produce, the stamp bears comparison to the advisory label affixed to produce. The privilege of circulation purchased vis-à-vis the stamp is impure, subject to the same police technologies as the fruit body—x-ray, irradiation, and customs. At a larger scale, the familiar USDA food pyramid circumscribes desire and proscribes ethical consumption based in nutritional science through posters in schools and medical offices, as a PDF on the internet, and in occasional media coverage. Harmonized bodies and diets effect a national subjectivity, wherein caloric and nutrient values extend beyond shelf life into the arena of political capital. The security of the national food supply is transparent to the American way of life.

Michel Serres's anecdote regarding movement implies an overturning of event and place: assume that the ball leads the game; teams arrange themselves accordingly. By analogy, contested territory is sited in the vegetal body, rather than in the land in which it is grown. This abstraction is both conceptual and literal, bringing into relief questions of legibility and traceability. The movements of fruits and vegetable, their exchange in fetishized, denatured, mediated, and whole forms, privileges realms of classification, tabulation, monitoring and storage, over spaces of production.

The domestic market demands the image of pure fruits and vegetables, unblemished skins that glow with the promise of health. The market also demands quality control, legal circulation, and added value through brand recognition. The site of accord is the produce label, a two-dimensional space of legislation. The 2002 Farm Security and Rural Investment Act required that all produce bear

^ Produce labeled with Natural Light Labeling System®. Courtesy of Durand-Wayland, Inc.

COOL (country of origin) labels by the fall of 2006. While the PLU is specific to the point-of-sale, and the trademark logo ensures economic gains to a larger network of actors, COOL invokes food and territorial insecurity rather than economic efficacy.[2,3] Consumer choice—as between produce "Made in the USA," of foreign provenance, or of mixed origin, is reaffirmed as a right and rite of democratic society. This additional layer of traceback information is minimally transparent to the global circulation of commodities, superficially certifying safety as legibility, in so far as various nations are imagined as variously secure. The skin of fruits and vegetables is increasingly a surface of registration, yet also persists as a physical barrier to more specific forms of information.

"Track and trace" natural light labeling and laser branding lay bare the expanse of vegetal skin as a palimpsest for inscription. Replacing adhesive labels blown onto produce in the tumble through automated sorting machines, these technologies instrumentalize specific wavelengths of light, treating the

^ "Crops of the Americas" stamp series, 2006. Courtesy of United States Postal Service.

surface as venerately as the tradition of still life oil painting. This labeling is subtractive, rather than additive, implying the pixelation of vegetal skin and the selective removal of pixels with bursts of micrometer light. Future industrial use tends towards the saturation of surface with information—registering country of origin, quality control

1. Steve Buchanan created the designs from research in the late 1970s on indigenous agricultural methods in the Southwest. The crops—corn, chili peppers, beans, squashes, and sunflowers—had been cultivated in the Americas for centuries when Europeans first arrived in the New World. www.usps.com

2. The PLU identifies produce by a four or five digit code—the first identifying conventional produce grown with chemical inputs, the latter adding a 9 to indicate organic or an 8 to indicate GMO.

3. The trademark is the earliest and least specific means of asserting property rights over natural products, followed by the Plant Variety Protection Act certificate in 1930, legitimized by technologies guaranteeing statistical purity in asexually propagated plants; and the naturalized industrial patent, adjudicated through a series of Supreme Court cases from 1980-1987.

testing, and metrics, while offering a new medium for forms of graphic salesmanship. All this on an edible skin, consumed at the domestic table: "A pear is just a pear, except when it is also a laser-coded information delivery system with advanced security clearance."[4]

Techniques of the Electro-Magnetic Spectrum

John Maeda's "food" still life series (2002) is created from scanned produce, reconfigured, multiplied, and aestheticized as an abstracted color palette of informationally equivalent pixels. The volumetric mass of fruits and vegetables is transfigured through photic technologies that flatten and fetishize the surface. Olivier Babin's "Art for the Very Last People" (2005) and "Towards Infinite Freshness" (2005) serve as counterpoint. Using traditional artistic media, Babin explores the opacity of skin and the fictions attached to these liminal surfaces. Five bronze watermelons juxtapose image to materiality, the organic to the inorganic. A sectional painting demarcates the secret fruit body hidden within concentric frames of rind and skin—invisible to the eye but for the destruction of the integral object. These works bears comparison to an older USDA poster extolling Americans to "Catch a Rainbow Everyday," in which a dietetic flora of color coded fruits and vegetables is arranged in imitation of a rainbow, a natural phenomenon whose colors delimit the extent of the visible light spectrum. Produce is already a simulacrum of nature, a collage of ontological fruit and vegetable units. Maeda suggests that the image is further reducible to information, primed for manipulation through a series of technological interfaces, a position shared by natural light labeling.

The primacy of *natural light* is unparalleled in its catalysis of plant and fruit maturation. This same light effects optical perspective, heats greenhouses, causes leaf pigments to photosynthesize ATP, and reflects pigmentation as color. The extension of light technologies into the electromagnetic spectrum, in addition to surface enrichment, overcomes the aporia of vision in its inability to gain increasingly specific knowledge, monitors circulation, and creates new subjectivities with respect to cultural conditions of mediation. Electromagnetic technologies effect a virtual delamination, less violent than Babin's representational cutting, through

^ **John Maeda: *GreenZucchini, BluePurpleTopTurnip*, and *RedBeet* from the *Smoothies* series (right, top to bottom), 2002, cibachrome prints; *Traffic*, 2002 / cibachrome print (left). Courtesy of Cristinerose / Josee Bienvenu Gallery.**

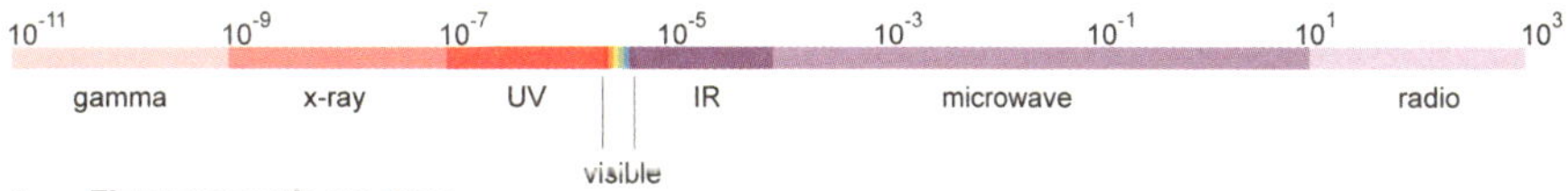

^ **Electromagnetic spectrum.**

non-destructive quality testing. X-rays scan the fruit body, the seed, and bulk shipments of the former, imaging at once vegetal injury and disease, seed viability, and illegal circulation—including human smuggling. In other words: lifespan, the possibility of life, and licit and illicit forms of life—reproducing almost exactly Foucault's notion of biopower. Light that governs temporal cycles of planting, gestation, and harvesting, competes with 24 hour, year round electromagnetic surveillance concerned with multiple standards of normal and pathological.

Techniques of Classification and Circulation

Circulation is predetermined by classification. The corollary also holds in that resistance to circulation—natural, legislative, or otherwise, problematizes existing terms of classification. *Classify* first appears in French in 1799, from the Latin *classis*, one of six orders in the Roman tax system; as originally "the people of Rome under arms," *classis* is related to *calare*, "to call to arms," and then *calendar*. US Army usage circa 1944 refers to the degree to which a document of an intelligence nature may be circulated.

Post-harvest produce is indexed by microclimates monitored in transit—between sovereign territories, *oikonomia* and *oikologia*,[5] classification and public consumption, nature and its representation. Transit conditions reveal instrumental relationships to various epistemes: free market liberalism; life as intellectual property; security through technology—rather than by transparent producer-consumer relationships or diplomacy with one's enemies. Border ports of entry are obligatory passage points wherein the legislative space of the label meets the legislative space of the land. Systems of classification and exchange are the lingua franca of a global world, but have never been neutral representations of equivalence.

4. Moskin, July (2005) "Tattooed Fruit is on Its Way," *New York Times*, July 19, 2005.

5. Greek roots of "economy" and "ecology" as the management of household; thrift, administration, arrangement; and the study of the house, dwelling place, and habitation of God's creatures.

^ "Catch a Rainbow" poster, 2004. Courtesy of United States Department of Agriculture.

They produce nature as artifact and fact vis-à-vis scientific and capital legitimation made public by law. The opaque discourse of legislation elides the consequences of defining actionable information from within this naturalized economic taxonomy. Instead, we have depthless guarantees of purity, security, and the invocation of patriotism by diet and other means. Produce is thus an event, a milieu, and an aporia; an icon of fixity and flux.

In our contemporary era, seed banking shifts the balance of distribution and collection by collapsing desires of agribusiness, environmentalists, pharmaceutical companies, and developing countries into a singular spatial product. Storage technologies that minimize genetic drift[6] refuse the principles of evolution through replication of identical copies in perpetuity. Classification according to patent utility: industrial, medicinal, agricultural, ornamental, facilitates simultaneous bioprospecting and biodiversity protection. Seed banks initially follow preexisting forms of organization: the USDA's National Plant Germplasm System maps onto existing experimental, collection, and security sites; the Millennium Seed Bank Project continues the tradition of economic botany at the Kew Botanic Gardens;[7] then evolve speculative forms of altruism and profit margin. Witness the reduplication of the cold war bunker by subterranean storage in the Norwegian Seed Bank, low humidity, sub-zero fortification inside a sandstone mountain on a frozen Arctic island, and the further subjugation of natural products to shareholder concerns in Monsanto's 2005 billion dollar bid for Seminis, Inc., the world's largest fruit and vegetable seed company.

Further, fruit genome databases collapse Foucault's classificatory grid of difference onto classes of information—public domain, password protected, intellectual property, and classified, recapitulating the dialectic between technologic and legislative discourses of legitimation. Private firms subjugate the five kingdoms of life to the Roman alphabet in Array Ready Oligo Sets, obsequiously called Grape, Peach, Rice, or Tomato Version 1.0. The crisis of taxonomy that follows the proliferation of regimes of signification has catalyzed several projects of recuperation: the Berlin Taxonomic Information Model, ZooBank, BioCode, uBio, and Wikispecies[8] compete to normalize the standards of classification. Natural selection responds to venture capital and search algorithm, as well as competition from opaque databases of classified or patented information. Password protection is as much an obligatory point of passage as the checkpoints through which natural products travel.

Nationalism and Chlorophilia

The USDA's production forecasts drive speculation in international produce prices and guide farmer sales strategies. The National Agricultural Statistical Service compiles estimates of planting and harvesting acreage through random sampling of farms by automated telephone system, post, and limited live contact. Farm survey data is corrected for assumed conservatism, compared with on-site probabilistic sampling, and adjudicated through board consensus before encryption and "lock-up."[9] Liberation from temporary classification, at precisely 3:00 PM on the 9th day of the month,[10] is followed by the furious movement of economically sensitive information and electronic payments through communication infrastructures. Agriculture securities are secured by redundant market protections, ranging from timed release to options on futures. The insurance of risk hedging carries within itself the seeds of market disaster. As RAND reports, "an adversary could attack for financial gain, if they have invested in agricultural futures."[11]

The produce economy partakes of every infrastructure—material and immaterial; every sector—private, public, government; every natural resource—water, soil, climate, gasoline, and labor, from sea to shining sea. The integrity of this distributed system is robust, absorbing illegal circulation of commodity and labor; refusal of circulation by import, tariff, phytosanitary regulation, or biodiversity protection; and socialization or habits that localize the economy in farmer's markets and food co-ops. Into this game, war enters as a curveball. The subtle and obvious ways in which war taxes, usurps, blocks, or diverts flows, paths, and rhetoric creates a system in disequilibrium. The evolution of the system under stress, whether by creation of novel forms or by self-replication and error correction, reveals underlying ideologies.

World Wars I and II occasioned a home-grown produce effort in the face of redirected domestic resources and crippling food shortages across the Atlantic. The National War Garden Commission exhorted citizens, school children, and soldiers not yet deployed to sow Liberty and Victory Gardens. The "city farmer" would till "idle soil," recuperating "slacker lands" and lost productivity for the war effort. The Commission, founded in 1917 as an ensemble of conservationists, university administrators, governors, a prominent plant breeder, and a retired state agriculture director, was a non-legislative body associated with the American Forestry Association.[12] The mission was spread through posters, free pamphlets, newspapers, and town hall meetings, with USDA participation in the dissemination of preservative technology once produce yields exceeded demand. The return of domestic, urban, and government lands to sustenance farming re-equilibrated the national agricultural system during war, conserving labor and infrastructure, and freeing valuable meat and grain for the overseas "doughboy diet."

Rather than a distributed diagram of self-sufficiency with heterodoxical units of

6. Cryogenic technologies suspend orthodox germplasm in a pre-animate state; recalcitrant germplasm is continually replanted and harvested in patrolled, barbed wire-enclosed experimental stations.

7. The record-breaking germination of a 2000-year-old date palm seed adds to Kew Garden's success stories, including acclimatization of North American cash crops to colonial environments and construction of the 19th century's largest greenhouse.

8. Yoon, Carol Kaesuk (2005) "In the Classification Kingdom, Only the Fittest Survive," *New York Times*, 10/11/2005.

9. Understanding USDA Corn and Soybean Production Forecasts," *Agmas*, February, 2006.

10. Didier, Emmanuel (2005) "Releasing Market Statistics," *Making Things Public*, eds. Bruno Latour and Peter Weibel, Cambrdige: MIT Press, 638-641.

11. Gilmore Commission Minutes, RAND, 9/22/1999.

12. Pack, Charles Lathrop (1919) "The War Garden Victorious," Lippincott Company: Philadelphia.

localized production, food security is now defined through control of vulnerable points in an expanded "garden to table continuum." The food security projects during the world wars stirred and appeased fears of hunger. Food security after 9/11 projects resurrected isolationist anxieties onto fears of network attack from accidentally or intentionally introduced pathogens,[13] democratizing courtly presentiments of intrigue and poisoning. In many ways, the stakes are incommensurate. The world wars caused major disruptions of global circulation, whereas unilateral action creates selectively punitive no-fly, no-go, and no-receipt zones. Technologies of the 1917 *Manual of Home Storing, Pickling, Fermentation, and Salting of Vegetables* survive as artisanal crafts, while industrial cold storage, ripening, and the selection of transport genes enable the circulation, rather than preservation, of surplus. Rhetoric creates paranoia through the overdetermination of terms. Contemporary "food security" collapses the definition of *safety*, from the Latin *salvus*, as lack of injury or harm, with *security*, from *se cura*, which comes to mean, almost simultaneously, in the 16th century: free from care and insurance from risk.

The terms of food security enacted through diversion of agricultural labor and land have been replaced by redistribution of bureaucratic labor and office space. Before a historical farm population of 58%, Lincoln could declare the USDA the "people's department." With only meat, poultry, and eggs under its direct supervision, the USDA is now the nominal government of rural America—managing pests, prices, and marketing; monitoring weather and climate; conducting vulnerability assessment for the country's 788,995 farmers or farm managers (US Census 2000). It remains the department of people's bodies in its management of malnutrition and obesity; of people's natural resources in its stewardship of forests, rangelands, seed banks and phyto-databases. With a current farm population of less than 1%, the FDA, as a consumer protection agency, has acquired the more representative duties: produce labeling, customs enforcement, criminal investigations, advisories, recalls, sanctions, and supply line standards of safety and purity. With the 2002 Bioterrorism Act, the FDA now certifies all produce imports and registers all facilities that procure, process, pack or hold produce. The further reorganization of the USDA under the DHS equates safety with vertical, rather than horizontal, information systems. The ethics of suspicion are legitimized through threat management as a legislative strategy and an industry to end all industries. The garden to table division of labor is fixed by DHS-private sector councils: pre-farm, farmer, processor, transporter, warehouser, wholesaler, and retailer. The destabilized image of wholesome produce, its fraught investiture with ideals of nationhood, prompts the sublimation of extracted and abstracted natural products into the idea of homeland. The USDA has lost research facilities to the DHS, which prefers academic research in "agro-security [measures] related to post-harvest food protection;" but acquired offices of emergency preparedness and surveillance. After 9/11, the USDA remains the people's department as the interface for the National Consumer Complaint Monitoring System, a complement to its instrumental role in farm surveys—information collection, adjustment, adjudication, and temporary classification.

Evidence of security is no longer harvested, but mediated through artificial interfaces that supplement the natural interface of skin. Epistemological crises of taxonomy and classification are brought into relief, then overcome, by the promise of x-ray and RFID to effect the simple binary of safe and unsafe. Supply threats are mediated and publicized as a color change in the non-specific Homeland Security Advisory System—the integrity of the food pyramid immured in the visible light spectrum. The neological Biosurveillance Integration System and Biodefense Analysis & Countermeasures Center position the DHS between the FDA's enforcement and the USDA's public relations, surveilling both by access to and creation of classified information. The precision of statistical sampling is increased through bureaucratic redundancy—the EPA, CDC, Departments of Commerce, Transportation, and Justice, local governments, FBI, and USPS have additional monitoring duties.

The unstable Crops of the Americas are reflected overseas in the infelicitous image of occupied territories in need of a "green revolution" and through "seeds of democracy" rhetoric. Environmentally sensitized cash crops harvested from annexed or leased territories once entered a colonial maritime economy. By the end of the last presidency, export plans included introducing US food stocks, produce, and hybrid F1 sterile and GMO seeds to territories abroad as an opportunistic refusal of Mesopotamia's 8,000 year-old agricultural tradition. These plans follow from Bremer Executive Order 81: "Patent, Industrial Design, Undisclosed Information, Integrated Circuit and Plant Variety Law," in which nature, with no remaining barriers to capital penetration, circulates, or does not, in the same way as other forms of intellectual property. USAID, in a perverse nod to the 18th century USPS,[14] has "deliver[ed] thousands of tons of US-origin 'high-quality, certified [corporate rather than national] seed' for subsidized, initially near cost-free distribution."[15] In 2006, there may have been no national border in the world like Iraq's post-war border: tariff, duty, inspection, and tax free. Labels affixed to relief supplies need not conform to this absent legislative space, but qualify acts of aggression against the sovereignty of native land, ecosystems, and seed. Replacement products of nature are accompanied by spatial typologies and chemical ingredients that serve to make space generic: the experimental station is collapsed with the trade show booth in "wheat extension demonstration sites"[16] in northern Iraq; pesticides, herbicides and fungicides follow GMO seed into fertilized microclimates of precise O_2 and N concentrations, as if to sanitize the space of violence. The invasion destroyed Iraq's national seed bank at Abu Ghraib, with its 1,400 crop varieties, but the recovery of native biodiversity and food crops may yet occur via a representative 200 seed deposit at Syria's ICARDA in 1996.[17]

The imagined agricultural future is staged in military installations outside of US and Iraqi domestic jurisdiction. Rather than oppose combat diets of protein and

13. Overseas contests of sovereignty in WWII were already translated into domestic fears of foreign contaminants and microbial pests. Colomina, Beatriz (1999) "The Lawn at War: 1941-1961," *The American Lawn*, ed. Georges Teyssot, New York: Princeton Architectural Press, 134-153.

14. Since its establishment in1862, as a formalization of the collection activities of the US Navy and an evolution of the Patent Office, the USDA has relied on the USPS to circulate—via ship, rail, pony, and air, natural products and information under its purview. Through the early 20th century, this amounted to mass distribution of free seed, deposition of type plants at USDA experimental stations and land grant universities, and monthly crop surveys.

15. Engdahl, F. William (2005) "Iraq and Washington's 'Seeds of Democracy,'" *Current Concerns*, 6/5/2005.

16. Ibid.

17. Clarke, Tom (2003) "Seed Bank Raises Hope of Iraqi Crop Comeback," *Nature*, v. 424, no. 242, 7/17/2003.

^ Z Backscatter ® X-ray scan of cargo truck. Courtesy of American Science and Engineering, Inc.

non-perishables to home diets based in produce, deployment obtains to a mirror of domesticity. High priority DHL shipments of irradiated, RFID tagged produce arrive at desert airfields twice a week.[18] Class I receiving facilities and organic truck platoons modify and construct spatial precursors to the market and rehearse infrastructural solutions to produce distribution. As feasibility studies, non-combat military operations are an extension of the war game that prepares combat, by war as a game that prepares capital. Additionally, soldiers, monitored for changes in blood lipids, zinc, magnesium, and other metabolic data, serve as an experimental group for nutritional research conducted by Military Nutrition Research Task Teams. The military body is a statistical average of diverse bodies subject to identical environmental conditions. The MRDA (military recommended daily allowance) qualifies the nutrient requirements of the military body against variations in climate, duration, humidity, transit and situational stress; a body fed by produce similarly indexed. The MRDA is transparent to USDA dietary regulations, which proscribe nutritional content through metrics with non-identical standard deviation. These twin projects of the American diet differ in degree rather than in kind. The citizen-soldier diet is further collapsed in the picture-perfect Thanksgiving mess-hall dinner, an image protected and guaranteed by the Bremer wall and Executive Order, broadcast live to the domestic interior.

Unfettered access to public, private, and classified data promises actionable information of the highest specificity. Cultural impediments to free circulation are materialized in the failure of food logistics, an imagistic negative to the successful Thanksgiving dinner. As reported to Congress in 2005, KBR, under contract to the Armed Forces, systematically contravened Military Preventive Medicine sanitation rules; over billed rations for up to 5,000 phantom troops; tripled produce costs by air-lifting American produce from Philadelphia; and fed soldiers food items that were "expired as much as a year," while fresh supplies spoiled for lack of refrigeration.[19] These examples of literal and economic waste in a rhetorical wasteland, make visible what is in excess of, or escapes from, supply chain management through accidentally or intentionally introduced miscalculations. When juxtaposed to surrounding no-go zones, where order is replaced by the confusion of (violent) fact as law, the military camp oddly manifests as the inverse of Agamben's state of exception.

The agricultural project strives towards the perfection of a free trade fantasy, but settles for imperfect reproductions of itself—legislative redundancies, problematic export markets, and exponentially increasing quantities of data as products of immaterial evolution. Produce futures rarefy climate and environment as forecasts to be hedged against, while seed banks and databases replace in-situ conservation in National Parks, Forests, or Wetlands. Manifest destiny relies on the reconfiguration of paths and flows to exclude, bypass, or absorb errors in the supply line and overcome remaining legislative or natural barriers to capital penetration. The culture of cultivation that predates the Farmer's Almanac is replaced by horticultural management software.Supply side strategies are indexed to market closings, password protection, the perpetual time of surveillance and seed bank technologies, the correction of standard deviation, and the vagary of shock-and-awe. Agriculture resembles a produce BIOS more than cultural production of bios. The disequilibrium caused by war is shifted throughout, replacing canned or pickled storage of war garden excess with pathologies of waste at peripheral sites—the shadow of supermarket overabundance and choice. The citizen-farmer is not mobilized. However the citizen-consumer can phone in suspicious food activity, thus participating in a democratic and anonymous "real time, early warning system." The analogue to 1917's War Garden Commission is 2003's ARDI (US Agricultural Reconstruction and Development Program for Iraq). This legislative body, led by a single individual who is, at once, a former USDA official, a former grain executive, and a former trade treaty author,[20] exports the redundant bureaucracies of America's food security team. In the war on (t)error, fruits and vegetables are no longer the home front diet that preserves resources for war-time troops, so much as accessories of war itself—the war garden sublimated, denatured, patented, coded, scanned and exported at will.

>>> **Jennifer W. Leung is an architect and critic based in Brooklyn, NY. She currently teaches at the Yale School of Architecture.**

18. In 2002, General Tommy Franks ordered all supplies to Afghanistan, including produce, be RFID tagged for "total asset visibility." Overby, Stephanie (2004), "Inside an Agile Transformation," *CIO Magazine*, 8/15/2004.

19. "Statement of Rory Mayberry, Former Halliburton Employee," *Senate Democratic Policy Committee*, 6/13/2005.

20. Engdahl, F. William (2005) "Iraq and Washington's 'Seeds of Democracy,'" *Current Concerns*, 6/5/2005.

nomadic allotments: london's farming future

heather ring

The advent of farming was significant in the transition from a nomadic lifestyle to a settlement culture, and with that, land and property divisions, urban development, and social stratifications. Today, with all of London's land divided into plots, or slated for development in vacant lots, and with communities displaced and histories erased, allotments have taken to roaming—nomads stopping between transient homes.

^ 2012 Olympic redevelopment area. The top circle is the Velodrome, the middle green is the Manor Garden Allotments and the bottom circle is the Stadium.

Olympic Plots

Like many sites slated for regeneration, the myth propagated about the grounds designated for the 2012 London Olympics is one of a post-industrial wasteland—an area of disuse and blight. Situated in Hackney Wick along the Lea Valley Corridor, most Londoners—excited by the prospect of hosting the games—have never ventured this far into East London, and so readily accept a narrative that fails to address the many communities and habitats uprooted by the development. Hackney Wick is, in fact, a challenging place to explore, though there is a compelling disorientation in navigating the industrial maze of artist-squatted warehouses and untamed marshlands along the canal. The Manor Garden allotments, which have occupied the site for a century, have now become a symbol of the communities displaced by the Olympic development scheme. Allotment gardens, or land leased from public or private owners and distributed to families or individuals for the purpose of cultivating food, have been around since the 19th century as country retreats. They were later re-appropriated in the early 20th century as a wartime strategy to encourage citizens on the home front to contribute to the domestic food supply. Interest in these initiatives subsided after World War II, and London's subsequent development schemes did not set aside new land for urban agriculture. In recent years, there has been a renewed interest in changing this through the grow-your-own movement, but the land shortage resulted in waiting-lists for plots as the surviving allotments are passed down through generations of families.

When the Olympic development committee announced plans to develop this site for the games, a few vocal activists spearheaded a campaign to save the Manor Garden allotments, but they were unable to mobilize the majority of the original plot-holders so as to present a unified front. Unlike the collective efforts that groups of people dedicate towards community gardens, which simultaneously function as local public spaces, allotments are divided into adjacent private plots—protected by fences and planning approvals—often miles away from the homes of their owners, each with their own idiosyncratic storage sheds. The plots are occupied by people of diverse ethnicities, ages, and backgrounds, which results in differing approaches to how their gardens are tended. (Traditionally, the overgrown romantic patches of the middle class sit adjacent to the tidy functional plots of the working class.) This also challenges the perception of whether the allotment gardens are viewed as a shared responsibility or a set of individual concerns. Those who attempted to speak on behalf of the group in protest brought significant visibility and press to the negative impact of Olympic development, and made a case for preserving the allotments as an international symbol of London's commitment to the urban agriculture movement. The Olympics marks a critical opportunity to bring international visibility to the urban food systems of London, but the Manor Garden allotments were slated to be swept to the side, like homeless cleared from the streets before a televised parade. They did however manage to secure an arrangement in which the allotments would temporarily be moved to another site, and then re-established in the vicinity of their original site, after the games.

What does it mean to move an allotment out of range and back again? In 2008, the plot-holders were barred from entry by the iconic blue fence that encircled the Olympic construction site. After this eviction, few of the original gardeners chose to establish gardens in the new (temporary) site at Marsh Lane

^ **The Manor Garden Allotments on the Olympic site. Photo © Gesche Würfel.**

^ **The interim site for the Manor Garden Allotments. Photo © Gesche Würfel.**

^ **VACANT LOT (2007) by What If: projects Ltd.**

Field. In fact, they were not particularly welcomed: Marsh Lane Field was registered as common land, unfenced and set aside for public use, including the grazing of cattle and the foraging of fruits. The local community that defended the field as a Commons was not interested to have the site subdivided into privately cultivated plots.

Unlike the original allotments, which enjoyed an intimate and secluded setting, the proxy-allotments are compromised by a sense of exposure and traffic noise from the bordering road. After a hundred years of passed-down knowledge of the land, the gardeners now are tasked to cultivate immature ground with inadequate soil. Tested for contaminants, but not growing potential (which turned out to be poor), the soil was inadvertently compacted by construction vehicles trampling the site, leading to flooding in the front half of the allotments and plot holders sinking up to their knees in the sludge. For its second season, French drains are being installed, and perhaps over time, the land will be learned and a community will build around this provisional site, making it once again difficult to move the allotments—and its corresponding community—back from the interim site to the original.

There is talk about shifting the typology of the allotments, once they finally return home, from individual growing plots to community gardens, presumably an amenity for the influx of residents of the new Olympic Village development. Whichever form they take, they'll be received by an altogether different community, one that has undergone radical gentrification and a recent history that dominates—and potentially erases—the narratives of the place that existed before the Olympics. As the allotments become dissociated from their context, uprooted from the soil and unanchored from their community, they become nomadic—a mobile landscape identity separated from a physical reality. And, after the games, the allotments will return from their period of exile to an unrecognizable home.

Vacant Lot(s)

With the 2008 economic collapse, a frustrated opportunity emerged between the land shortage and the growing number of vacancies (as shops closed) and vacant lots (as construction projects stopped). Private development in London, initially spurred by the promise of financial stimulus from the Olympic Games, was put on hold—motionless cranes in mid-air, yet buildings demolished. These transitional spaces have been addressed by the art and architecture practice, What if: projects Ltd, who explored the potential for vacant lots to function as nomadic inner-city allotments. Their project, VACANT LOT (2007), introduced a field of growing plots that could be moved between temporary spaces without disruption to either the soil or the site. Initiated in 2007, the first vacant lot occupied was a formerly run-down and inaccessible plot of housing estate land in East London. London's council estates often have extensive grounds of worn turf—land that is neither productive nor recreational—failing to give these disorienting and Brutalist complexes a sense of place. There have been recent council funding initiatives offered to residents of these estates to introduce urban agriculture schemes, community gardens, and allotments—and some interesting projects have emerged through this support.

Drawing from the urban practice of growing potatoes in small bags of soil on fire escapes and balconies as well as the *organopónicos* developed in Cuba to address the impact of the US embargo on food supply, the What if: projects Ltd utilized seventy 1/2-ton bags of soil in rows to transform a vacant lot into a vacant al-*lot*-ment. In the prototype, low brick walls cast a hazy red glow over the lot, which was once a children's playground. The grow-bags have been distributed both to the long-standing families that overlook the lot from the balconies of the surrounding council tower blocks, as well as to the transient young professionals that populate the nearby Hoxton square, a gentrified area full of boutiques, galleries and clubs. As a small, shared space that holds individually contained plots,

^ **Wayward Plants at the Market Estate. Photo © Effie Fotaki on behalf of the Wayward Plant Registry.**

VACANT LOT operates somewhere in between a community garden and an allotment. Alongside the rows of bags, there are compost bins, benches, a shared rainwater tap, and a barbeque. These shared resources require cooperation and invite social exchange. As more long-standing gardeners revisit their bags over seasons, or if decisions and organizational responsibilities are handed to the community, there may be growing evidence of local investment in the lot.

Although it's unlikely this council property will be slated for development in the near future, it's possible that the plotholders might resist the relocation of this plot just as they had with the Olympic site. But unlike the Manor Garden allotments, which were symbolically preserved but physically abandoned for a new site with different soil, the grow bag allotments can be lifted, soil and plants intact, and moved to another space without disrupting the individual plots. The group of bags could be separated, rearranged or disseminated without compromising their contents. And as they are lifted, the vacant lot will remain intact, with the ground unbroken, and evidence swept up. The site becomes a temporary stage for a mobile production, testing the importance of context to a community.

Halfway Homes

When planning permissions are acquired and regeneration schemes move forward, not all gardens have exit-strategies like the Manor Garden allotments. The Wayward Plant Registry has been salvaging the left-behind plants from private and communal gardens and allotments of estates undergoing demolition. This includes The Market Estate, a 1960s modernist estate in Islington, and the monolithic Heygate Estate that will be torn down as part of the regeneration scheme for Elephant & Castle. These Brutalist estates were considered a great failure in social housing, with open passages, communal spaces and informal gathering areas that were seen as an invitation for crime and other "anti-social" behaviors. The new housing developments designed to replace them turn their back on these once utopian elements, creating wholly privatized units that deny any opportunity for the communal cultivation of land.

While the residents of the estates are displaced from their homes, their gardens are stranded, waiting to be buried in the rubble. The Registry takes in those plants that uproot easily, while accepting propagation cuttings from willful trees that refuse to be moved. These plants, along with other unwanted, displaced, abandoned, and salvaged species, come together to form "halfway homes for wayward plants"—temporary gardens that might settle comfortably into these intermediary spaces, or last just a moment—marking the point of their exchange or adoption. As the plants travel to new homes, seeking fresh pastures, the histories embedded in the estates are carried with them.

A New Nomadism

The new nomadism of allotments marks a certain resiliency in the face of regeneration, and resourcefulness when addressing London's land shortage. This separation of farming from settlements also reflects the disconnect many Londoners feel between their food and the land it is grown on—through abstract production methods, a dependency on oil and an unsustainable reliance on the transport and distribution systems of the food supply.

One encouraging effort is a new London initiative called Landshare, which matches private homeowners that have neglected their back gardens with eager food growers who wish to cultivate their land and share the produce. While thus far the program has attracted more interest from stray gardeners than from generous landowners, it requires a reframing of the relationships between neighbors, testing the boundaries of personal space and productive partnerships.

To feed the city, there is a pressing need for an infrastructure of urban agriculture in London's development plans, as well as continued innovations on underutilized land. Allotments can no longer wait to be given homes, they'll need to occupy any space that they can find.

crop rotation / sequencing

David Newton
MANIFOLD Design
cityLAB and Roger Sherman Architecture + Urban Design
Moving Cities and Adrian Blackwell
Corine Vermeulen
Carey Clouse

performative landscapes

david newton

^ Hylomorphism vs. Self-Organization. Traditional levy and pier design (left) and soft infrastructural system (right).

"Performative Landscapes" examines and strategizes the production of artificial wetlands, port facilities, and recreational and commercial fishing areas by utilizing the excess dredge material generated by the Houston Ship Channel. The proposal establishes a developmental tool kit and set of logics that will guide the growth of new wetlands and foster the production of integrated programming along Galveston Bay, Texas.

From the Discreet to the Field

At the edge of Houston's dense urban mat, suburban and industrial clusters flake off and fragment into the undeveloped countryside. Each cluster secures a moment of partial autonomy until the dense mat, driven by market forces, can eventually catch-up, absorb these forward deployments, and dispatch yet others further ahead. Like viral particles hop-scotching through a body, these clusters advance, projecting the city's material and spatializing logics into the dense ecological webs of the coastal wetlands, replacing multi-performative ecological infrastructures with the mono-performative mineralized tissues of the modern city. In this exchange, soft systems are traded for hard. The spongy wetland floor meshworks are replaced with hard mineralized strata such as asphalt, concrete, and compacted earth. Speed is exchanged for slowness, and with that erosion, flood, and pollution protection is compromised, as well as scenic beauty and wildlife habitat. Field logics are supplanted with exclusionary fragmentation. In other words, the ground condition moves from a logic of gradients and blends toward one of hard boundaries and homogeneity.

This clash has lead to an environmental, social, and economic crisis in and along Galveston Bay. In the last 30 years alone the bay area has lost over 30,000 acres of wetland from a multitude of human activities in and along the bay. With this sobering reduction in wetlands, pollution in the Bay has risen to dangerous levels as stormwater run-off, no longer filtered or detained by wetland meshworks, is desposited directly into the Bay by the city's hardened surfaces. This process has tended to increase the amount of fresh water and therefore disturb the salinity levels within the Bay, causing valuable species such as clams to have high mortality rates.

Soft Infrastructures – The Cell

Another fascinating dynamic in the coastal areas, especially along the Gulf Coast, is the production of excess clays, silts, sands, and shells, from the dredging of the shipping channels of commerce. A significant question remaining once the material is harvested from the ocean floor is how to dispose of it. In Galveston Bay, the Port of Houston has developed a federally funded program to use the material to build artificial wetlands and recreational areas in order to replace some of the wetlands lost over the previous thirty years. The Beneficial Use Project, as it is called, seeks to construct more than 8,000 acres of wetlands over the next 50 years. The project is running out of funds from federal cutbacks and is now desperately seeking ways in which the project might fund itself, while unfortunately operating within the same exclusionary organizational diagram that has been so disastrous to these coastal cultural and natural ecosystems. "Performative Landscapes" seeks to appropriate this man-made flow of geology towards the production of a hybrid landscape of both cultural and natural performances. This is engendered with the organizational schemas endemic to the material logics of wetland meshworks in order to develop an inclusive coastal design and planning strategy. The project begins with the development of a soft infrastructural component for dredge retention, erosion protection, wetland generation, and foundational support needs.

The port of Houston's Beneficial Use Plan currently goes about its task by constructing 140-acre containment cells, which are then used to contain and consolidate dredge that is emptied into them. The cell walls of the levees are constructed mostly of imported stone and are continuous except for infrequent drainage pipes that allow the contaminated water from the dredge to drain into the bay. The cells have three stages in their development: cells are filled with dredge (12 months); cells are left to consolidate (1-2 years); cell walls are broken down in random locations to allow for greater water permeability and emergent channels linking the Bay to the new wetland. This process is ineffectual for three reasons. First, it ignores the potentials of utilizing accumulation and silting self-organizational processes,

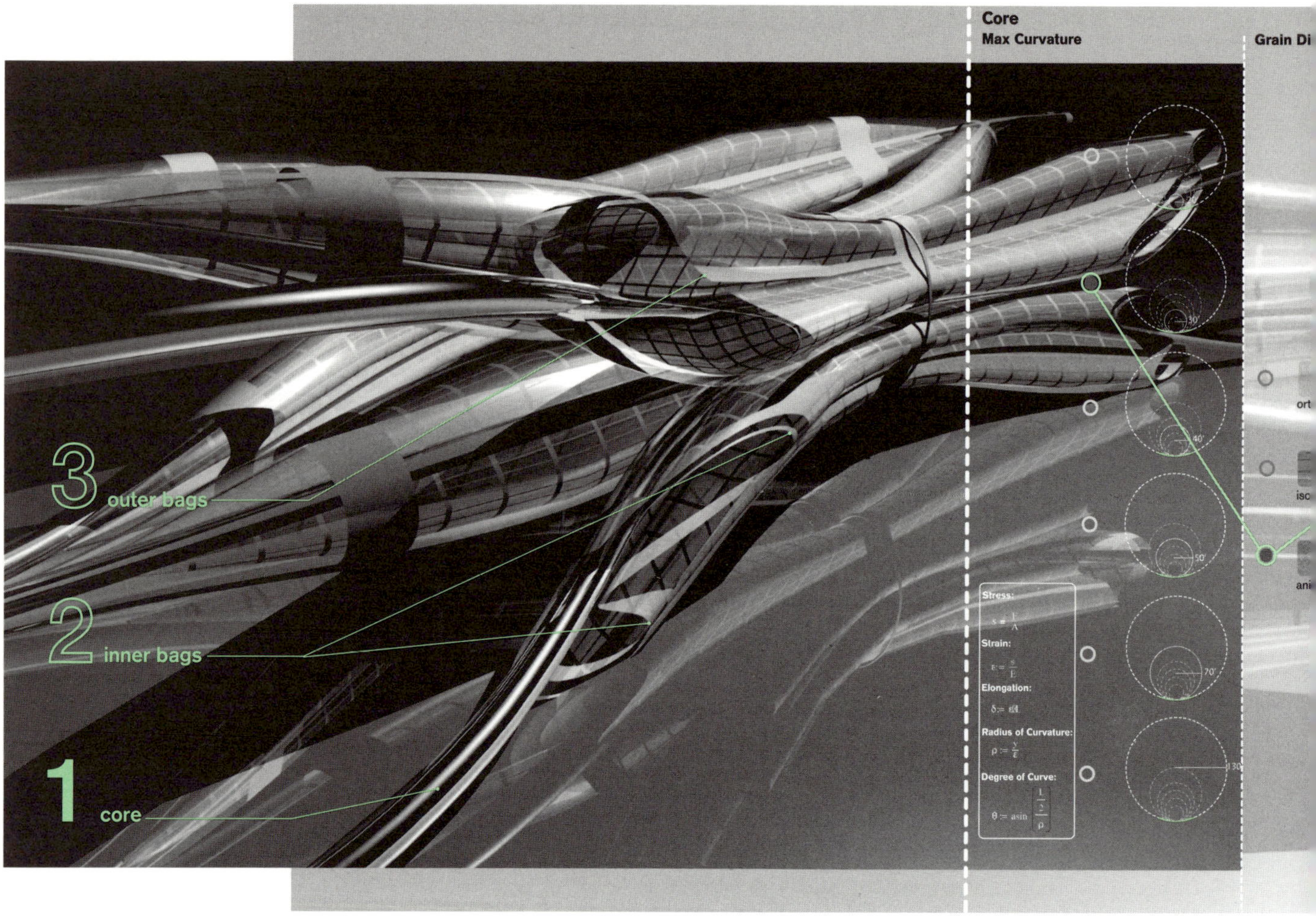

^ **Material phylogram: core, actuators, actuation.**

pursuing instead a top-down hylomorphism consisting of filling up a mold. This top-down approach consequently requires that the cell walls be taken down at some point in order to allow for emergent morphology. Second, stone must be imported from out of state to construct the cell wall levees because hard clays are not abundant enough locally to supply the project. But, could there be a way to use soft silts and clays for the job? Third, this process cannot adequately allow for foundational support for building projects unless a piling system is added to it. Therefore, this process, for all the reasons above, cannot fund itself and needs to rethink its material logic.

The quest for an alternate material system started with the goals of utilizing self-organizational silting and accumulation processes in the Bay for the production of wetlands, and a system that could provide erosion protection and load-bearing performances without the need to import stone or construct pilings. The design of the local cell of the system takes the hydraulic structural performances of the Acetabularia as an operational dynamic, and employs the use of water, air, or liquefied dredge pressure to effect a geometric adaptational deformation of its rigid elements. The cell is composed of four rigid hollow spine elements, flanked and attached to a bag system capable of being filled with water, air, or liquefied dredge material. The system can consequently use dredge material to inflate volumes that protect against erosion. Therefore, importing stone for cell walls is no longer necessary. The system, when proliferated, also acts as a silting collection mat in which cell sizes inform self-organized silting and accumulation densities.

The Difference that Makes the Difference

The system deployed in a mat configuration takes advantage of self-organizing silting and accumulation dynamics by varying its density, elevational profiles, global/regional curvatures, and porosity. In doing so, the system obtains a soft control of the location of elevational profiles, island morphology, water channel densities, pond densities, and cove densities. All these factors of soft control are fundamental in encouraging various types of flora/fauna ecosystems in the Bay, which may perform in ways that can inform human habitation and environmental sustainability. The "difference that makes a difference," as Gregory Bateson writes in *Steps to an Ecology of the Mind*, turns out to be an elevational difference of about 50 centimeters along the Bay, which produces several different types of wetlands. For instance, at around 25 cm above the mean tide level a type of wetland dominates that is highly productive for fostering blue crab and various shrimp species. This wetland type is called *Spartina alterniflora* edge condition and is estimated to be worth about $45.65 per square meter in shrimp and blue crab alone, based on current prices and bay studies on the number of individuals usually found per square meter in this type of wetland. At the same time, this type of wetland has been found to be able to remediate by-product salts, petroleum, and hydrocarbons in polluted water. These are all pollutants that Galveston Bay has a serious problem with.

By tuning the mats in terms of density and profile height, land morphologies and elevations can be encouraged and specific emergent wetland types are produced that respond to any needed performance. The major performances this proposal elicits are chemical remediation potentials to decontaminate

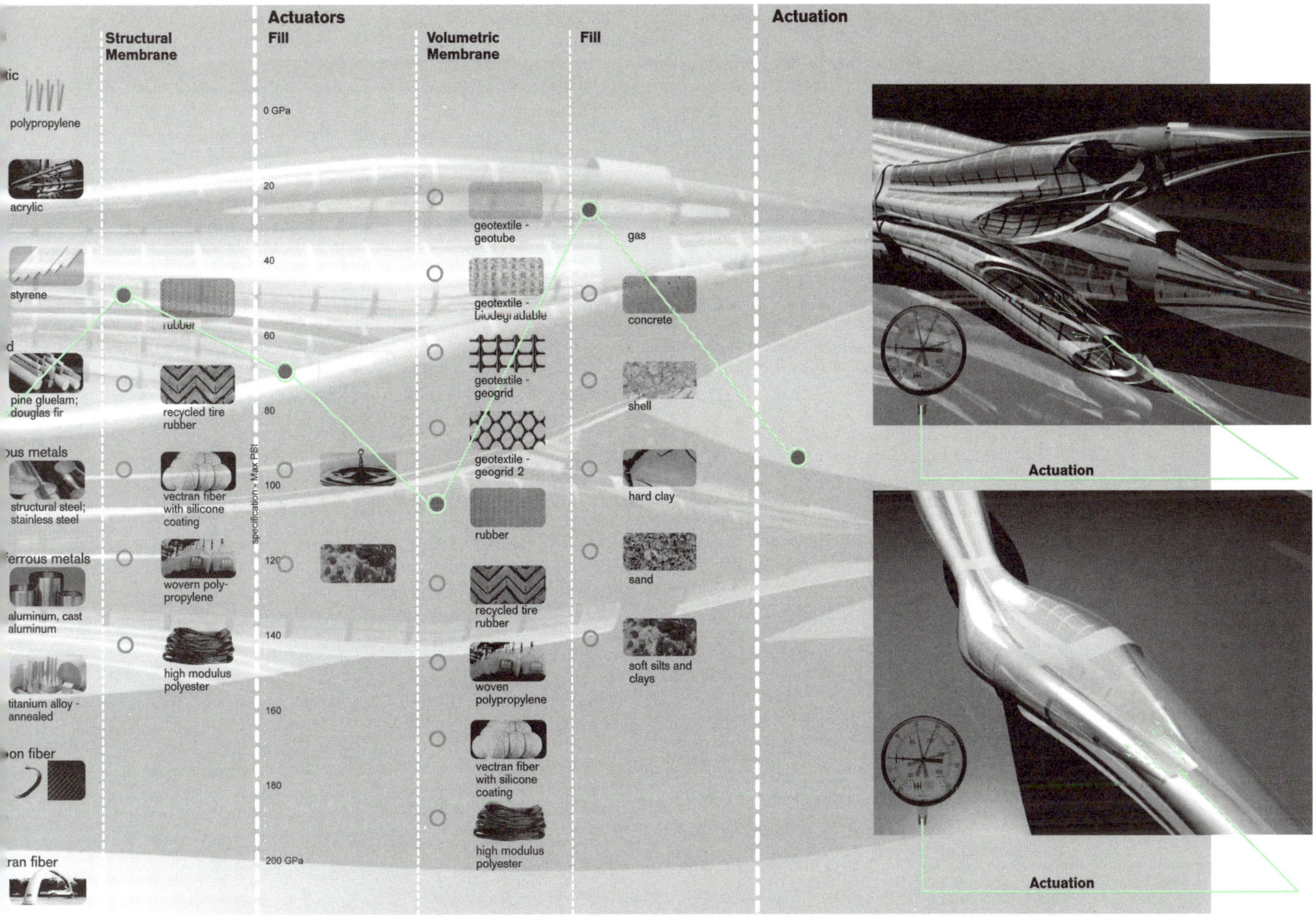

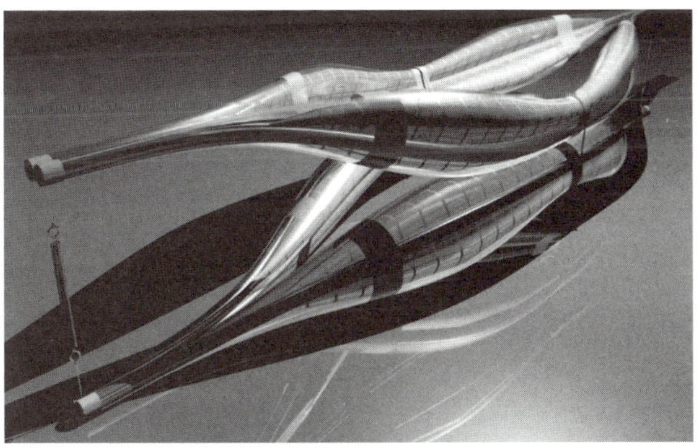

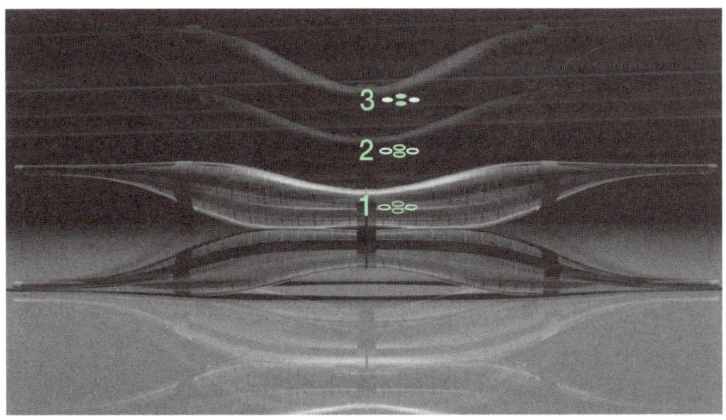

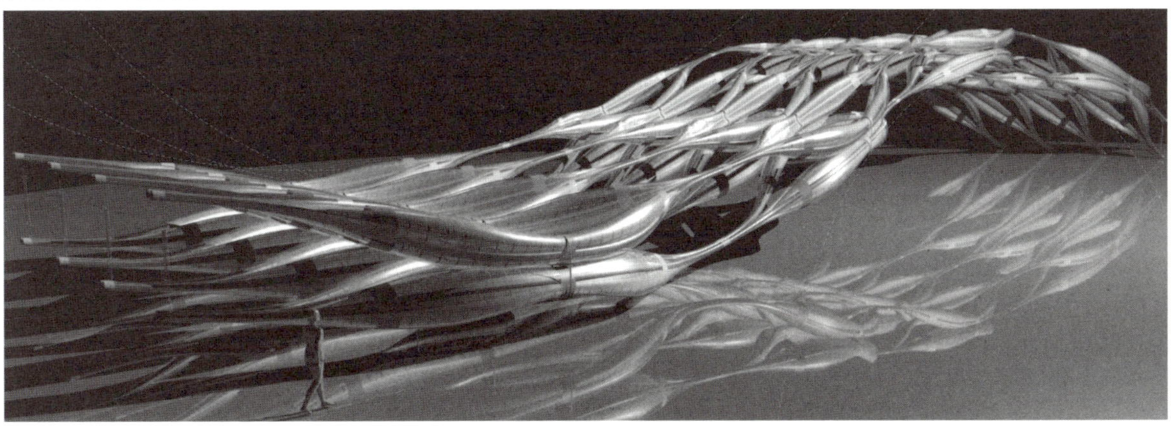

^ **Soft infrastructures – The Cell.**

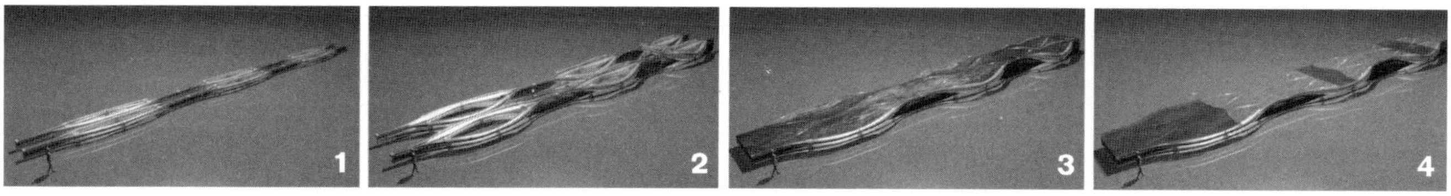

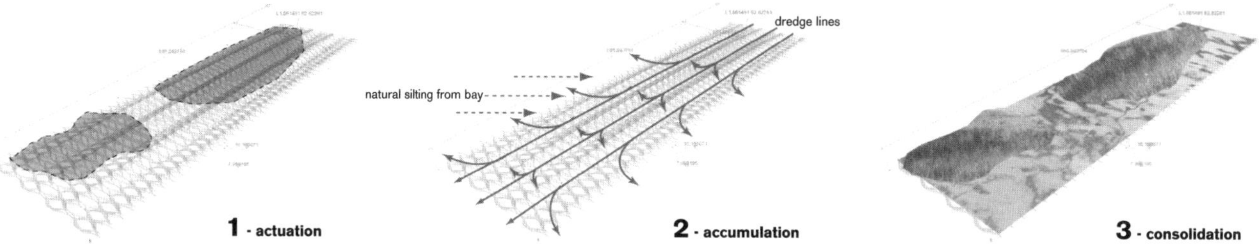

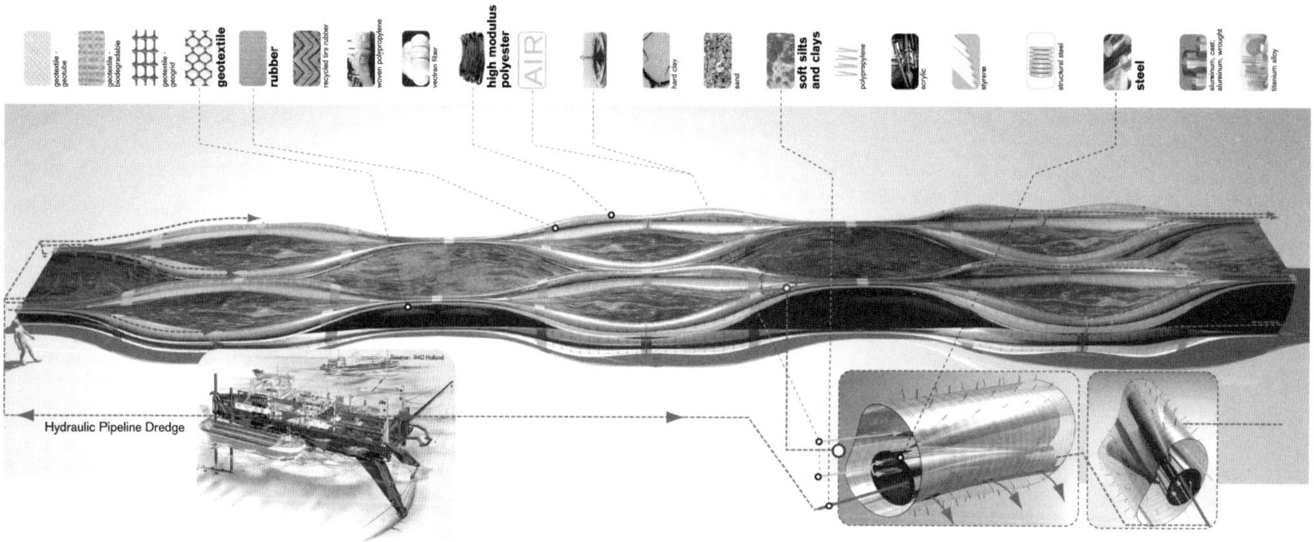

^ **Cell stages: actutation, accumulation, consolidation. Self-organizational silting and accumulation processes.**

dredge material and the Bay, aquacultural productive wetlands for commercial and recreational fishery species, and an endangered bird habitat that can also attract eco-tourism and bird watching enthusiasts. The system may also provide foundational support depending on the density of the elements. The system is capable of producing varying degrees of enclosure and hybrid performative spaces between remediation strata, recreational wetland strata, and habitable spaces, thereby presenting us with an inhabitable metabolic thickness, or perhaps what Stan Allen has referred to as the "thick 2-D."

The proliferated mats have several possible conditions; the mats may rest on the bay floor and create topography by actuating to various heights from that datum, or they may be actuated with air producing a floating meshwork of wetlands and port facilities from the datum of the water line. The mats can be filled by pumping dredge through the hollow spines or collected from a single release point remotely.

Global Deployments

At the global and regional scales, what is at stake in the project is a proposal for a comprehensive and inclusive design strategy and developmental tool kit that can integrate the production of wetland habitats, remediation wetland complexes, recreational areas, and port facilities. The developmental tool kit allows this multiplicity of performances to be strategized and produced in various modes through a single component system, displacing the hylomorphic and clumsy deployment of traditional levee and pier construction in wetland/port production. This system, designed from the cell and proliferated in metabolic mats, also suggests a method for global deployment that is, arguably, superior to the current strategies employed in the Beneficial Use Plan presented by the Port of Houston Authority.

There are two port facility design paradigms, which "Performative Landscapes" attempts to subvert and rethink. The first strategy is that of the consolidation of all port facilities into a discreet closed figure, a "Super-harbor" as the former design firm PLOT referred to it. This bounded figure resists growth of its boundaries, while arranging program into discreet boxes and planes one stacked on the other. Here we see the exclusionary and homogenizing logics of the reductionist paradigm perhaps at its most extreme. The Port of Houston Authority, on the other hand, seems content to strategize wetland production and port development in an exclusionary strategy, deploying ports in discreet points along valuable natural wetland and community shoreline. Performative Landscapes seeks to radicalize the boundary condition and integrate port production through a metabolic complex of wetland and port facilities that has the ability to grow and decay as the market requires. This allows for the boundary of culture and natural ecology to be in constant play as port facilities once used will dissolve into recreational wetlands and vice versa.

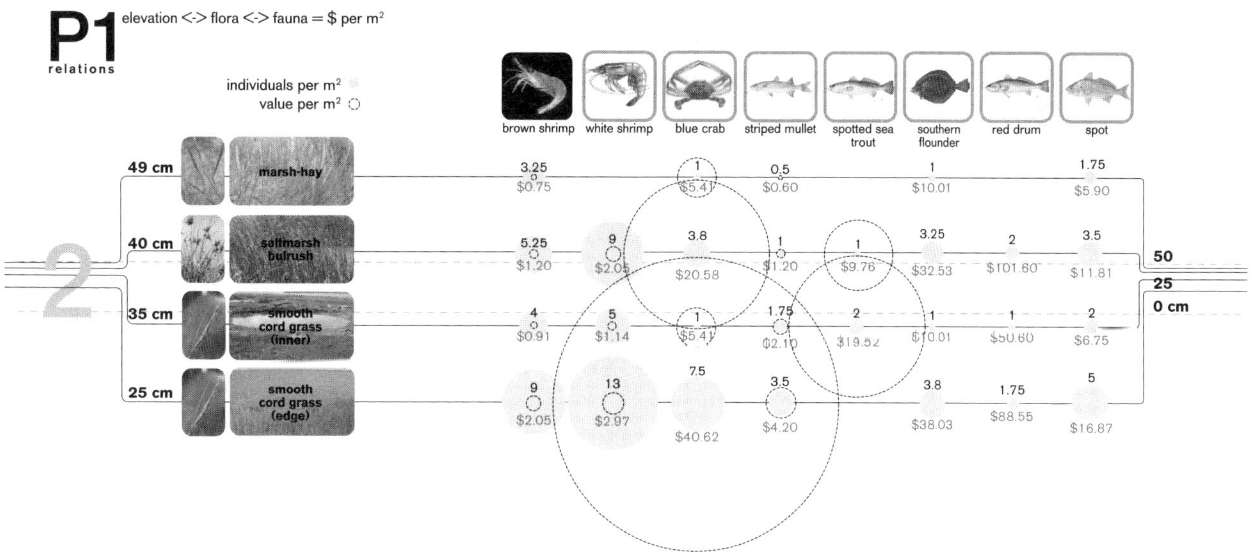

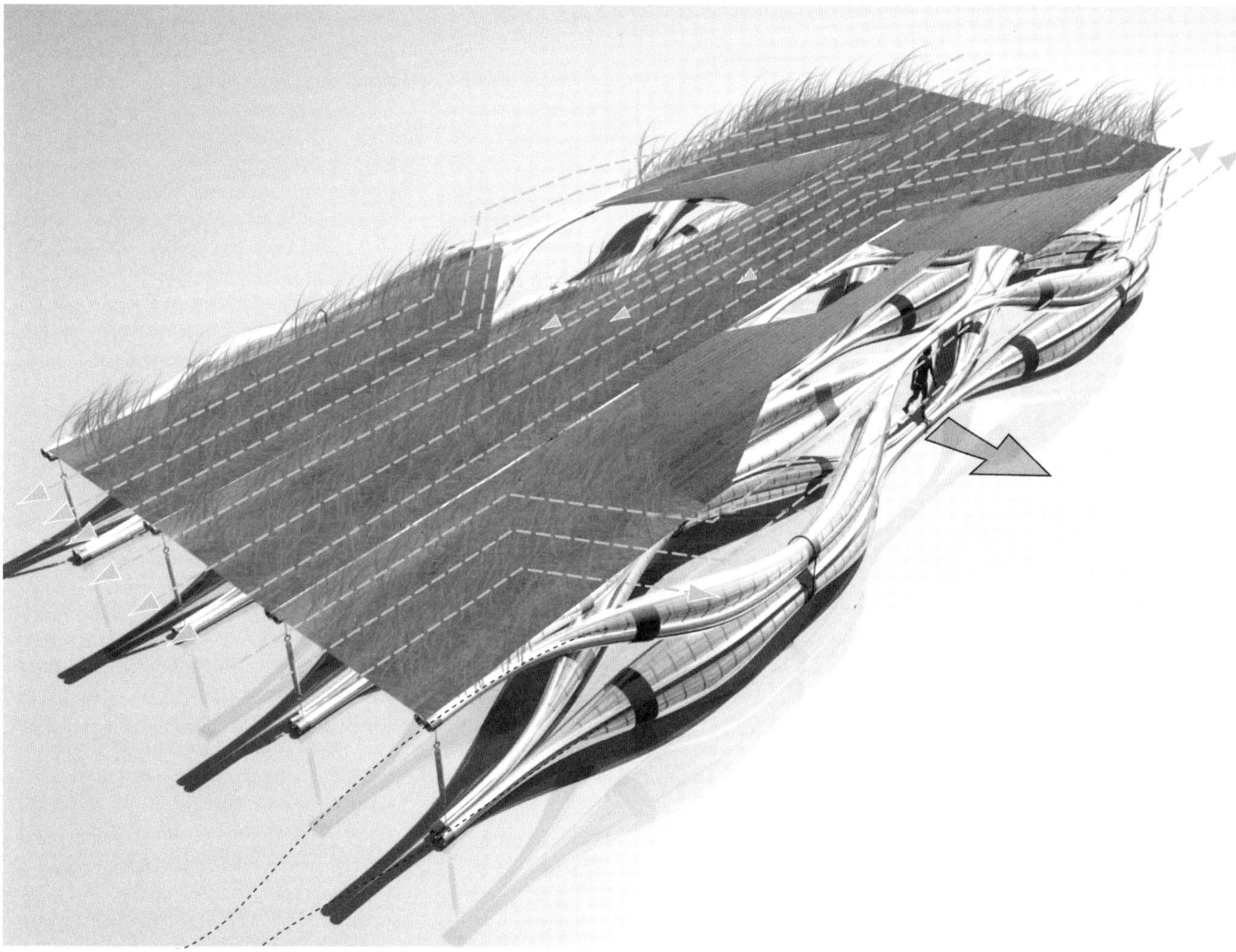

^^ Mat configurations: The difference that makes the difference. A small range of elevational difference of about 50 centimeters along the bay produces several different types of wetlands, which all perform differently.

^ Habitable deployments. Remediation strata and pedestrian circulation.

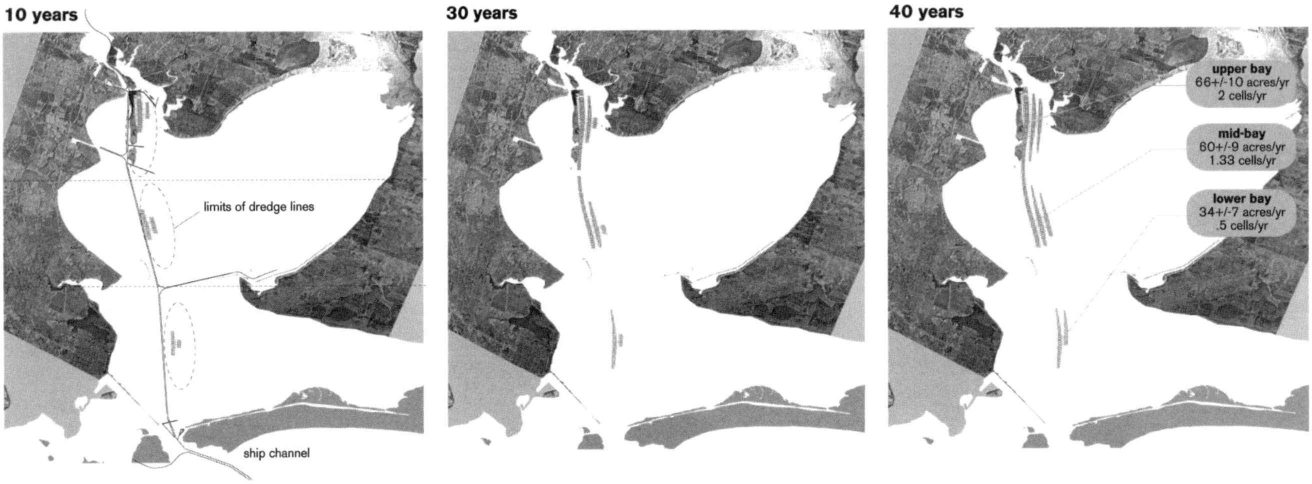

^ 50-year plan: Performative Landscapes unfolds and develops itself according to the flow rates of dredge material.

Flows of Earth: Calculating the Growth Rate

The flows of earth generated by dredging the Houston ship channel in the next 50 years are estimated to be able to produce around 8,000 acres of wetland. This flow of earth has a yearly rate, and provides the growth rate for the system. In greatest abundance is the soft silts and clays at 150 million cubic yards (MCY) of generated material in a 50-year span. The levee grade material of hard clays, sands, and shells are in much smaller supply with an estimated 75 MCY over the same duration. This dynamic underscores the value of the presented component that can utilize the soft silts and clays for erosion protection without having to import stone.

Dredging takes place in three main regions of the Galveston Bay: upper bay, mid-bay, and lower bay. It is divided based on dredging schedules and the varied characteristics of each segment. Another constraint on deploying dredge is the distance the dredge can be pumped from its excavation point in the channel. With these constraints, the boundaries and points of growth for the wetland / port / recreational meshwork becomes clear. There will be three seed locations in each section of the bay with a possible width constrained by the dredge pump constraints. Through analysis of projected dredge amounts, the upper bay was found to have the fastest growth rate at 66 (+/-10) acres per year or 2.0 cells per year; mid-bay had the second fastest rate at 60 (+/- 9) acres per year, or 1.33 cells per year; and the lower bay had the slowest rate of dredge production at 34 (+/- 7) acres per year, or 0.5 cells per year.

Tuning the Bay

"Performative Landscapes" unfolds and develops according to the flow rates of dredge material, however, it is important to note that the flow of material is variable and can change the phenotype presented in the 50-year plan as cells expand or shrink to accommodate varying rates of dredge caused by extreme events. Along with varying dredge rates, the phenotype is also informed by the regimes of water flow across the Bay and the corresponding salinity and pollution concentrations. Through flow analysis presented in several studies on the environmental state of the Bay, several important dynamics emerged. The salinity in the west bay was too low, increasing flora and fauna mortality, due to fresh water inflow from the urbanized wetland areas and a lack of cross current flow. Opening up the cross flow area would allow bay currents to circulate and balance salinity levels, while at the same time flushing the pollution concentrations through a strategic field where the remediation meshworks could be deployed. The global phenotype then works through the differentiated permeability of the component system to allow a cross flow and filtering to occur, therefore tuning the salinity and pollution levels of the Bay with its spongy meshworks.

Regimes of Flow

The 50-year phenotype situates itself over the central flow of ocean currents in the Bay, tuning the Bay's salinity levels to a balance and directing polluted currents over the remediation wetland meshworks. The geometric and material logics of the proposed component system allow varied and differentiated cross flow. The components are proliferated into mats that form local sections with a maximum length of 3,100 feet, a minimum length of 1,100 feet, a maximum surface area of 20 acres, and a minimum surface area of 6 acres. When this local section is linked to other local sections a regional complex is produced, which invites a networked capacity to initiate and control regional flows. The regional geometry of the networked cells is constrained in terms of its maximum and minimum surface areas and curvature by the material properties assigned at the beginning of the process to the sub-local cells, in this case structural steel tubes with a maximum curvature of 30 feet at the sub local level and a maximum curvature of 90 feet in a local configuration. In this way, material resistances and constraints are embedded in the possible geometries of the phenotype. Cell sizes in these regional complexes are also sized in relation to one another. The first two cells are sized for the 30 acres of cargo space needed for a 2-berth port, the wetland cells are then sized to accommodate the required stormwater retention and remediation of a cargo facility of this size required by the port of Houston. The third cell is sized to accept stormwater retention, but its wetland type is designed to remediate a much reduced concentration of pollutants as a secondary catchment cell while acting as a habitat for blue and brown shrimp.

Performative Meshworks in the Phenotype

Departing from the flow characteristics of the global phenotype, I will now address the performance that this actuated deployment hopes to foster and entangle in its differentiated lattice mat. In other words, this is a shift from the skeleton to the meshwork of natural and cultural performances accumulating on it. The design of the single unit of the phenotype, the local cell, is designed into a toolkit of performative generics:

50 years

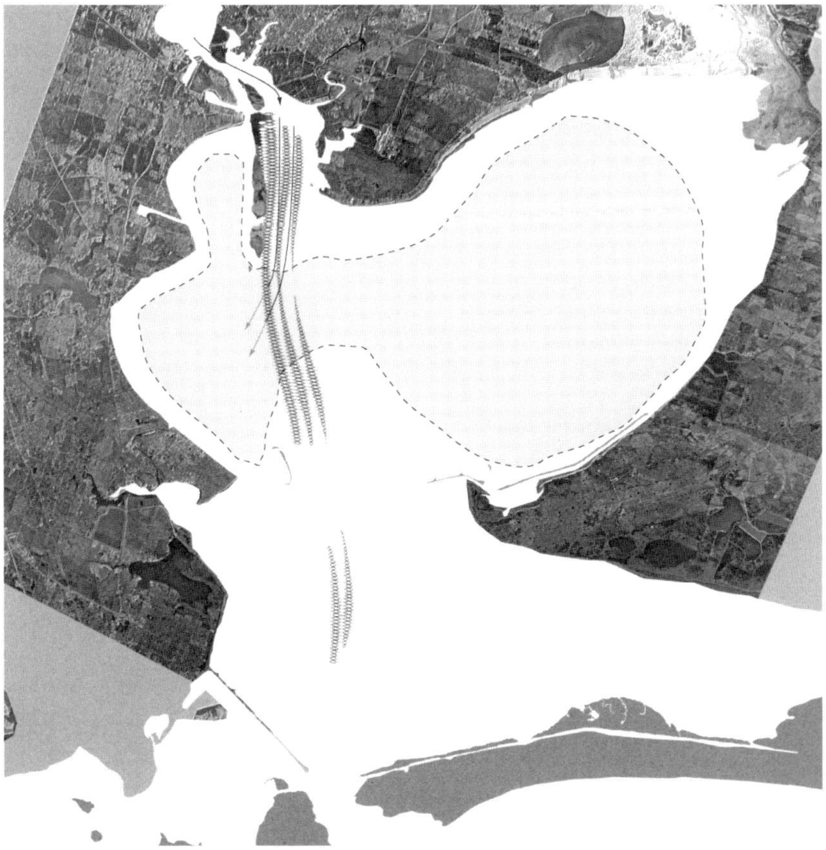

^^ In 50 years Performative Landscapes allows current cross flow.

^ It also operates to balance salinity and acts as a pollution filter.

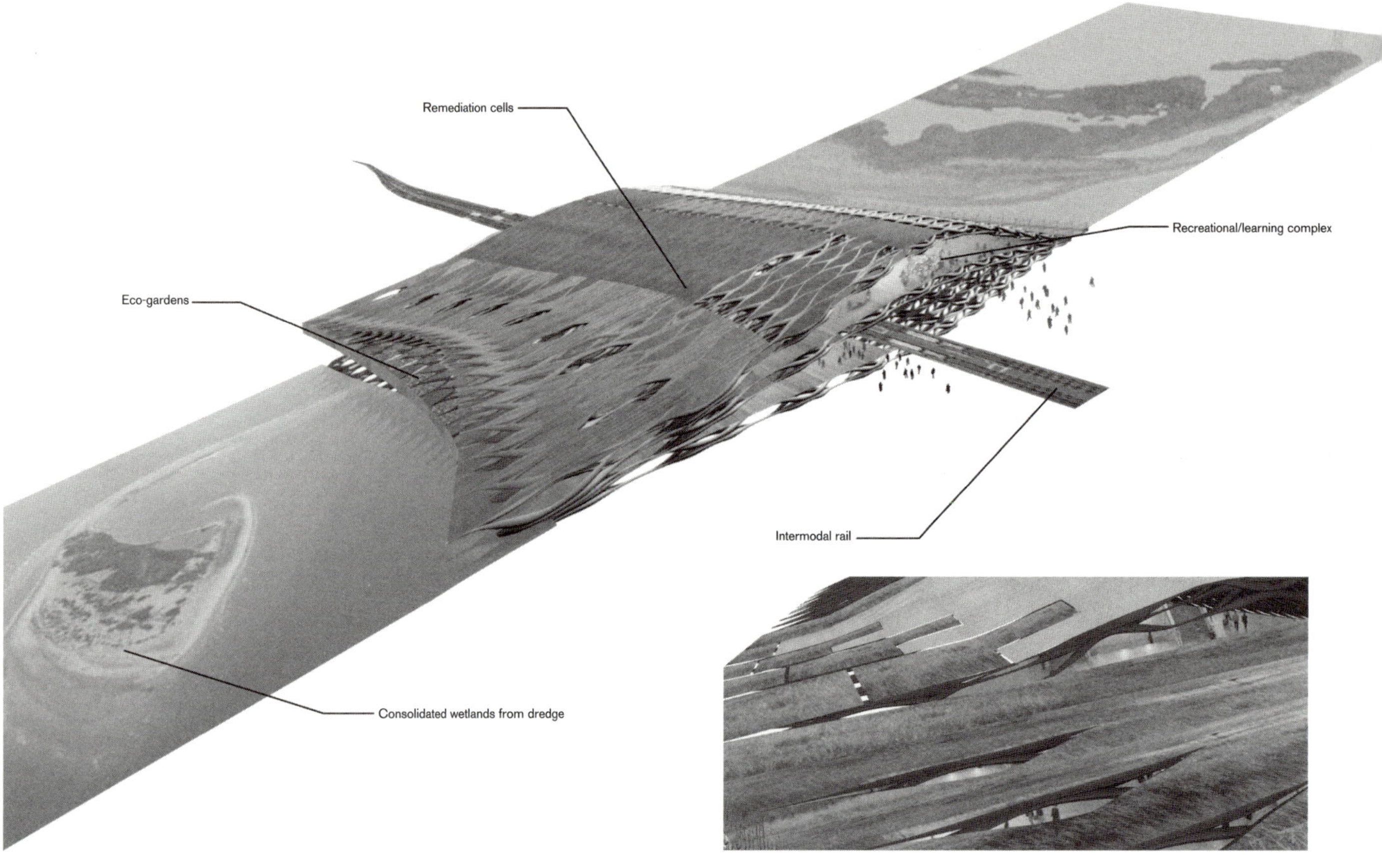

^ Wetland infrastructural meshworks: A breathable and catalytic boundary.

remediation, bird habitat, recreational/commercial fishing, port facilities, and dredge consolidation cells. Each generic cell, through specific actuation and/or subtraction of sub-local components, is tuned to target ranges of water channel, pond, and cove densities, as well as specific elevational ranges that have been identified to encourage specific wetland types. In controlling the percentage of a particular wetland type, one can also encourage and garner recreational, eco-touristic, and commercial fishing value and performance from these wetland types.

The 50-year plan is but one phenotype amidst a multitude, and is geared specifically towards increasing economic fishing value through shrimp and blue crab species, while at the same time acting as a tissue of remediation for the Bay. In this scheme the far west line of cells are a dredge consolidation and remediation band, functioning as a spillway for newly dredged material to be dumped and the excess water and liquefied dredge to flow from that consolidation cell into the next line of cells for further remediation. The furthest easterly line of cells is then populated with cell qualities that foster *Spartina alterniflora* edge condition wetlands for maximized shrimp and crab species as well as some qualities that foster large bird habitat and therefore birdwatching, and finally some recreational fishing cells.

Rendering contribution: Micah Stanley

>>> David Newton is a Phoenix-based architect and lecturer at Arizona State University. His work focuses on the intersection of computation, biomimcry, and design.

reforestation at greenwood farm: an emergent landscape and intervention

manifold design (karen lemmert & david naill)

Three adjacent farms, located in Carroll County, Maryland, have been functioning dairy and chicken farms producing feed crops since 1779. While, the three 90-acre farms are defined as part of the Chesapeake Bay Watershed, physical and economic pressures from suburban development have eroded the agricultural landscape. In an effort to give rest to both farmers and land, and to preserve the parcels in their entirety, the owners placed the farms into the Conservation Reserve Enhancement Program (CREP). This program, sponsored by the State of Maryland, was created specifically to preserve wetlands and riparian areas adjacent to streams and other bodies of water. These sensitive land areas have a tremendous impact on water quality and provide critical wildlife habitat.

The CREP program takes 'sensitive' land out of agricultural production and stabilizes waterway buffer areas through the planting of grasses and native trees. The program has a specific set of parameters directed at stabilizing soil conditions through the performative qualities of the plants. However, most of the reforestation projects, implemented throughout Maryland, seem disconnected from the agricultural terrain and associated landscape.

The Reforestation at Greenwood Farm project recognizes the plantings of 20,000 trees, of varying native species in varying soil conditions, as a man-made landscape within the traditions of agriculture. An opportunity emerged to design a system for organizing species and for implementing that system across many acres. The design incorporates variations in leaf colors, tree heights, and branch densities.

Through a 16-tree module with alternate placeholders, the location of a single tree could be established within the weave of tree species. Hand cards placed with the module were used for planting in the field. Reference points established through counting notation provided enough strucuture for the adaptable plan to develop and spread. Alternate positions in the modules allowed for tree species to be exchanged as determined by local soil conditions.

The system oscillates between the establishment of a systematic structure of the landscape and the specific conditions of each planted region.

>>> **Karen Lemmert and David Naill are principals of the architecture practice MANIFOLD design, and teach at the Maryland Institute College of Art.**

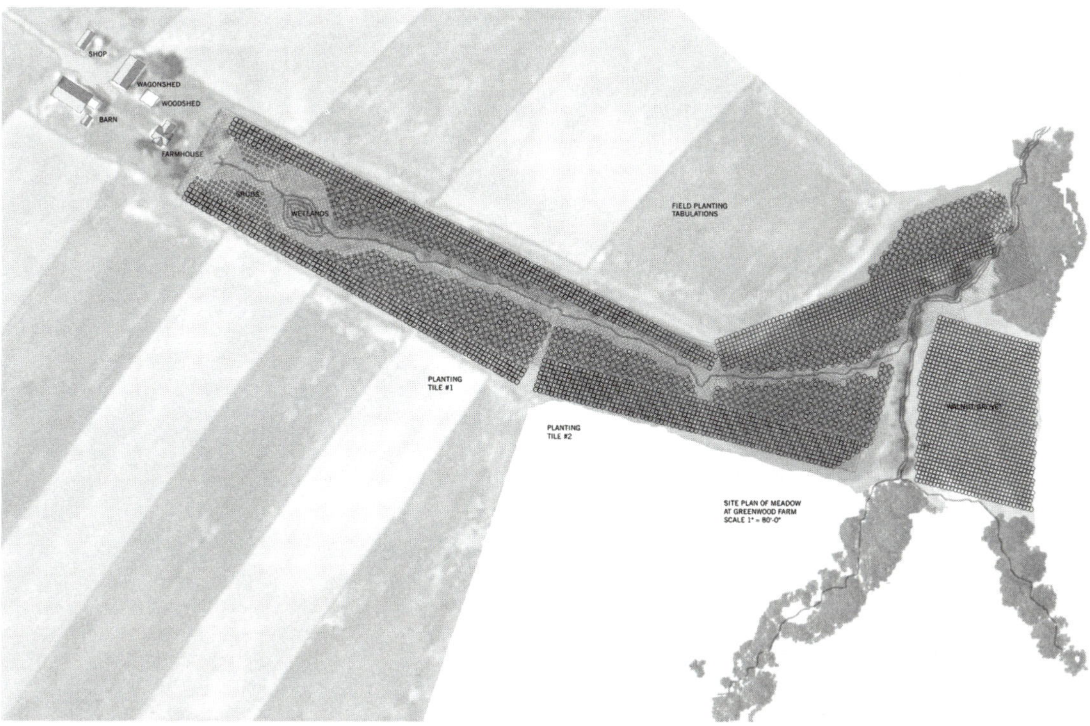

^ **Greenwood Farm plan.**

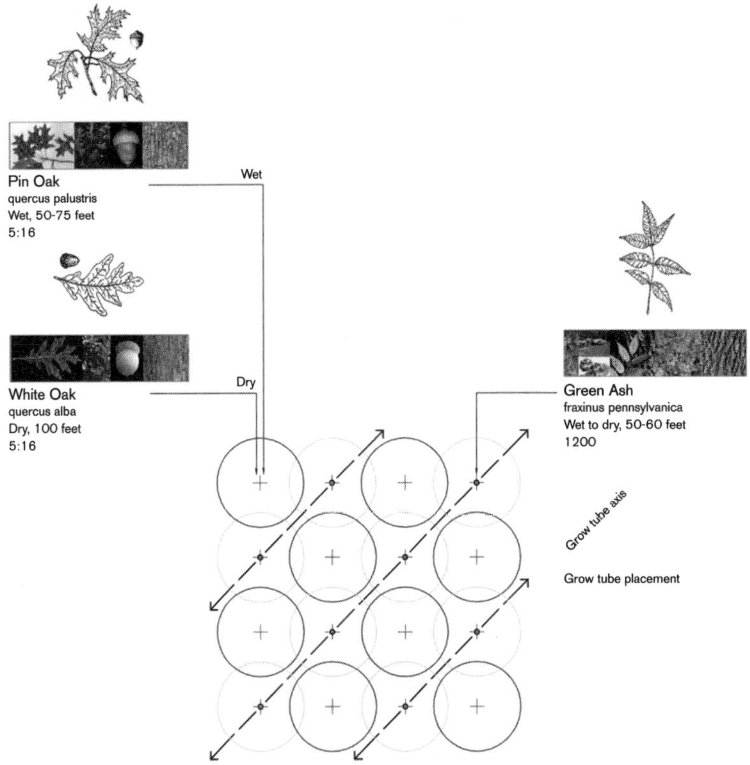

^ **Planting module #1.**

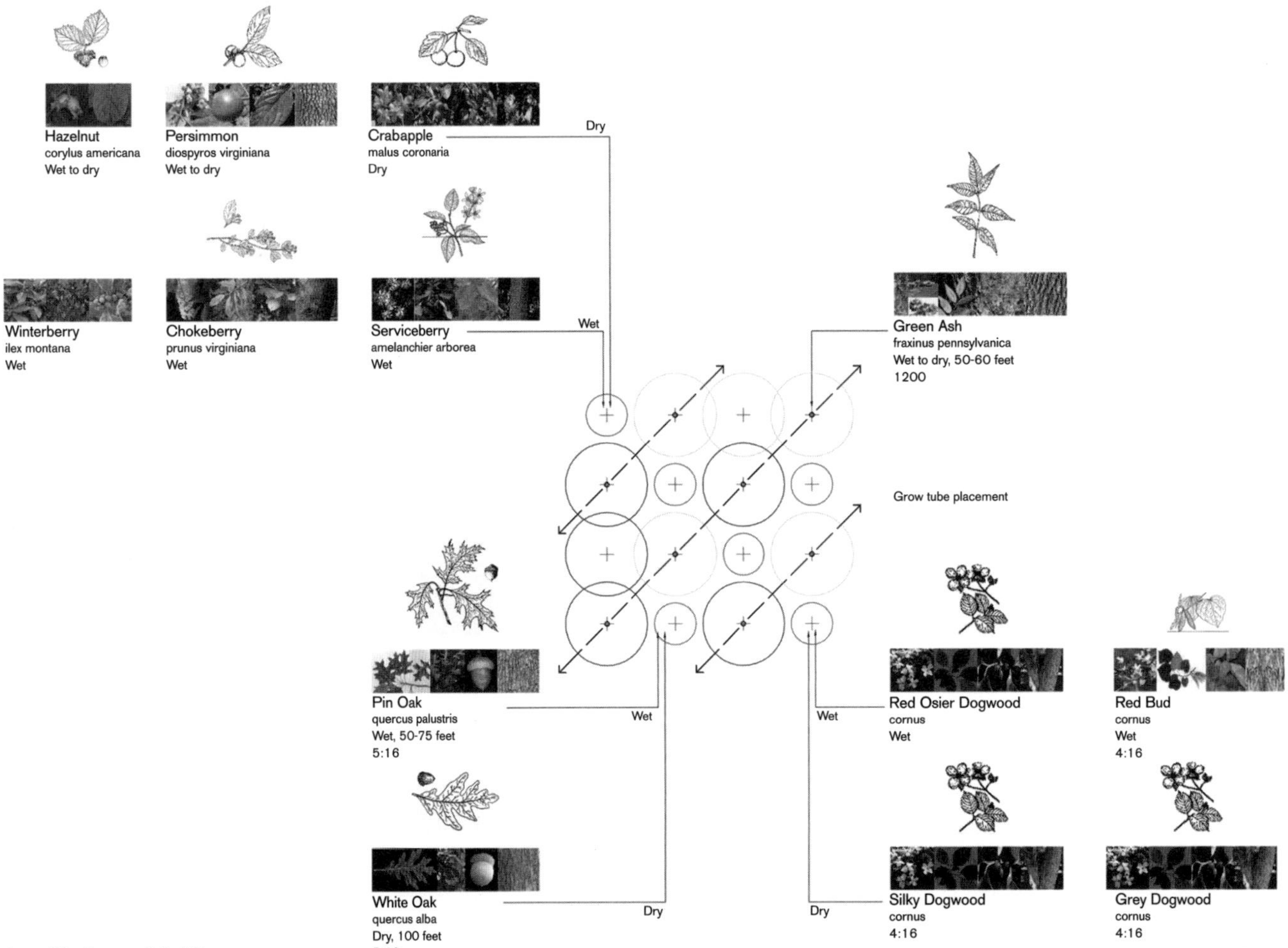

^ **Planting module #2.**

^ Greenwood Farm, Fall 2004.

^ Implementation, Spring 2005.

chia mesa

citylab & roger sherman architecture + urban design

"Chia Mesa" is a proposal to transform the strip mall into a prototype and a recovery strategy for Phoenix, Arizona. Rapid urbanization of the Metro area—its unbridled horizontal expansion into the outlying landscape—has destroyed not only its potential agricultural productivity but its identity as well. With Chia Mesa, a new imageability emerges for the city as a whole—one whose morphology ironically recollects that of the landscape it is rapidly consuming. Instead of being anonymously absorbed into the endless and undifferentiated urban fabric (Phoenix's own version of Banham's "Plains of Id"), Chia Mesa envisions the next generation of strip malls as climactically and socially 'cool hotspots' in an otherwise arid field. They will become the new landmarks of Phoenix's urban future, as the indistinct identity of this city develops an imageability as singular and recognizable as Dubai's Palm Islands, New York City's skyline, or—closer to home—Vegas' neon glow. The repurposing of the strip mall becomes the rebranding of Phoenix—from land of malls to land of mesas.

Chia Mesa brings life to the defunct strip mall, literally and figuratively. A new 3.5-acre hydroponic farm wraps its top and sides, giving it a unique exterior identity and interior ambience that will transform it into a bonafide destination. Occupying a zone between three and nine feet thick, the farm is comprised of water-fed plant trays and attendant service walkways covered by operable polycarbonate louvers. Three stories of additional rentable space and parking are accessed by bridges that span across a pedestrian 'gorge' from one to the other. Meanwhile, the all-important visual presence of businesses from the boulevard is built into the north and west faces of the project. Here, the polycarbonate envelope is perforated by 'cliff-sellings:' large backlit signs that identify the tenants within. A drive-thru biodiesel station, in the repurposed existing building, offers a direct link between the farm and the consumer.

Chia Mesa offers a strategy for land-constrained cities, where farmland is no longer available. It offers residents fresh, locally-grown produce, sparing the economic and environmental costs of importing from California and beyond. It re-strengthens—in a new way—the ever-diminishing connection between Phoenix and its landscape context, while also generating the critical mass of economic and social activity necessary to transform the strip mall into a public destination. It literally and figuratively grows green: the color-striated lettuces, vegetables, flowers, and algae that wrap the shops and offices represent a new desert ecology of cars, consumers, and productivity. Capitalizing upon the extreme productivity and profitability of hydroponic farming, Chia Mesa offers a business plan with the financial robustness to pay for the cost of transforming the existing strip mall model by gaining leverage from an industry whose long-term prospects are more secure than most retail and commercial tenants. Profits from hydroponic farming in this project are estimated at over $2.8 million per year.

The hydroponic farm is also critical to Chia Mesa's marketability. Its interface with the everyday operation of the commercial spaces and attendant parking is complementary, providing an oasis-like ambience and microclimate that redefines the shopping experience and public spaces associated with it, that will attract tenants ranging from natural food stores, juice bars, gyms and urban spas—as well as a weekly farmer's market selling produce harvested on-site. It also lends viability to other types of retail and office tenants, such as medical, dental, and law practices, for whom the association with the farm provides collateral benefits ranging from a soothing backdrop to an association with sustainable practices. Finally, the same hydroponic technology also serves as a natural air conditioner that oxygenates the building's microclimate, generates moisture and precipitates that clean the high quantities of particulate matter in such an arid climate, and through conduction, naturally cools the air.

CityLAB / Roger Sherman Architecture and Urban Design: Roger Sherman / Dana Cuff, Team Leaders; Team: Miguel Alvarez, Sergio Figuereido, Jennifer Gilman, Mira Henry, Quyen Luong, Brendan Muha, Travis Russett, Magdalena Stolarczyk; Jessica Campion, model; Tom Bonner, photography

>>> **Dana Cuff is Director of the think tank cityLAB and a Professor of Architecture and Urban Design at UCLA.**

>>> **Roger Sherman is a Los Angeles-based architect and urban designer, and Adjunct Professor at UCLA.**

^ From malls to mesas: The repurposing of the strip mall becomes the rebranding of Phoenix.

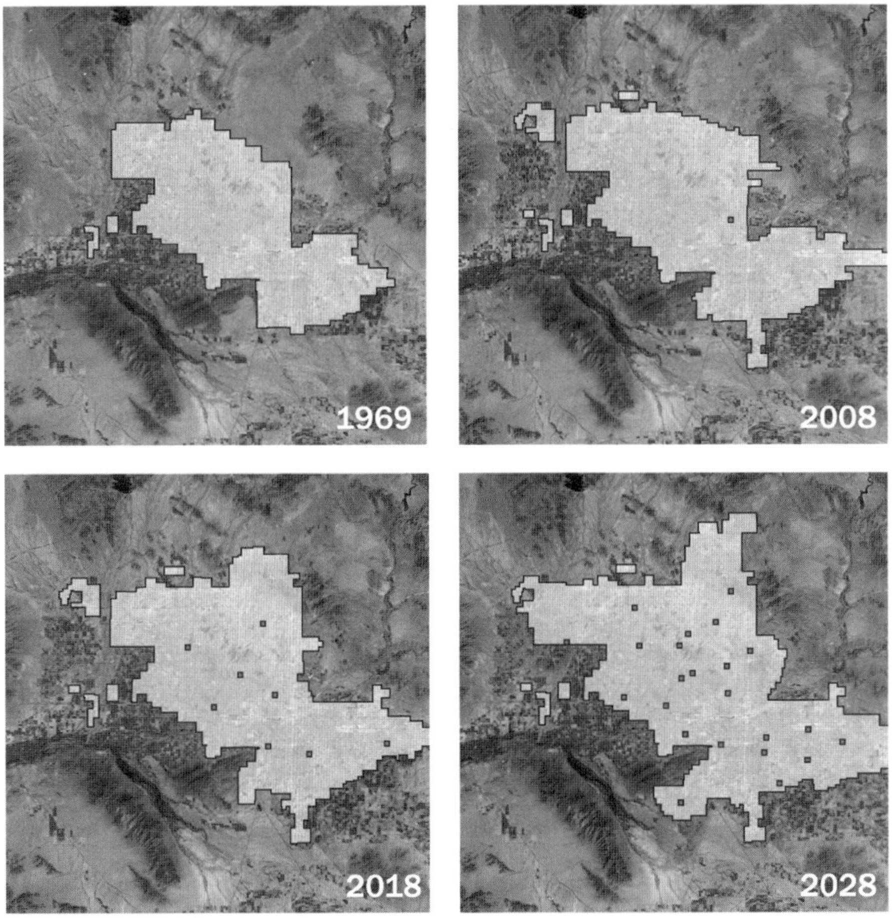

^ **As Phoenix grows, mesas show: As a recovery strategy, Chia Mesa enables the city to expand agricultural production even as it consumes more land.**

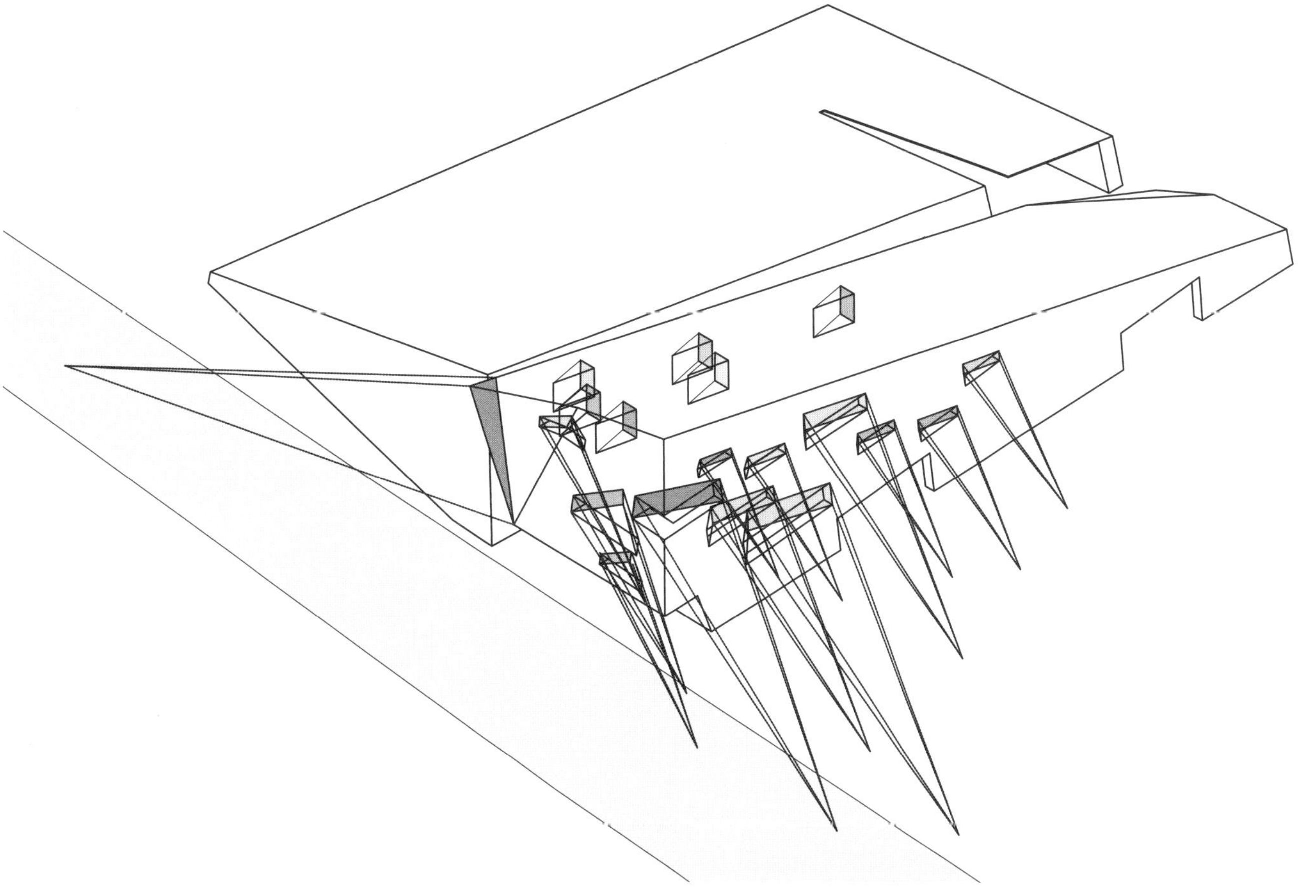

^ **Cliff selling: With a skin of operable louvers and advertising signage, Chia Mesa brings vitality to the street. Image by Mike Amaya.**

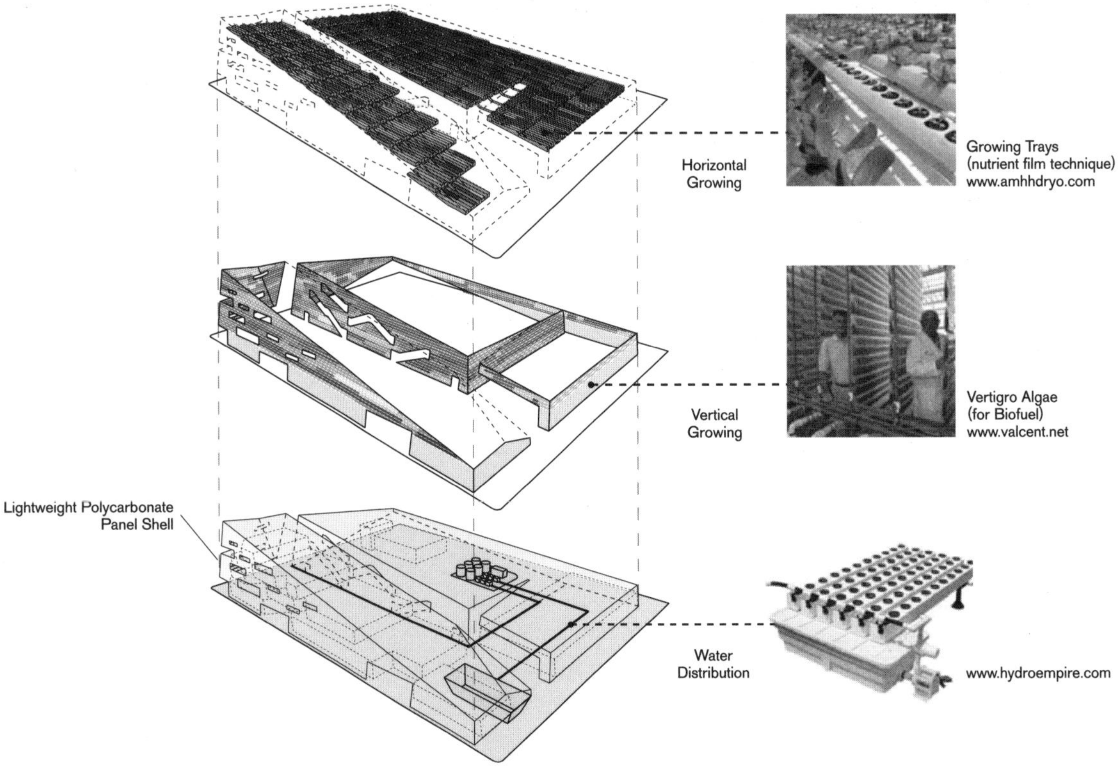

^^ Hydroponic system.

^ Sectional perspective.

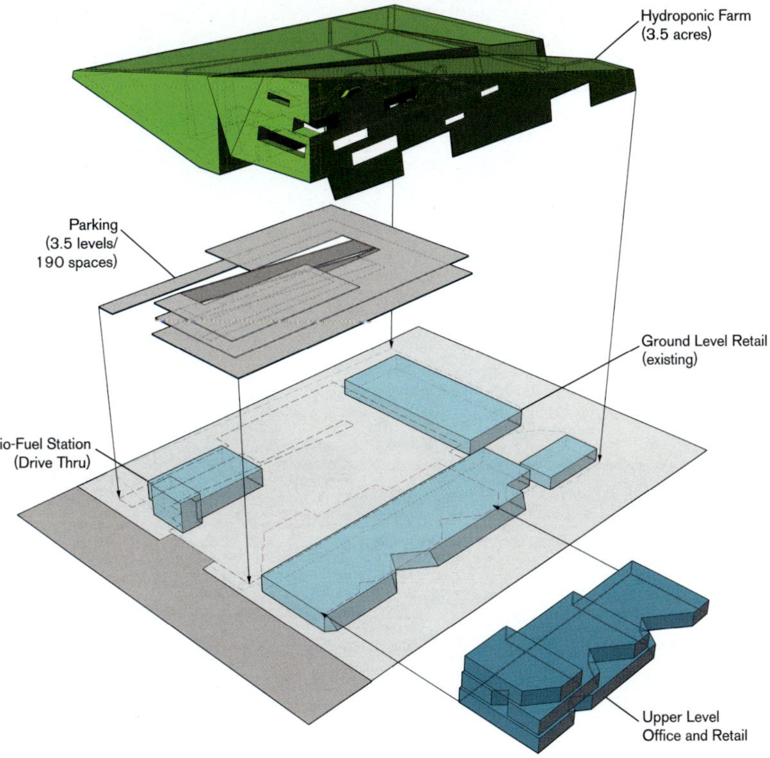

^^ Land use.

^ Greenback: Chia Mesa envisions the next generation of strip malls as climatically and socially cool oases in an arid urban field.

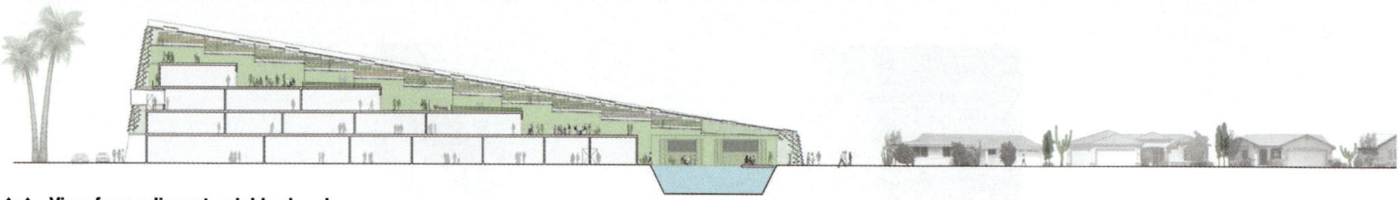

^^ View from adjacent neighborhood.

^ Longitudinal section.

line 13 – superlinearity

moving cities (bert de muynck & mónica carriço) & adrian blackwell

> By placing all its components in continuous variation, music itself becomes a superlinear system and enters the service of a virtual cosmic continuum of which even holes, silences, ruptures, and breaks are a part.
> —Gilles Deleuze and Felix Guattari, *A Thousand Plateaus*

Beijing's Line 13 was built in the late 1990s as a speculative infrastructure in order to stimulate the development of the countryside north of the city. At the time it was built, only Line 1 and Line 2 existed. So Line 13, as unlikely a route as its number suggests, sat as an anomalous form, designed to irrigate the periphery for real estate speculation. This process has allowed strange new cities like Huilongguan, a commuter suburb of monotonous slabs, to grow up surrounding the line, creating zones of uneven development where the messy vitality of existing uses meets the overbearing coherence of new subdivisions. Despite its urban character as infrastructure, the areas that Line 13 traverses are in between sites; housing blocks, universities, office, and business parks share space with golf courses, farms, villages, and desolate wastelands.

However, the unevenness of the Line also opens up potentials in the gaps between overdeveloped lands. "Superlinearity" examined the northern (10 km long) leg of Line 13, the area between Shangdi and Beiyuan, studying three systems to see how each might be redesigned to create opportunities within this socially polarized, ecologically degraded, and culturally empty landscape. These interventions work to render the peripheral system more complex, breaking the linear and uni-directional system of Line 13 into a superlinear composition, creating multiple connections between its parts.

This design proposal was produced collectively by 12 architecture students from the University of Toronto. The students worked in groups related to the three themes of transport, landscape, and labor, and in groups related to their respective stops along the route.

Transportation Networks

Beijing is a city notoriously difficult to traverse. It is made of over-scaled blocks and streets. Line 13 connects different parts of the city, yet it cuts through large plots of land and diverse programs. It suggests a form of transit-oriented development. However, after examining the formal and informal transportation networks surrounding Line 13, it is clear that the current model does not efficiently connect housing, retail, and employment to realize this potential.

This project proposes new infrastructures that move laterally across the grain of the existing line connecting the urban periphery to the city as well as the distant periphery through a set of loops.

Productive Landscapes

Beijing has one of the most intensive systems of peri-urban agriculture in the world. Its periphery is heavily urbanized and densely inhabited by international standards, and yet it is also very intensively farmed. Migrant farmers leave their own farms in distant provinces to live on the urban

^ Photographs of productive landscapes—existing unproductive landscape uses.

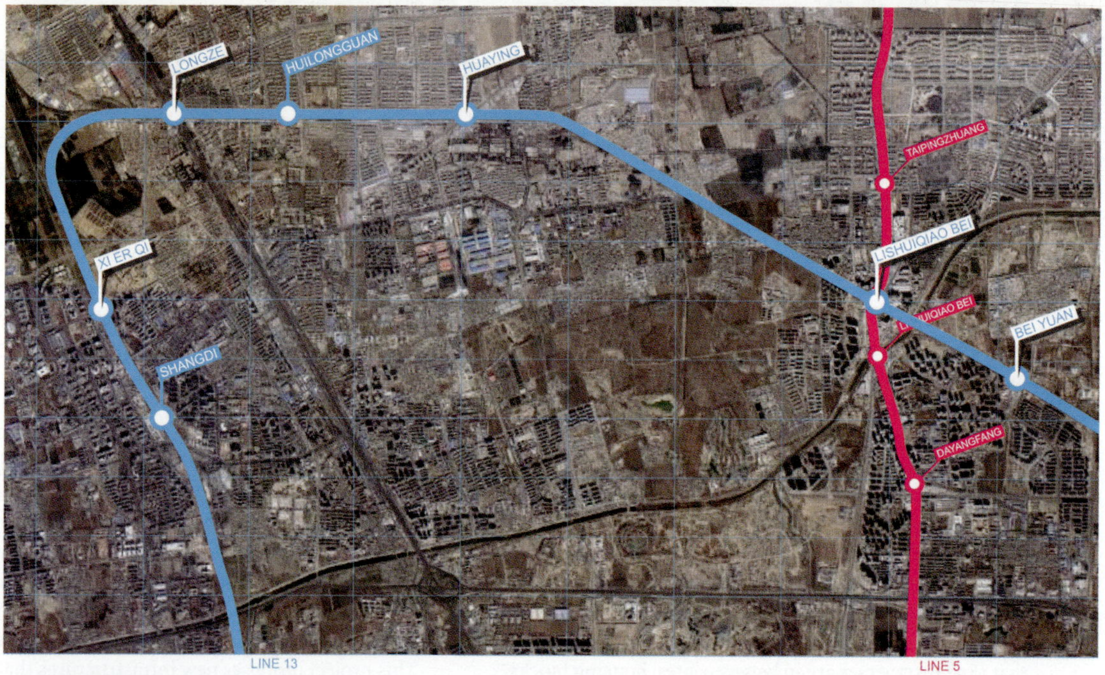

^ **Plan of the northern section of Line 13 from Shangdi to Beiyuan.**

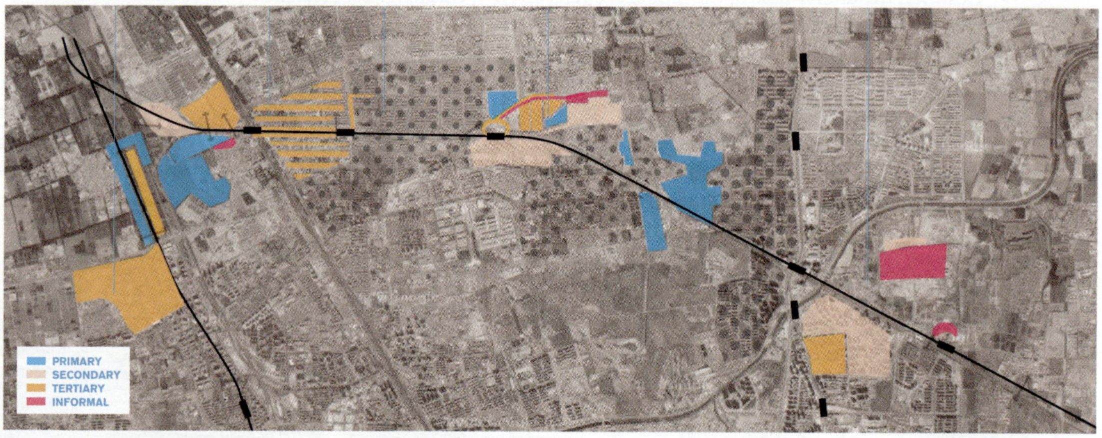

^ **Plan of the new forms of work along Line 13.**

^ **Diagram of traditional Beijing concentric urban development 14th C – 20th C; Diagram of Beijing concentric urban development 21st C; Diagram of bypassing Beijing's urban development (left to right).**

edge, rent land from local farmers, and grow vegetables where higher profits can be made from farming. This local food production is good for the city. It opens possibilities for a better understanding of food cycles, reduces transportation costs, offers employment, diversifies planting, and greens urban spaces. This project expands existing farming to create an agricultural spine along Line 13.

New Forms of Work

Investigating Line 13 from bottom up, the objective is to identify the potentials for new forms of creative labour. A patchwork of programs, fixed and floating economies, is active in these spaces. Within them, we identify what Marx called the 'general intellect' at work in the diversified local work systems of Chinese and foreign migrants, and in the communicative networks linking them back to their home spaces. The Beijing periphery already contains crucial aspects of the city's creative industries—artist villages, galleries, foreign immigrants, software developers and development in Zhongguancun. These creative industries are founded on the basis of cheap migrant labour that is also housed on the urban periphery. This proposal focuses on fostering the potential of these new industries to create emergent local economies.

The University of Toronto's John H. Daniels Faculty of Architecture, Landscape, and Design and Moving Cities' five-day workshop at Theatre in Motion, Beijing. Assisted by: Xu Jian, Sichuan Art College. Students: Gary Chien, Maya Desai, Hayley Imerman, Holly Jordan, Safora Khoylou, Esmond Lee, Timothy Lee, Antoine Morris, Mariangela Piccione, Matthew Spremulli, Sando Thordarson, Sandy Wong, Joseph Yau.

>>> **Moving Cities is a Shanghai-based think-thank investigating the role that architecture and urbanism play in shaping the contemporary city. Established in Beijing, 2007, by Bert de Muynck (BE) and Mónica Carriço (PT), Moving Cities publishes, collaborates, research, interacts, talks and walks, and operate as embedded architects.**

>>> **Adrian Blackwell is an artist and urbanist who teaches architecture at the University of Toronto.**

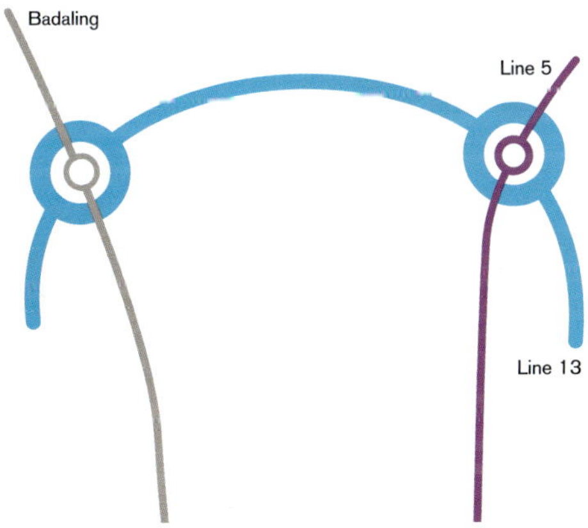

^ Diagram of North Section of Line 13 showing its intersections with the Badaling expressway and Line 5.

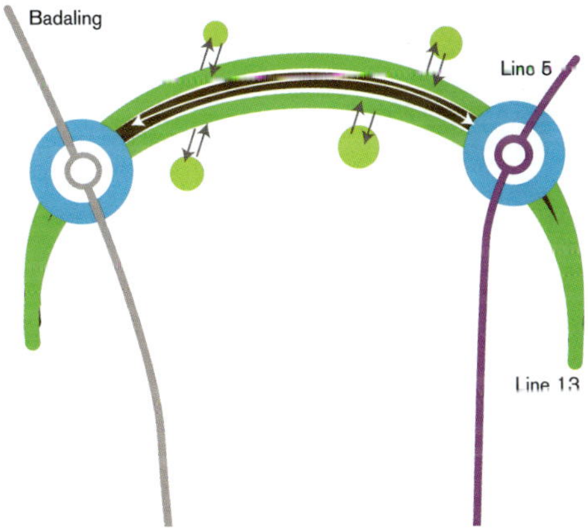

^ Diagram of productive landscapes: the travelling landscape.

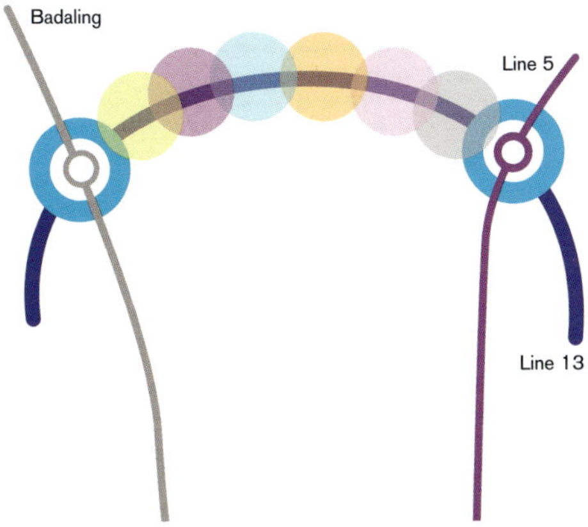

^ Diagram of new forms of work: hybridization.

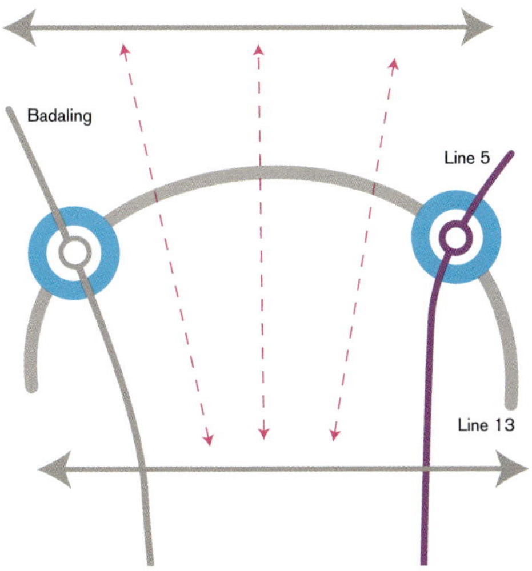

^ Diagram of transportation: connective infrastructure.

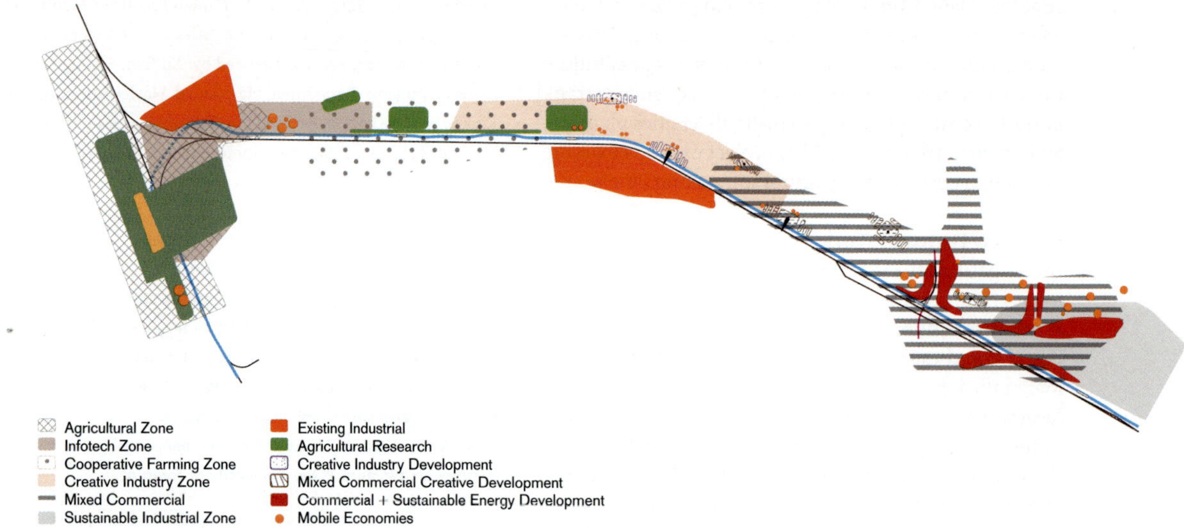

^ Plan of Line 13 showing proposed new employment uses.

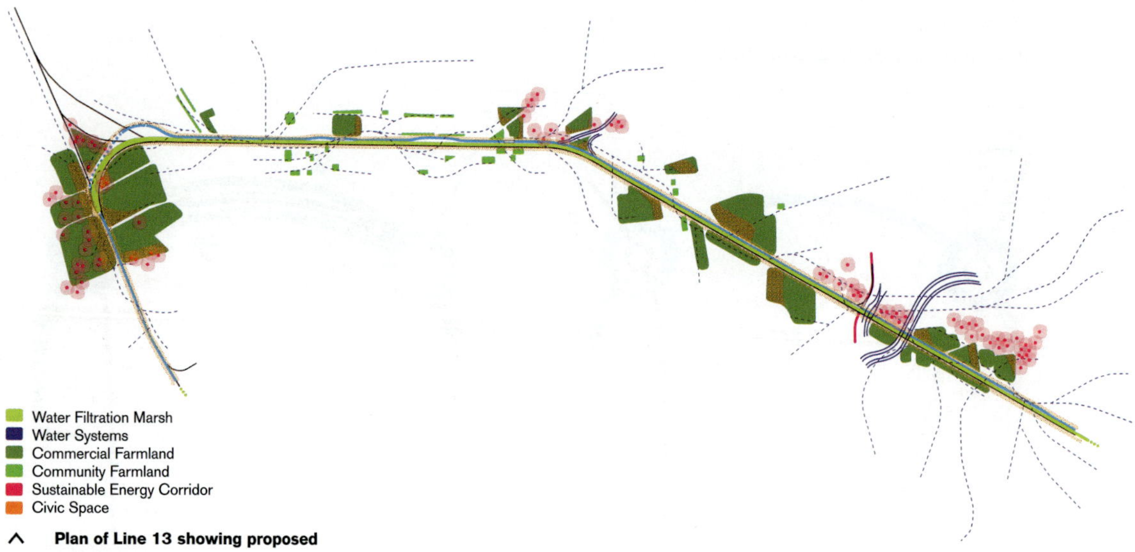

^ Plan of Line 13 showing proposed productive landscapes.

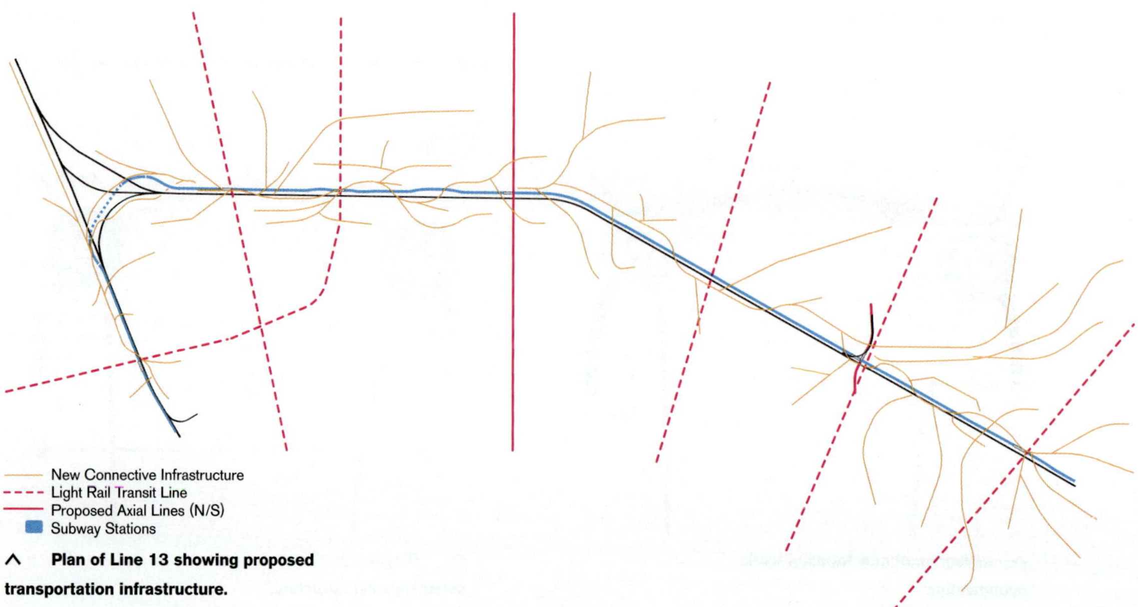

^ Plan of Line 13 showing proposed transportation infrastructure.

^ View 1: Wetland water remediation along Line 13 (Timothy Lee).

^ View 2: Superlinear garden hub, showing bike and pedestrian paths, ramps and bridges, and a linear farm between the tracks (Joseph Yau).

^ View 3: New high-rise development around wind farms in the hydro corridor (Holly Jordan).

your town tomorrow

corine vermeulen

^ Skyline, Atwater Street, Detroit, 2007.

It has been almost three-hundred years since Antoine de la Mothe Cadillac sent word back to Paris describing Detroit's landscape as "so temperate, so fertile, and so beautiful that it may justly be called the earthly paradise of North America."[1] Receiving Cadillac's recommendation, it was Count Pontchartrain's countrymen laid out "ribbon farms" that billowed from the river in strips up to a mile inland. It was these farms that were paved over and built upon once the industrial age came—the age of Henry Ford and his automobile. But despite the concrete and the trampling of people, the land never went away; it simply lay dormant, and given the opportunity it shows itself again. It creeps through the crevices to reveal its resilience. And people once again see it's beauty, and find a use for its rich soil.

The most important thing to understand about Detroit—as you see it today, with its abandoned buildings both large and small, and its empty lots, where buildings once stood, scattered about—is that half its population left. Nearly one million people, in a matter of a few decades, were no longer using the infrastructure that was built to accommodate all two million of them. Henry Miller's prediction had come true. Miller, referring in *The Air-Conditioned Nightmare* to Detroit in the 1940s, wrote: "The capital of the new planet—the one, I mean, which will kill itself off—is of course Detroit. I realized that the moment I arrived."[2]

Detroit, at one time, produced half the world's automobiles. Jobs were plentiful and anyone, from anywhere in the world, could come to Detroit and get one. The city evolved into the 20th century's prime example of industrial prosperity. But, like Miller foresaw, it wouldn't last. There are many reasons why Detroit is in the condition it is in today. These reasons have been debated and reiterated for years. Regardless, Detroit is here to stay, in whatever guise it may now evolve into. Coleman A. Young, mayor of Detroit from 1974 to 1993, summed it up well in his autobiography: "In the evolutionary urban order, Detroit today has always been your town tomorrow. ...Detroit remains a surpassingly purposeful place, as important to the nation right now as it has ever been—maybe more so, because right now it is telling us that the cities are in trouble. Detroit is the advance warning system—the flashing red light and siren—for what could be a catastrophic urban meltdown, and the country had damn well better pay attention."[3]

More than a decade after Young wrote this, the city proves to be the harbinger of which he speaks. Life in Detroit can be harsh. Those who have stayed, either by choice or, more likely, because their economic situation left them no option, have been forced to adapt, while dealing with the poverty and crime that have resulted from the loss of the once plentiful jobs. It may surprise many people, however, that a place so synonymous with urbanism is actually getting more and more agrarian. The definition of what a city is, is gradually being changed within the borders of Detroit.

Postscript: The gradual change witnessed here in 2007 has been further accelerated by the recent economic recession. Detroit now seems more barren than ever and its plight appears to be the world's fascination. The current mayor Dave Bing has announced plans to relocate people living in "desolate areas" saying "If we don't do it, you know this whole city is going to go down."[4] Rest assured, the natural world will have a response.

Project and photography: Your Town Tomorrow by Corine Vermeulen; All images © Corine Vermeulen.
Text: Matt Casadonte, Detroit.

>>> Corine Vermeulen is a Dutch photographer who continues to focus her lens on Detroit's shifting social and geographic ecologies.

1. Helen Frances Gilbert, *Tonquish Tales* (Plymouth, MI: Pilgrim Heritage Press, 1984), 121.

2. Henry Miller, *The Air-Conditioned Nightmare* (New York: New Directions, 1981), 41.

3. Coleman A. Young and Lonnie Wheeler, *Hard Stuff: The Autobiography of Coleman Young* (Viking Adult, 1994), xxv.

4. Christine MacDonald, "Detroit Mayor Bing emphasizes need to shrink city," Detroit News, February 25, 2010.

^ **Urban Prairie with Peregrine Falcons, Detroit, 2007.**

^ Emlee, Detroit, 2007.

^ Kurt, Detroit, 2007.

^ Emlee's House, Detroit, 2007.

^ Kurt's Car, Detroit, 2007.

^ Xavier, Birdtown Community Garden, Detroit, 2007.

^ Leah, Birdtown Community Garden, Detroit, 2007.

^ **Old Train Station, Detroit, 2007.**

^ **Peacemakers International Gardens, Detroit, 2007.**

^ **Geodesic Domes, Detroit, 2007.**

BFGoodrich

rethinking urbanism in the shrinking city of new orleans

carey clouse

^ Lower 9th ward in New Orleans, LA. (left). In the Lower 9th ward, vacant and unoccupied lots (fill) outnumber occupied lots (right). Both map bases courtesy of the City of New Orleans.

New Orleans is a ragged city. In the aftermath of Hurricane Katrina, this metropolis has rebounded in fits and starts, with the gap-toothed development of new housing flanking broken buildings, and swaths of activity ringing large vacant sections of the city. While Frederick Jackson Turner asserted in the 1890s that the "frontier has gone,"[1] and Joel Garreau declared Edge Cities "the final frontier,"[2] the shrinking city of New Orleans offers up a fresh model of untested territory. In New Orleans, the last frontier may actually be found in the resurrection of urban wilderness.

It is expected that the recovery of New Orleans will engage the same narrative that served to legitimize the settlement of the New World, incorporating capitalism, technology and the heavy footprint of human development.[3] The spirit of the city, however, is much better suited to a radical grassroots approach. Beyond the unconventional urban forms traced by New Orleans' canals and neutral ground, lie equally unique place-based issues, such as high soil toxicity, extreme wealth disparity and the threat of flooding from a sunken topography. New Orleans shoulders such a magnitude, multitude and curious assortment of problems that the city defies conventional solutions.

The recovery process doesn't have to involve restraint, discipline, or the neat frames of civilization. Because New Orleanians are comfortable with the chaos that has historically amplified this city, there is room for radical innovation in the rebuilding effort. New Orleans is an ideal place for creative and messy solutions, progressive and strange methodologies; it isn't too precious, and it isn't in control.

This untamed environment promotes rugged individualism, ripens the ground for innovation, and cultivates the creative renewal of a people's relationship to place. However, the City of New Orleans, FEMA and even local non-profit aid organizations have been slow to respond to this region's rebuilding needs. Worse, these institutionalized planning efforts lack imagination and a progressive sustainability agenda. In contrast, the following design interventions provide simple and speedy improvements for greening the urban frontier of New Orleans.

1. Frederick Jackson Turner, *The Frontier in American History* (New York: Henry Holt, 1920), 37-38.

2. Joel Garreau, *Edge City, Life on the New Frontier* (New York: Anchor Books, 1991), 14.

3. Carolyn Merchant, "Reinventing Eden," *Uncommon Ground*, (New York: W.W. Norton & Company, 1996), 137.

Forty Acres and a Mule

One hundred thousand buildings stand vacant in New Orleans today, under threat of demolition due to associated arson, blight, and maintenance.[4] The City of New Orleans could immediately restore its shrinking population and also bring life to these buildings by implementing squatter's rights, immigration reform and an open call for homesteaders. Like the many prior waves of U.S. immigration, these new groups would boost multiculturalism in New Orleans, while reducing the rampant homelessness and exorbitant rental rates in the city, both of which have more than doubled since 2005.[5]

Relax Density

Vacant parcels of land, some thirty thousand already seized by the New Orleans Recovery Authority,[6] are logical opportunities for the extension of greenspace in the city, rather than new built development. In practice, this might manifest itself in the expansion of single residential lots that seize neighboring plots, or a new urban fabric of pocket parks and community gardens. Either way, this growing patchwork of natural spaces affirms an intention to invite wilderness in the city, while embracing a reduction in density.

4. Information accessed at: http://www.noraworks.org/about_nora.htm

5. The Brookings Institution Metropolitan Policy Program & Greater New Orleans Community Data Center, "The New Orleans Index," January 2009, 12.

6. Information accessed at: http://www.noraworks.org/about_nora.htm

Detoxify Land

In the weeks following Hurricane Katrina, New Orleans held a body of contaminated sludge that democratically seeped over lot lines and across neighborhood boundaries. Bioremediation offers an opportunity to cure soil toxicity by trapping heavy metals in natural plant growth over time.[7] This process could begin with guerrilla-style phytoremediative seed bombs launched from the street, and grow into a more thorough and widespread planting effort. Because bioremediation takes several years to cleanse contaminated soil, and doesn't have many negative development effects, every lot could immediately begin this process, without regard for future development plans. In the interim, these sites would benefit from the creation of new natural habitat, and an overall reduction in maintenance, urban heat sink issues, and stormwater runoff.

Link Scattered Sites

Beyond greenspace, vacant lots in New Orleans offer cheap and useful space for the generation of local fuel sources, food production and other grown products. While this land exists in separate parcels scattered throughout the city, they can be coordinated by a single entity and managed for synchronized biofuel production, tree nurseries for free tree programs, urban gardens. If New Orleans's population ever rebounds, this reduction in density proves temporary, as these uses readily support future development.

7. Artist Mel Chin has already developed a strategy to bioremediate New Orleans, but lacks the funding to carry out his vision.

Ban Mowing

Post-Katrina New Orleans has adopted a special fondness for the lawn mower. In this verdant city, green shoots split through brick sidewalks, plants overtake whole houses, and thick jungles appear overnight in empty lots. Teams of volunteers from well-meaning mission groups descend upon the city to neaten empty lots and trim voluminous yards. The city would become more sustainable simply by halting this activity; volunteers could be more effective change agents for meaningful work, and natural areas would thrive within the city, thereby boosting urban habitat, cleaning the air, and improving soil health. More importantly, there would be an overall reduction in the use of fossil fuels.

Collect Water

The sunken topography of New Orleans collects rainwater: this runoff feeds into geriatric pumping stations and is lifted out into Lake Ponchatrain, at the rate of 1.5 inches per hour. This system has two major flaws: toxic urban runoff moves directly out to the lake, and the pumps are woefully undersized for their task. Water pools in streets and yards, occasionally flooding cars, before it gets expunged from the city. Instead, vacant lots in New Orleans can act as temporary holding areas for this runoff. Old swimming pools and other depressions instinctively act like cisterns on these sites, providing immediate storage capacity and reducing the threat of flooding. Eventually, these pools could become living machines to filter that toxic water, and perhaps even provide the space for small-scale urban catfish and crawfish farms.

Seize Potholes

20,000 potholes interrupt the streets of post-Katrina New Orleans.[8] Caused by sinking soils and the high water content of this marshy land, potholes reach soil below the street, and can be both dangerous to pedestrians and hard on cars. A fast alternative to road resurfacing would be to empower locals to fill these depressions with a plant, tree or bush. A single vertical planting would serve to slow traffic, clearly identify potholes, and add greenery to the urban streetscape. These planted roads would help to absorb water runoff, while serving as a physical reminder that the city is at once verdant and invincible.

Share Livestock

Spaces that require mowing in New Orleans need a neighborhood goat. These friendly animals could be tethered at playgrounds, available for check-out to trim yards or to prune back growth on empty lots. Enterprising locals would have access to their milk for cheese production. Meanwhile, chicken tractors could occupy the space of the empty neutral ground in New Orleans, where the birds would be free to eat insects, fertilize the city's median grasses, and provide eggs for the entire neighborhood.

8. Information accessed at: http://www.cityofno.com/portal.aspx?portal=1&load=~/PortalModules/ViewPressRelease.ascx&itemid=3858

Hijack Canals

The canal network in New Orleans operates like an American Southwest wash system. During periods of heavy rain, canals come to life with roiling waters and swept-up debris. Otherwise, these skeletal concrete corridors support tenuous emergent plant life and a constant, if lean, shelf of water. In an effort to promote this green growth, while pairing opportunities for urban hiking and biking, canals could feature an elevated bike path and fissures for volunteer plants. Similar to a reverse New York High Line, this trail system could involve minimal initial infrastructure, starting with a series of stiles that would act as entry and egress points into the canal system, and then phasing in grassroots-driven signage and mapping, graffiti murals, an open-source guidebook, aeries for birds, and finally, an elevated path.

Pirate Public Space

In New Orleans, buildings typically fill an entire lot, often with minimal front, side and rear yard setbacks. As a result, many residents lack access to their own land for gardening. However, the city's consistent swath of greenspace between the street and the sidewalk in front of each house makes an ideal urban garden. This land typically benefits from good solar access, and accommodates both produce and fruit trees. Like Ulf Matthiesen's Space Pioneers,[9] this pirating of public space also signifies a step towards the empowerment of urban subcultures.

9. Philipp Oswalt, "Space Pioneers," *Shrinking Cities* (New York: D.A.P., Distributed Art Partners, 2006), 380-381.

Redefine Native

The greening of New Orleans' empty urban spaces involves an inclusive and unrestrained approach. Emergent ecologies often include non-native plants, and these volunteers occasionally introduce harmful environmental responses. On the other hand, volunteers are hardy, well-suited to the hot and humid climate and harsh urban environment, provide habitat and a density of foliage on each site, and fix both the soil and the air. More importantly, non-native volunteers are a relevant local phenomenon. Like the carpetbaggers that have descended upon post-Katrina New Orleans to aid in the redevelopment of this city, and the fresh ideas that these people initiate, volunteer plants thrust New Orleans forward.

Thanks to Zachary Lamb, Culum Osborne and Michael Keller for contributing to this piece.

>>> **Carey Clouse is an architect and educator working to resurrect the city of New Orleans.**

on farming — editorial board

Fritz Haeg works between his occasional architecture and design practice, the ecology initiatives of Gardenlab (including Edible Estates and Animal Estates), the happenings and gatherings of Sundown Salon (now Sundown Schoolhouse), and various combinations of building, curating, dancing, designing, exhibiting, gardening, organizing, talking, teaching, and writing. His home base since 2001 is a geodesic dome in the hills of Los Angeles. Haeg studied architecture in Italy at the Istituto Universitario di Architettura di Venezia and Carnegie Mellon University. He has variously taught in architecture, design, and fine art programs at Cal Arts, Art Center College of Design, Parsons School of Design, and the University of Southern California. Haeg has produced projects and exhibited work at Tate Modern; the Whitney Museum of American Art; San Francisco MoMA; Casco, Utrecht; Mass MoCA; the Institute of Contemporary Art, Philadelphia; the Netherlands Architecture Institute, Maastricht; The Indianapolis Museum of Art, among others. *The Sundown Salon Unfolding Archive* (Evil Twin Publications) was released in 2009, and the expanded second edition of *Edible Estates: Attack on the Front Lawn* (Metropolis Books) in 2010.

Maya Przybylski is a Toronto-based designer and educator. She is a graduate of the Faculty of Architecture, Landscape and Design at the University of Toronto (M.Arch. 2007) where she was awarded the Royal Architectural Institute of Canada Medal for her thesis work on post-oil opportunities for the Caspian Sea. Maya previously earned a degree with a specialization in Software Engineering at the Department of Computer Science at the University of Toronto (B.Sc.Hon. 2003). Maya has taught at the University of Toronto, Ryerson University and the School of Architecture at the University of Waterloo. Through her own practice, Maya has collaborated with a range of design practices including: RVTR, Lateral Office, WilliamsonWilliamson, and Bruce Mau Design. In 2008, Maya became a director of InfraNet Lab, a non-profit research collective probing the spatial byproducts of contemporary resource logistics. The Lab's research into urban infrastructures will be published in *Pamphlet Architecture 30* (2010). Maya is co-editor of *Bracket 1* (Actar, 2010).

Heather Ring is a landscape architect and artist practicing in London. She received her Masters in Landscape Architecture from The University of Pennsylvania, and has been working with Gustafson Porter and Martha Schwartz Partners on the design of large-scale public spaces while developing a socially-engaged art practice. She is the founder of the Wayward Plant Registry, developing adoption events for unwanted plants, and the Wayward Land Trust, a strategic design practice for orphaned land and interim spaces. Her work has been included in exhibitions at the Barbican London; the Canadian Centre for Architecture, Montreal; The Graham Foundation, Chicago; and the London Festival of Architecture. She is a Senior Editor at Archinect.com.

Michael Speaks is Dean of the College of Design and Professor of Architecture at the University of Kentucky. Former Director of the Graduate Program at the Southern California Institute of Architecture in Los Angeles, Speaks has also taught in the graphic design department at the Yale School of Art, and in the architecture schools at Harvard University, Columbia University, The University of Michigan, UCLA, Art Center College of Design, and the Berlage Institute and TU Delft, in the Netherlands. Speaks is a founding editor of the cultural journal *Polygraph*, and is former Senior Editor at *Any* in New York, where he also edited the book series, "Writing Architecture," published by MIT Press. In addition, Speaks served for many years on the editorial advisory board of *a+u* and as a contributing editor for *Architectural Record.*

Nathalie de Vries was educated at University of Technology Delft (Department of Architecture and Urban Planning), where she graduated with a Master in Architecture in 1990 with honorable mention. In 1993 she founded, together with Winy Maas and Jacob van Rijs, the office of MVRDV. They have realized architectural designs, urban designs, master plans and research. Currently, Nathalie de Vries is working on a residential masterplan for the city of Leiden, an apartment building in Amsterdam, and a development strategy for Airport Tegel, Berlin. In China, she is working on a water city in Xinjin, near Chengdu, which foresees the creation of new islands in the city's river. Nathalie de Vries lectures and teaches throughout the world and takes part in international juries. She has been guest professor at the TU Berlin (2002-2004) and was the 2005 Morgenstern Visiting Critic at the IIT in Chicago. She has also been teaching at the Berlage Institute in Rotterdam, the ABK in Arnhem, and the TU of Delft. Additionally, she was a board member of the Netherlands Architecture Fund (1999-2005); she was member of the Gestaltungsbeirat of Salzburg (Austria, 2003-06); and a foundation board member of the Dutch architectural journal *Oase* (2004-). Nathalie de Vries was National Railroad Architect on behalf of ProRail/NS (2005-08). She recently joined the supervisory board of the Graphic Design Museum, Breda.

Charles Waldheim is the John E. Irving Professor of Landscape Architecture and Chair of the Department of Landscape Architecture at Harvard University's Graduate School of Design. His teaching and research examine the relationships between landscape and contemporary urbanism. Waldheim coined the term "landscape urbanism" to describe the recent emergence of landscape as a medium of urban order for the contemporary city. Waldheim has authored numerous articles and chapters on the topic, and edited *The Landscape Urbanism Reader.* His writing on landscape and contemporary urbanism has appeared in *Landscape Journal, Topos, Log, Praxis, 306090, Canadian Architect,* and *Landscape Architecture Magazine.* Citing Detroit as the most legible example of urban industrial economy, Waldheim is editor of *CASE: Lafayette Park Detroit* and co-editor, with Georgia Daskalakis and Jason Young, of *Stalking Detroit.* Waldheim is a licensed architect and principal of Urban Agency, a multi-disciplinary consultancy in design and contemporary urbanism. Waldheim received the Master of Architecture from the University of Pennsylvania where he was awarded the Paul Cret Medal for Master's Thesis Prize and the Wil Melhorn Prize for work in architectural theory. He has taught and lectured on contemporary urbanism across North America, Europe, Australia, and Asia. Waldheim is recipient of the Rome Prize Fellowship from the American Academy in Rome; the Visiting Scholar Research Fellowship at the Study Centre of the Canadian Centre for Architecture; the Cullinan Chair at Rice University, and the Sanders Fellowship at the University of Michigan. He is presently writing the first book-length history of Chicago's O'Hare International Airport: *Chicago O'Hare: A Natural and Cultural History.*

Mason White is an architect, writer, and educator based in Toronto, Canada. He is Assistant Professor at the University of Toronto Daniels Faculty of Architecture, Landscape, and Design. He received a BArch from Virginia Tech and MArch II from Harvard Graduate School of Design. White is a founding partner of Lateral Office in 2003 with Lola Sheppard, and a founding director of InfraNet Lab, launched in 2008. Recent research pursues questions of the role of infrastructure and networks within contemporary spatial practice. White's design research exists at the intersection of architecture, landscape, and urbanism. It is often situated within sites where the systems and codes that determine these environments must be uncovered and reconsidered. Lateral Office has won or been shortlisted for competitions internationally, most recently the WPA 2.0 competition (2009) and *Pamphlet Architecture 30* (2009), a publication forthcoming from Princeton Architectural Press (2010). Lateral Office is the recipient of the 2010-11 Professional Prix de Rome from the Canada Council for the Arts; Young Architects Award (2005) from the Architectural League of New York; and the Lefevre Fellow (2004) from Ohio State University. Mason is an Arthur Wheelwright Fellow (2009) from Harvard GSD. At InfraNet Lab, he is a curator of *HYDROCity: Hydrology and Urbanism* (2009-10) and a co-editor of *Bracket 1* (Actar, 2010).

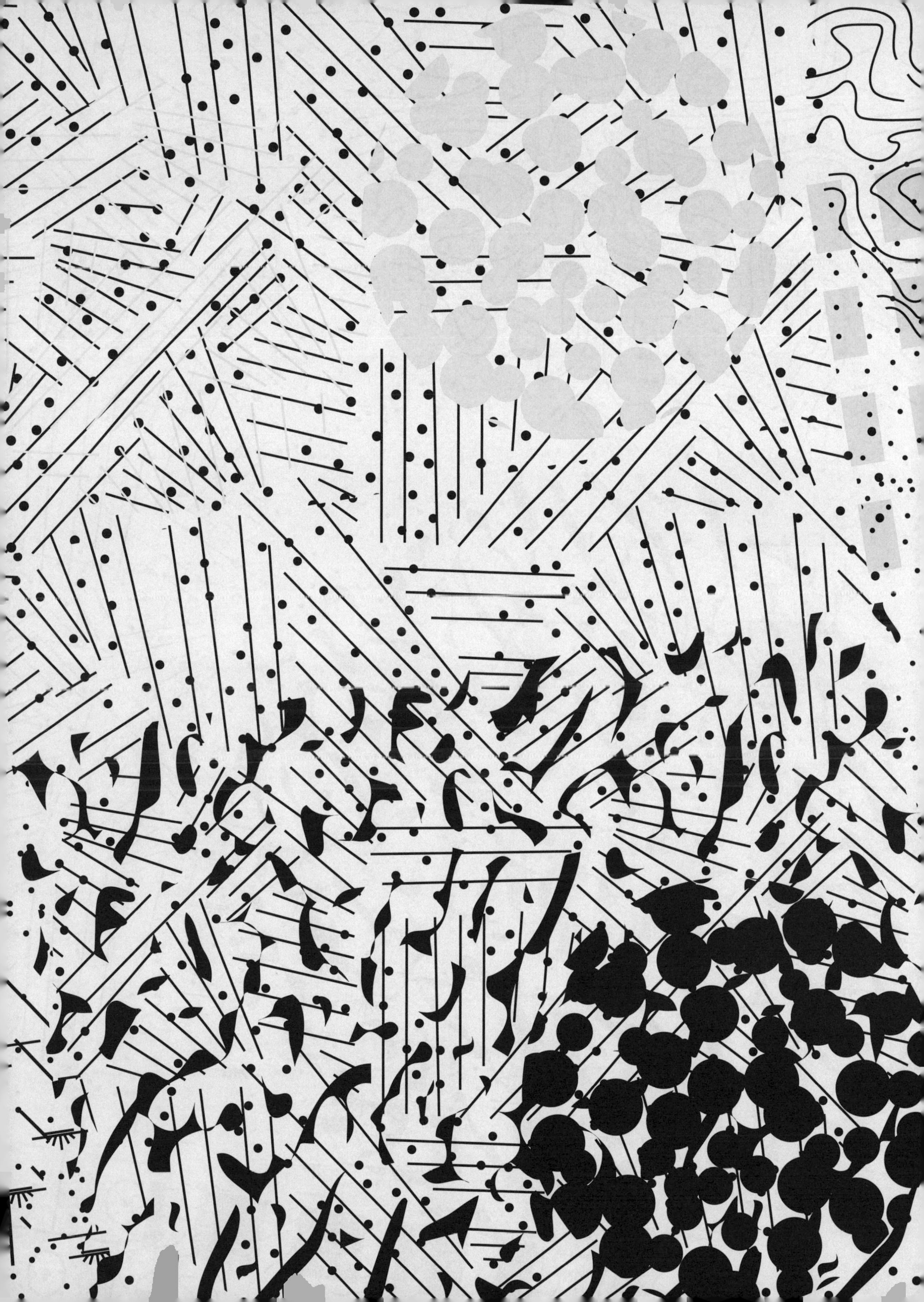

bracket — goes soft — almanac 2

call for entries —

Bracket 2 invites the submission of critical articles and unpublished design projects that investigate physical and virtual soft systems, as they pertain to infrastructure, ecologies, technologies, landscapes, environments, and networks. The role of soft systems has gained relevance more recently within a culture of declared crises—economic, ecological, and climatic. Bracket 2 seeks to critically position and define soft systems as a potential for new spatial networks, forms of architecture, urbanization and nature.

Bracket 2 seeks critical essays, research, and projects that investigate the role of soft systems within design. From software to soft powers, from soft spaces to fluid territories, Bracket 2 questions the use and role of responsive, adaptable, and immaterial systems in the built environment.

Please visit www.brkt.org for further information and schedule.

Bracket is an annual alamanc of architecture, environment, and digital culture founded by InfraNet Lab and Archinect.